AF595559

Advances in Sheet Metal Forming Processes of Lightweight Alloys

Advances in Sheet Metal Forming Processes of Lightweight Alloys

Editors

Mateusz Kopec
Denis Politis

Basel • Beijing • Wuhan • Barcelona • Belgrade • Novi Sad • Cluj • Manchester

Editors
Mateusz Kopec
Head of the Materials and Structures Testing Laboratory
Institute of Fundamental Technological Research Polish Academy of Sciences
Warsaw, Poland

Denis Politis
Department of Mechanical and Manufacturing Engineering
University of Cyprus
Nicosia, Cyprus

Editorial Office
MDPI
St. Alban-Anlage 66
4052 Basel, Switzerland

This is a reprint of articles from the Special Issue published online in the open access journal *Materials* (ISSN 1996-1944) (available at: https://www.mdpi.com/journal/materials/special_issues/sheet_metal_forming_alloy).

For citation purposes, cite each article independently as indicated on the article page online and as indicated below:

Lastname, A.A.; Lastname, B.B. Article Title. *Journal Name* **Year**, *Volume Number*, Page Range.

ISBN 978-3-0365-8916-9 (Hbk)
ISBN 978-3-0365-8917-6 (PDF)
doi.org/10.3390/books978-3-0365-8917-6

Contents

About the Editors

Mateusz Kopec

Dr Mateusz Kopec graduated from the Military University of Technology in Material Sciences (2015). He received his PhD in the Mechanical Engineering Department of Imperial College London in 2020. Since 2021, he has been the Head of the Laboratory for Materials and Structures Testing at the Institute of Fundamental Technological Research of the Polish Academy of Sciences (IPPT PAN). Since 2020, he has been a member of the Polish Materials Science Society (PTM), member of the Board of the Polish Society of Theoretical and Applied Mechanics (PTMTS) and council member of The European Society for Experimental Mechanics. He received over 30 awards for his scientific and didactic achievements including a scientific award from the Board of the Polish Materials Science Society for the best doctoral dissertation in the field of materials science in 2020, a scientific award of Division IV From Technical Sciences of the Polish Academy of Sciences and a scientific award from the Committee of Mechanics of the Polish Academy of Sciences. His main research interests include the hot stamping of titanium alloys, implants for biomedical applications, fatigue damage development in power engineering steels and thermal barrier coatings for nickel alloys.

Denis Politis

Dr Denis Politis is a Lecturer and Head of the Manufacturing and Materials Modelling laboratory in the Department of Mechanical and Manufacturing Engineering at the University of Cyprus. He has more than 10 years of research experience in sheet metal forming and forging technologies. He has worked on numerous research projects in collaboration with industrial companies that have been sponsored by European FP7 and H2020 grants.

Editorial

Advances in Sheet Metal Forming Processes of Lightweight Alloys

Mateusz Kopec [1] and Denis J. Politis [2,*]

[1] Institute of Fundamental Technological Research, Polish Academy of Sciences, Pawinskiego 5B, 02-106 Warsaw, Poland; mkopec@ippt.pan.pl

[2] Department of Mechanical and Manufacturing Engineering, University of Cyprus, Nicosia 1678, Cyprus

* Correspondence: politis.denis@ucy.ac.cy

Citation: Kopec, M.; Politis, D.J. Advances in Sheet Metal Forming Processes of Lightweight Alloys. *Materials* **2023**, *16*, 3293. https://doi.org/10.3390/ma16093293

Received: 5 April 2023
Accepted: 16 April 2023
Published: 22 April 2023

With the continuously growing need for more fuel-efficient and sustainable vehicles, the characterization and modeling of metal-forming processes have been indispensable in the development of new products. In the automotive and aviation sector, low-strength structural components are commonly produced from aluminum alloys, and higher-strength structural components are made from ultra-high-strength steels (UHSSs) and titanium alloys. The main issue experienced during the hot forming of complex-shaped components from difficult-to-form alloys is that they are time-, energy-, and cost-intensive. The aircraft industry currently uses methods such as superplastic forming (SPF), superplastic forming with diffusion bonding (SPF-DB), hot stretch forming, hot gas-pressure forming, and isothermal hot forming. Moreover, novel techniques have been developed to produce complex-shaped structural components including solution heat treatment, forming and in-die quenching (HFQ), quick-plastic forming, hot stamping using rapid heating, and fast light alloy stamping technology (FAST).

This Special Issue covered a wide range of topics, including novel materials (HSLA steel [1], titanium alloys [2–5], magnesium [6,7]), forming techniques (single-point incremental forming [8], magnetic pulse forming [9], rigid-flexible sequential loading forming [10]), and advanced predictive models [1,5,9,11] developed for such processes.

Behrens et al. [1] presented experimental and numerical investigations on HSLA steel friction drilling in which the temperature, strain rate, and rolling direction-dependent tensile tests of the HSLA HX220 were executed and used to parametrize the Johnson–Cook hardening and failure models. Further, the experiments were numerically modelled using different methods. Since the comparison of the simulations and the experiments showed a good agreement, it was assumed that the methods used for the material characterization and modelling were appropriate.

Dang et al. [2] investigated dynamic softening and hardening behavior and microstructure evolution of the TC31 titanium alloy during high-temperature tensile deformation. The authors conclude that the TC31 titanium alloy exhibited clear softening behavior during hot tensile deformation at a temperature of 850 °C and a strain rate of 0.001 s^{-1}~0.1 s^{-1}, with an increase in the deformation temperature to 950 °C~1000 °C and an increase in the strain rate to 0.1 s^{-1} discontinuous yielding occurred, and quasi-steady flow appeared at a temperature of 950 °C~1000 °C and a strain rate of 0.01 s^{-1}, with a decrease in the strain rate to 0.001 s^{-1}, resulting in a slight dynamic hardening phenomenon. Furthermore, the authors observed significant microstructural changes when the deformation temperature increased from 850 °C to 950 °C. It was found that the volume fraction of the β phase increased from 20% to 41% after it deformed to a strain of 0.7 with a strain rate of 0.01 s^{-1}, whereas the volume fraction of voids was significantly reduced from 11.2% to less than 1%. Since the increased fraction of the β phase at higher temperatures improved the deformation compatibility and reduced the void damage, a relatively high deformation temperature was recommended for the forming of complex TC31 titanium alloy components to avoid void damage. On the other hand, when the TC31 titanium alloy was deformed at 950 °C, the grains grew at the strain rate of 0.001 s^{-1} and were refined at the strain rate from 0.01 s^{-1} to 0.1 s^{-1}. The refinement was more significant under the higher strain rate conditions. It was concluded that the appropriate strain rate should be approximately 0.01 s^{-1} during the forming of the TC31 titanium alloy sheet considering both the grain coarsening and uniform deformation.

Li et al. [3] conducted research on the effects of the forming temperature and heating rate on the maximum flow stress and elongation of TC4 by using hot tensile tests and microstructural analysis. It was observed that in the forming temperature range between 850 °C and 950 °C, both the maximum flow stress and elongation decreased with an increase in the forming temperature. When the heating rate was 10 °C/s, the flow stress was larger than that at a heating rate of 1 °C/s, while the elongation remained constant. During microstructural observations, it was found that the volume fraction of the β phase increased with an increase in the heating temperature and a decrease in the heating rate, the average grain size decreased with an increased heating rate and a higher volume fraction of the β phase, and finer grains improved the material ductility. Based on the microstructure observation results, a model was established to predict the volume fraction of the β phase under different heat treatment conditions. The prediction error of the model was 5.17%, which would contribute to a qualitative analysis of the mechanical properties of TC4 titanium alloy under high-temperature deformation conditions.

McPhillimy et al. [4] generated a laser metal deposition tailored preform with a variable thickness to mitigate thinning, a common defect in the room temperature single-point incremental forming (SPIF) of titanium parts with high angled walls. An initial material study of a laser metal deposition (LMD) tailored CP-Ti50A sheet with localized thickening was performed. Subsequently, single-point incremental forming was performed on a LMD tailored CP-Ti50A preform sheet. To facilitate the hybrid LMD + SPIF process, a modular fixture was designed to constrain the tailored titanium sheet during LMD, post-processing, and SPIF.

Su et al. [5] developed an FE method based on the thermal–elastic–viscoplastic macroscopic model to predict the shrinkage, deformation, relative density, and crack of injection-molded Ti-6Al-4V after sintering, using commercially available Simufact software. In the authors' research, experiments were simultaneously performed to justify the accuracy of the sintering model and simulation method. The results exhibited a good agreement between the experimental measurements and numerical simulations with a 3% error. The slightly larger sintered density and shrinkage in the experiment than those in the simulation were found due to additional thermal convection and conduction during the sintering experiments. It was concluded that the deformation was affected by gravity, friction, specimen shape, and support mode.

Ullmann et al. [6] studied the orientation-dependent flow behavior of the ZAX210 magnesium alloy to provide basic yield model data for the numerical simulation. The authors found that the plane anisotropy Δr was between 0 and −0.2 at all tested temperatures, which indicate a slight anisotropic behavior in the sheet plane. The obtained r-values are a direct result of the crystallographic texture present in the ZAX210 alloy, combined with the relative resolved shear strengths of the slip and twinning systems. The authors conclude that the in-plane material flow behavior can be identified as orthotropic, with decreasing anisotropy at elevated temperatures.

He et al. [7] investigated an unusual phenomenon of strain neutral layer (SNL) spreading revealed during the V-bending test. It was found that the SNL on the middle of the symmetrical surface perpendicular to the transverse direction extends to the compression region with a mound-like boundary. The SNL in the side position was distributed with a parallel band feature. This difference in SNL distribution was mainly attributed to the difference in three-dimensional stress distributions between the side position and the middle position of the bending sample. Finally, it was concluded that the three-dimensional compressive stresses in the compressed region were responsible for the SNL spreading phenomenon.

Yan et al. [8] introduced and optimized a two-stage forming strategy in SPIF to reduce the geometrical deviation and the processing time compared to those manufactured using a single forming tooling. A simulation model of the SPIF has been developed and solved using an explicit finite element analysis to study the optimal tool path for a truncated cone. The design of experiments using a response surface method was used to optimize the proposed two-stage forming strategy. The simulation results showed that the two-stage forming technique could significantly reduce both the geometrical deviation and the forming time. The forming time and part geometric deviation were reduced by 56% and 25%, respectively. In addition, the part thickness distribution was found to be more uniform after optimization, and the minimal thickness decreased by 1.6%.

Mahmoud et al. [9] presented an efficient approach for the simulation of the magnetic pulse forming process of thin sheet metals. The model was developed by combining an electromagnetic solver, relying on Maxwell's equations, with a mechanical solver, based on the conservation of momentum equations. The overall results showed that the accuracy obtained with a low-resolution SHB approach (a small number of elements) was comparable to that of a high-resolution MINI-element-based technique (a large number of elements). The SHB element was shown to be less stiff than the MINI element. Finally, a computational cost study was carried out, and this demonstrated a higher computational efficiency for the SHB element since a smaller number of elements could be used while maintaining a comparable accuracy to that of the MINI element.

Zhang et al. [10] studied the spring back behavior of large complex multi-feature parts (aluminum alloy inner panel) in the rigid–flexible sequential forming process. Based on the theoretical prediction and experimental results, the spring back compensation of the complex inner panel was carried out. The authors showed that the hardening model has a greater impact on the accuracy of spring back prediction than the yield criterion, and the prediction accuracy of Barlat'89 + Yoshida–Uemori mixed hardening model is the greatest. Finally, the optimized loading locus of hydraulic pressure was obtained, and the accurate results from the compensated parts were used to verify the accuracy of the analysis model.

Liu et al. [11] presented an analytical formability model for the sandwich panel and demonstrated its capability by predicting the critical failure of a sandwich panel consisting of two skin AA5754 layers and a core PVDF layer as a case study. It was found that the developed FLD model overcame the limitation of traditional FLD models and was capable of predicting the formability of sandwich panels made from composite materials. The results presented in this study provided a safety evaluation and theoretical guidance on the plastic-forming and critical failure of the composite sandwich panels for lightweight sealing and insulating component applications.

Funding: This research received no external funding.

Institutional Review Board Statement: Not applicable.

Informed Consent Statement: Not applicable.

Data Availability Statement: Not applicable.

Conflicts of Interest: The authors declare no conflict of interest.

References

1. Behrens, B.-A.; Dröder, K.; Hürkamp, A.; Droß, M.; Wester, H.; Stockburger, E. Finite Element and Finite Volume Modelling of Friction Drilling HSLA Steel under Experimental Comparison. *Materials* **2021**, *14*, 5997. [CrossRef] [PubMed]
2. Dang, K.; Wang, K.; Liu, G. Dynamic Softening and Hardening Behavior and the Micro-Mechanism of a TC31 High Temperature Titanium Alloy Sheet within Hot Deformation. *Materials* **2021**, *14*, 6515. [CrossRef] [PubMed]
3. Li, D.; Zhu, H.; Qu, S.; Lin, J.; Ming, M.; Chen, G.; Zheng, K.; Liu, X. Effect of Heating on Hot Deformation and Microstructural Evolution of Ti-6Al-4V Titanium Alloy. *Materials* **2023**, *16*, 810. [CrossRef] [PubMed]

4. McPhillimy, M.; Yakushina, E.; Blackwell, P. Tailoring Titanium Sheet Metal Using Laser Metal Deposition to Improve Room Temperature Single-Point Incremental Forming. *Materials* **2022**, *15*, 5985. [CrossRef] [PubMed]
5. Su, S.; Hong, Z.; Huang, Y.; Wang, P.; Li, X.; Wu, J.; Wu, Y. Integrated Numerical Simulations and Experimental Measurements for the Sintering Process of Injection-Molded Ti-6Al-4V Alloy. *Materials* **2022**, *15*, 8109. [CrossRef] [PubMed]
6. Ullmann, M.; Kaden, C.; Kittner, K.; Prahl, U. Orthotropic Behavior of Twin-Roll-Cast and Hot-Rolled Magnesium ZAX210 Sheet. *Materials* **2022**, *15*, 6437. [CrossRef] [PubMed]
7. He, C.; Liu, L.; Bai, S.; Jiang, B.; Teng, H.; Huang, G.; Zhang, D.; Pan, F. Unusual Spreading of Strain Neutral Layer in AZ31 Magnesium Alloy Sheet during Bending. *Materials* **2022**, *15*, 8880. [CrossRef] [PubMed]
8. Yan, Z.; Hassanin, H.; El-Sayed, M.A.; Eldessouky, H.M.; Djuansjah, J.; A. Alsaleh, N.; Essa, K.; Ahmadein, M. Multistage Tool Path Optimisation of Single-Point Incremental Forming Process. *Materials* **2021**, *14*, 6794. [CrossRef] [PubMed]
9. Mahmoud, M.; Bay, F.; Muñoz, D.P. An Efficient Computational Model for Magnetic Pulse Forming of Thin Structures. *Materials* **2021**, *14*, 7645. [CrossRef]
10. Zhang, Y.; Lang, L.; Wang, Y.; Chen, H.; Du, J.; Jiao, Z.; Wang, L. Spring Back Behavior of Large Multi-Feature Thin-Walled Part in Rigid-Flexible Sequential Loading Forming Process. *Materials* **2022**, *15*, 2608. [CrossRef]
11. Liu, X.; Di, B.; Yu, X.; Liu, H.; Dhawan, S.; Politis, D.J.; Kopec, M.; Wang, L. Development of a Formability Prediction Model for Aluminium Sandwich Panels with Polymer Core. *Materials* **2022**, *15*, 4140. [CrossRef] [PubMed]

Article

Finite Element and Finite Volume Modelling of Friction Drilling HSLA Steel under Experimental Comparison

Bernd-Arno Behrens [1], Klaus Dröder [2], André Hürkamp [2], Marcel Droß [2], Hendrik Wester [1] and Eugen Stockburger [1,*]

[1] Institute of Metal Forming and Forming Machines, Leibniz Universität Hannover, 30823 Garbsen, Germany; behrens@ifum.uni-hannover.de (B.-A.B.); wester@ifum.uni-hannover.de (H.W.)
[2] Institute of Machine Tools and Production Technology, Technische Universität Braunschweig, 38106 Braunschweig, Germany; k.droeder@tu-braunschweig.de (K.D.); a.huerkamp@tu-braunschweig.de (A.H.); m.dross@tu-braunschweig.de (M.D.)
* Correspondence: stockburger@ifum.uni-hannover.de

Abstract: Friction drilling is a widely used process to produce bushings in sheet materials, which are processed further by thread forming to create a connection port. Previous studies focused on the process parameters and did not pay detailed attention to the material flow of the bushing. In order to describe the material behaviour during a friction drilling process realistically, a detailed material characterisation was carried out. Temperature, strain rate, and rolling direction dependent tensile tests were performed. The results were used to parametrise the Johnson–Cook hardening and failure model. With the material data, numerical models of the friction drilling were created using the finite element method in 3D as well as 2D, and the finite volume method in 3D. Furthermore, friction drilling tests were carried out and analysed. The experimental results were compared with the numerical findings to evaluate which modelling method could describe the friction drilling process best. Highest imaging quality to reality was shown by the finite volume method in comparison to the experiments regarding the material flow and the geometry of the bushing.

Keywords: tensile testing; material characterisation; material modelling; friction drilling; high-strength low-alloy steel

Citation: Behrens, B.-A.; Dröder, K.; Hurkamp, A.; Droß, M.; Wester, H.; Stockburger, E. Finite Element and Finite Volume Modelling of Friction Drilling HSLA Steel under Experimental Comparison. *Materials* **2021**, *14*, 5997. https://doi.org/10.3390/ma14205997

Academic Editors: Denis Politis and Mateusz Kopec

Received: 28 September 2021
Accepted: 8 October 2021
Published: 12 October 2021

1. Introduction

Modern lightweight construction concepts for large-scale automotive production are increasingly characterised by the use of hybrid material composites of fibre-reinforced plastics (FRP) and metals. Resource-efficient lightweight construction aims to exploit material-specific potential through appropriate material combinations. Various processes are currently being used to produce hybrid material compounds, such as hybrid injection moulding. Industrially manufactured injection-moulded components use the metal insert as an element of load transfer from the metallic joining zone with other machine components into the plastic reinforcement. On some components, the bonding between the plastic and metal is realised with through-moulding points. One way to improve the low bond strength of these components is to insert friction-drilled bushings into the component before injection moulding. The interlocking effect between the plastic and the bushing can increase the load-bearing capacity of the hybrid material bonding in the component.

Previous studies mainly used the same modelling techniques and focused on the simulation of the friction drilling process parameters such as force or torque, temperature distribution, and development, as well as the validation of these values. So far, few publications have investigated the material flow and geometry of the bushing in, as well as against, the feed direction, since it is not necessary for common subsequent processes such as thread forming. Depending on the application, however, the correct digital mapping of the geometry is essential for a realistic computer-aided design of the process.

Therefore, this study focuses on proper digital mapping of the bushing formation while friction drilling. First, a material characterisation of the HX220 steel by means of temperature, strain rate, and rolling direction dependent tensile tests was performed. The parameters of the Johnson–Cook hardening and failure models were determined using the tensile tests. Second, experimental friction drilling tests of HX220 were conducted as well as analysed, and compared with friction drilling tests of HX420. Finally, the friction drilling tests of HX220 were modelled with different numerical approaches using the characterised material data and contrasted with the experiments. Investigated methods were the finite element (FE) in 2D and 3D. In addition, the finite volume (FV) method in 3D by means of the Coupled Eulerian–Lagrangian (CEL) method was applied and tested to model a friction drilling process.

1.1. Friction Drilling

Friction drilling is a noncutting manufacturing process for the production of bushings in sheet metal and hollow profiles. For this purpose, a rotating tungsten carbide friction drill is pressed onto the work piece under axial force. The frictional heat generated between the drill and the work piece reduces the strength of the material to such an extent that the drill can be pressed through the sheet metal with significantly reduced force. Material is shifted by the drill, and a bushing is formed outside of the sheet metal. The lower part of the bushing consists of approximately two-thirds of the displaced material volume. Upwards against the feed direction, the remaining third is displaced and can be formed into a collar [1]. In combination with thread forming, which is also chip-free, highly resistant threads can be formed in these bushings. Thus, the process offers ideal conditions for the production of detachable connections in thin-walled work pieces. In order to exploit further areas of application and to optimise the processes, computer-aided modelling is indispensable for time- and cost-saving reasons [2]. FE modelling allows the efficient investigation of most important process variables and their effect, especially for parameters that are not experimentally determinable [3].

1.2. Friction Drilling Simulalation

The heat transfer of a friction drilling process was modelled in 2D using the thermal FE method in Ansys by Miller et al. [4]. Obtained from experiment and simulation, the trend of temperature increase matched well. Abaqus/CAE(Simulia (Dassault Systèmes), Johnston, Rhode Island, US) was used by Miller et al. in [5] to model friction drilling of AA 6061-T6 in 3D with FE using eight-node hexagonal elements, adaptive meshing, and element deletion. The criterion to delete an element was based on the value of critical plastic strain. A threshold value of 2.2 was required to be reached before an element was deleted. A comparison of simulated and measured force, torque, and temperature showed a good quality of the model. Since the simulation showed no bushing against the feed direction, it was concluded that the element deletion removed too many elements. It was suggested to improve the simulation by developing a model without using the element deletion. The results in [6] by Krasauskas et al. also showed no bushing against the feed direction in a similar simulation model in Abaqus/CAE. Dehghan et al. used a 3D FE model to analyse the stress, plastic strain, and temperature distribution in friction drilling of AA 6061 [7]. The model was also created in Abaqus/CAE with hexahedral C3D8RT elements with a 0.1 mm element size, an adaptive mesh technique, and a Johnson–Cook failure model. The modelling showed reasonable results regarding the process parameters, but a comparison of the cross-section view of the experimental and simulated friction drilled hole showed a difference in both bushings. The simulation had a shorter bushing, most likely due to many elements deleted by the failure model. The heat generation and bushing formation caused by friction in friction drilling were studied by Dehghan et al. in [8]. A FE model in Abaqus/CAE was used, and examined the materials were AISI 304, Ti-6Al-4 V, and Inconel718. The model consisted of the Johnson–Cook failure model, and the damage evolution law was modified to include the Hillerborg's fracture energy. The authors were able to represent the formation of the bushing well. Nevertheless, the bushing in the direction of feed was sometimes too short, and the bushing against the direction of feed was always too small.

Oezkaya et al. presented another 3D FE simulation model of a friction drilling in [9] for AlSi10Mg using the software DEFORM-3D, tetrahedral elements, and an adaptive mesh technique. A failure criterion was not mentioned. The experimental and numerical force, torque, and temperature showed a good agreement. A comparison of the bushing for experiment and simulation was not shown, but the simulation indicated the formation of a too-short bushing. The authors in [10–12] used a similar approach to model friction drilling processes: 3D FE models were created in DEFORM-3D and consisted of tetrahedral elements with an adaptive mesh technique. The study did not describe which failure models were implemented. Hynes et al. analysed friction drilling of copper Cu2C numerically and validated it through experiments with regard to force and torque [10]. The numerical temperature was not validated, and the bushing geometry was not analysed. Srilatha et al. analysed the aluminium AA 7075 numerically in terms of stress, strain, and temperature at different rotational speeds and feed rates, but did not validate the studies [11]. Bilgin et al. considered the AISI 1020 for friction drilling [12]. The numerically calculated temperature, torque, and force matched well with the experimental values for different rotational speeds. Unfortunately, no figures for the final geometry of the friction drilling simulation were presented, so the bushing could not be evaluated.

Other promising approaches to modelling friction drilling are mesh-free computational approaches. Those approaches are developed to deal with large deformation-induced material failure in destructive manufacturing. A common approach is the element-free Galerkin (EFG) method. To treat material failure more physically in destructive manufacturing processes, a genuine mesh-free method, the smoothed particle Galerkin (SPG) method, was developed more recently. Wu et al. simulated friction drilling of an AISI 304 steel with the SPG method in LS-DYNA [13]. The work piece was discretised by SPG particles with a nodal distance of 0.2 mm. The Johnson–Cook hardening model was applied to describe the flow behaviour of the work piece. However, the material failure was not described by the Johnson–Cook failure model. The failure process was modelled by the SPG bond failure mechanism. The numerical results showed a good match with experimental values regarding force and torque without deleting parts of the discretisation, as in the FE simulation [14].

2. Materials and Methods

2.1. Material Characterisation

To investigate the flow behaviour of the high-strength low-alloy (HSLA) HX220 steel, the miniaturised tensile test specimens shown in Figure 1a were used, as in [15]. The specimens were taken from a 1.2 mm thick sheet metal at 0°, 45°, and 90° to the rolling direction using water-jet cutting. Uniaxial tensile tests were carried out at elevated temperatures on a DIL 805A/D+T quenching(TA Instruments; New Castle; PA; USA) and forming dilatometer from TA Instruments. The test equipment of the forming dilatometer is shown in Figure 1b. Before testing, a stochastic pattern was applied, and a thermocouple was welded onto the specimen to regulate the temperature. It was then placed in the dilatometer, heated to the respective forming temperature via a heating coil with an integrated cooling system, and held for 20 s. Afterwards, the forming force was applied via the hydraulic system until failure of the specimen occurred. During the forming of the specimen, the change in length and width of the test area was measured with an ARAMIS (GOM GmbH, 38122 Braunschweig, Germany) optical measuring system from GOM GmbH. The forming temperature was investigated at 20 °C, 250 °C, and 500 °C, and the strain rate at 0.01 s^{-1}, 0.1 s^{-1}, and 1 s^{-1}. Every parameter combination was performed five times to ensure statistical coverage. The occurring force and the measured length change were transferred into uniaxial flow curves according to the state of the art [16].

Further, temperature-dependent anisotropy coefficients were derived for the strain rate of 0.01 s^{-1} according to the common procedure [17].

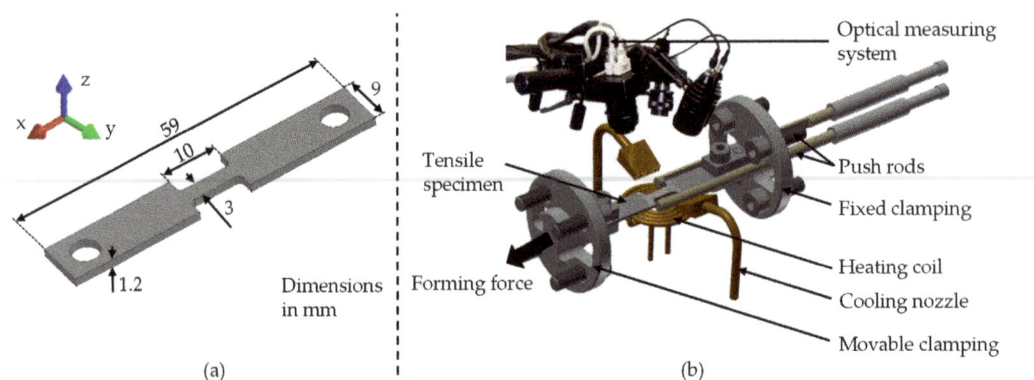

Figure 1. Miniature tensile test specimen geometry (**a**); and the experimental setup of the quenching and forming dilatometer with an optical measuring system (**b**).

2.2. Modelling of the Flow and Failure Behaviour

Since forming simulations often show plastic strains that are higher than the plastic strain of the uniaxial tensile test, extrapolation approaches are used to extrapolate flow curves up to the increased plastic strains. Furthermore, the approximation of the flow curve by an analytical approach is advantageous, since the FE calculation no longer requires time-consuming interpolation between the supporting points of the flow curve, but can be calculated efficiently using the analytical approach. A commonly used extrapolation approach is the Johnson–Cook hardening model, as shown in Equation (1) [18]:

$$k_{f,J\text{-}C} = \left(A + B\,\varepsilon_{pl}^{n}\right) \times \left(1 + C \times \ln \dot{\varepsilon}_{norm}\right) \times \left(1 - (T_{norm})^{m}\right) \quad (1)$$

$$\dot{\varepsilon}_{norm} = \dot{\varepsilon}_{pl} / \dot{\varepsilon}_{pl,0} \quad (2)$$

$$T_{norm} = (T - T_{room})/(T_{melt} - T_{room}) \quad (3)$$

where $k_{f,J\text{-}C}$ is the flow stress; A, B, C, m, and n are model parameters; ε_{pl} is the plastic strain; and $\dot{\varepsilon}_{norm}$ and T_{norm} are the nondimensional plastic strain rate and the nondimensional temperature, respectively. $\dot{\varepsilon}_{norm}$ is a function of the plastic strain rate $\dot{\varepsilon}_{pl}$ and the reference plastic strain rate $\dot{\varepsilon}_{pl,\,0}$, while T_{norm} is a function of the temperature T, the room temperature T_{room}, and the melting temperature of the material T_{melt}. The model was parameterised using the experimental flow curves and the least squares method. For a comparison of the material flow behaviour and a first validation, the tensile tests were simulated. The 3D FE simulation model and the associated boundary conditions are shown in Figure 2a.

The geometry of the specimen, material data, and the boundary conditions were implemented in Simufact Forming 16.0 (simufact engineering gmbh, 21079 Hamburg, Germany) Only the middle area of the specimen, which was the plastic deformed area in the experiments, was modelled using hexahedral solid-shell elements with five integration points over the thickness and an element edge length of 0.1 mm. Adaptive remeshing was used when the plastic strain exceeded a value of 0.4. For the material properties, the Johnson–Cook hardening model and the anisotropy parameters by means of the Hill'48 yield criterion [19] were used. As boundary conditions, the left nodes were fixed in all directions, and the right nodes were freely movable only in the x-direction, in which a displacement was applied with the same speed as in the experiment. The end of the simulation was defined by the end of the displacement in the x-direction, which was the average value of five experiments for the displacement at failure $u_{x,fail}$, as shown in

Figure 2b. The problem was solved using an implicit solver. Finally, the force–displacement curves from the experiments and simulations were compared.

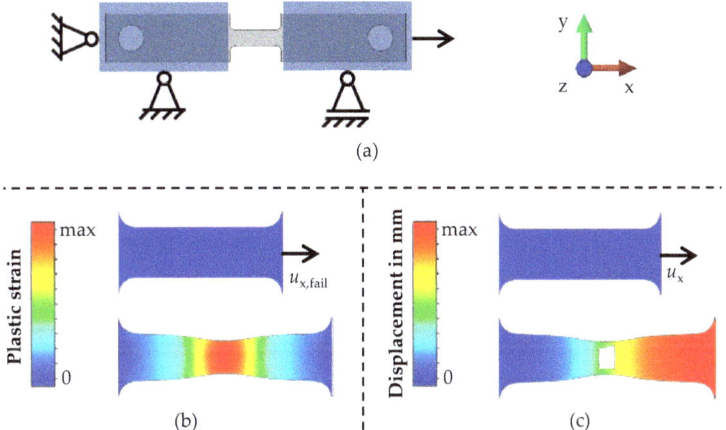

Figure 2. Geometry of the simulation model (**a**); and the boundary conditions for the comparison of the hardening model (**b**) and for the failure model (**c**).

To predict the material failure in sheet metals, usually forming limit curves are used [20]. Due to the deformation during friction drilling, however, a stress state is present, which is not represented by forming limit curves. Therefore, the Johnson–Cook failure model [18] was used as depicted in Equation (4):

$$\varepsilon_f = (D_1 \times D_2 \times \exp(-D_3 \times \eta)) \times \left(1 + D_4 \times \dot{\varepsilon}_{norm}\right) \times (1 + D_5 \times T_{norm}) \quad (4)$$

where ε_f is the failure strain, D_1 to D_5 are model parameters, and η is the triaxiality. In the prior FE simulation of the tensile tests, the triaxiality and the plastic strain were evaluated at the location and time of the material failure of the specimen, which both were determined in the experiments. From the plastic strain–triaxiality curve, the area-weighted triaxiality was calculated, and the maximum plastic strain was defined as the failure strain. To calculate the Johnson–Cook failure model parameters, both the area-weighted triaxiality and the failure strain were used when applying the least squares method. For another comparison of the material failure behaviour, the tensile tests were simulated using the prior 3D FE models including material damage. The Johnson–Cook failure model and element deletion were implemented. Further, the boundary conditions were changed as shown in Figure 2c. The displacement in the x-direction u_x was set to a high value, since the end of the simulation was determined by deletion of the first element. Finally, the force–displacement curves from the experiments and simulations were compared to evaluate the moment of failure prediction by the simulation.

2.3. Friction Drilling Experiments

For the experimental investigations, friction drilling tests were performed. The test setup was installed in a 5-axis DMU 100 Monoblock milling machine from DMG MORI(DMG MORI Aktiengesellschaft, 33689 Bielefeld, Germany), and consisted of a friction drill, a specimen with dimensions of 25 mm × 100 mm, and clamping and force-measuring plates. A schematic illustration of the experimental setup is shown in Figure 3a to clarify the resulting load path. For the experiments, a M4 friction drill and a separating agent from Flowdrill (Flowdrill Fließformwerkzeuge GmbH, 69469 Weinheim, Germany) were used. The tested rotational speed and feed rate were 5800 rpm and 50 mm/min. The resulting forming force while friction drilling was measured with the force-measuring plate. Measuring of the temperature on the bushing during the experiment was difficult due to the wide range of temperatures, the speed of the process, and the experimental setup. Therefore, thermocouples were welded on the specimens before testing to measure the temperature development. They were placed about 4 mm from the centre of the former. The tested specimens were made of the HX220 steel with a sheet thickness of 1.2 mm. To compare the influence of a higher sheet thickness and a higher material strength on the process, HX420 steel measuring 1.5 mm was also used in the friction drilling experiments. The forming was stopped at different strokes, such as 2 mm, 4 mm, 6 mm, and 8 mm. It was aimed to set the characteristic states entering (1) and breakthrough (2) of the conical part of the friction drill, as well as entering (3) and breakthrough (4) of the cylindrical part, as shown in Figure 3b. Therefore, it was possible to compare the material flow and resulting geometry of the bushing from the experiments to the simulation. Each experiment was performed five times to ensure statistical safety. Subsequently, the specimens were photographed from the outside with a VR-3200 3D profilometer from Keyence (KEYENCE DEUTSCHLAND GmbH, Neu-Isenburg, Germany). Further, a 3D profile of the upper and lower bushing was recorded with the 3D profilometer. Finally, the specimens were separated, mounted, grinded, and polished. The mounted separations also were microscopically examined with the 3D profilometer.

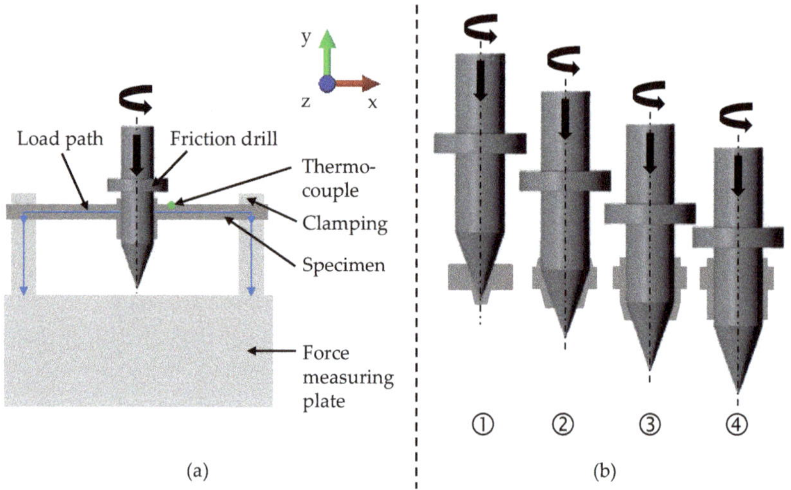

Figure 3. Schematic illustration of the experimental setup for friction drilling with force measurement (**a**) and states of the friction drilling (**b**): entering (**1**) and breakthrough (**2**) of the conical part of the friction drill; and entering (**3**) and breakthrough (**4**) of the cylindrical part.

2.4. Numerical Modelling of Friction Drilling

Due to the very complex physical phenomena of the friction drilling process, several assumptions were made in the numerical models. The contour of the friction drill with its forming studs was simplified to be rotationally symmetrical. Furthermore, the friction during the friction drilling process was a complex and changing condition, and the friction parameters were difficult to determine experimentally. Therefore, the friction was simplified using the Coulomb friction law. To create the numerical models, the effective part of the tools and the specimen were abstracted, prepared, and discretised with a finite element mesh. The boundary conditions were chosen according to real friction drilling tests. Therefore, an M4 friction drill with a rotational speed of 5800 rpm and feed rate of 50 mm/min was simulated. The contact between the specimen and the clamp tool was modelled as adhesive, and for friction between the specimen and the friction drill, a Coulomb friction coefficient of 0.3 was assumed. The characterised and modelled material behaviour of the specimen was used by implementing the Hill'48 yield criterion, the Johnson–Cook hardening model, and the failure model.

Lagrangian discretisation is a meshing method with a matter-fixed arrangement [21]. The model sections are divided as finite elements of the component. A movement is represented in the matter-fixed elements as a movement of those elements. Each element is a part of the model and moves with the model. Limitations of the range of motion are represented as contact surfaces, which may not be penetrated by the elements. Lagrangian discretisation is common for the FE method, and hence for applications such as solid-state computations and structural mechanics [22].

First in this study, a symmetrical 3D FE model of the friction drilling experiments was generated with the software Simufact Forming 16.0. Such 3D FE simulations are commonly used to model forming processes due to their accurate representation of reality [23]. With the high reality fidelity, the computation time increased massively. Figure 4a shows the geometry and the boundary conditions of the 3D FE model. The geometry consisted of the friction drill, the specimen, and a clamp tool. In preliminary simulations, the full geometry and different sections in 3D were investigated using axis symmetry to reduce the number of elements as much as possible. Finally, one-eighth of the geometry was modelled, because this offered the best ratio in terms of calculation time and result accuracy. The specimen was modelled as elastic-plastic, and the other parts as rigid bodies. Solid elements of the element type hexahedron with remeshing at plastic strains of 0.2 were used. The element edge length was 0.1 mm in the area of the bushing and 0.4 mm outside, also to reduce the computation time. In preliminary investigations, it was found that a coarser mesh in the area of the bushing led to a poorer reproduction of the friction drilling. An implicit scheme and an iterative solver were chosen for the computation.

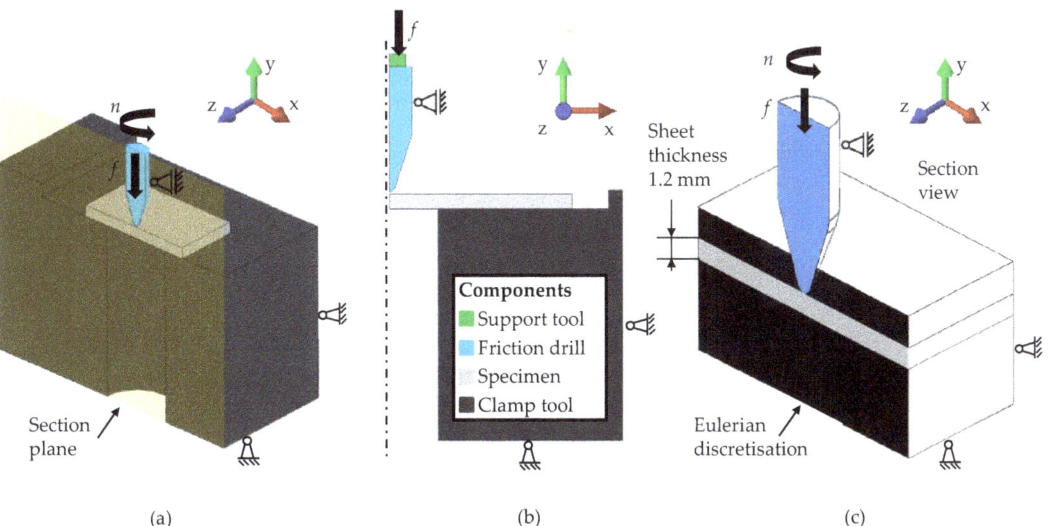

Figure 4. Geometry and boundary conditions of the numerical friction drilling models: section view of the 3D FE model (**a**); 2D FE model (**b**); and section view of the 3D FV model (**c**) with rotational speed n as well as feed rate f.

Second, a rotationally symmetrical 2D FE model of the friction drilling experiments was created with the software Simufact Forming 16.0. Using symmetry had the advantage of reducing the number of elements and shortening computation time, and is therefore also often used in the literature [24]. Hence, it was useful to study the suitability of the 2D FE simulation to reproduce friction drilling numerically. The geometry and the boundary conditions of the 2D FE model are depicted in Figure 4b. A further support tool was added in the simulation. The specimen was modelled as elastic-plastic, the clamp tool and the support tool as rigid, and the friction drill as an elastic deformable body. A quad element with an element edge length of 0.1 mm was used to discretise the specimen and the friction drill. The remesh criterion for the specimen was set to a plastic strain of 0.2. As in the 3D FE simulation, implicit calculation and an iterative solver were used. The friction welding module of Simufact Forming 16.0 was used to model the friction drilling. Due to the friction between the rotating bodies in friction welding processes, the joining zone is heated locally and the flow stress is reduced. Similar processes take place in friction drilling processes, and therefore the module was applied. In the friction welding module, the frictional heat was not calculated via the tool movement, but via a substitute model [25]. This allowed simulating such processes as 2D, since rotations are not describable in 2D. The substitute model described the generation of frictional heat q_{FR} as a function of conducted frictional work W_{FR} and its dissipation coefficient β_{FR}, as shown in Equation (5):

$$\mathrm{d}q_{FR(i)} = \beta_{FR} \times \mathrm{d}W_{FR(i)} = \beta_{FR} \times \tau_{R(i)} \times \omega \times r \times \mathrm{d}t \quad (5)$$

The incremental frictional work of a node i on the contacting surface can be described as a function of its axial distance from the rotation centre r, current frictional shear stress $\tau_{R(i)}$, rotational speed ω, and incremental time $\mathrm{d}t$.

Eulerian discretisation is a meshing method with a space-fixed arrangement [21]. The model sections are divided as a fixed network of elements in space. A movement is represented in the space-fixed elements as a balance of the quantity flowing in, the quantity flowing out and the quantity remaining in the elements. Limitations of the range of motion are represented as blocking of elements. A common method in Eulerian discretisation is the FV method, and therefore is used for applications such as fluid dynamics and flow processes. The CEL method has the possibility to couple an Eulerian formulation with a Lagrangian formulation to allow Eulerian and Lagrangian discretisation to interact within one model. The Lagrangian part can move through the Eulerian discretisation until it encounters an Eulerian material. Then, the contact between the different discretisations is calculated by a penalty contact method.

Third, in Abaqus/CAE, a 3D FV friction drilling model with CEL was built up. The geometry and the boundary conditions of the 3D FV model are illustrated in Figure 4c. It consisted of the friction drill, the specimen, and the Eulerian discretisation of the space around the specimen. The friction drill was modelled as elastic, and the specimen as elastic-plastic. For the Eulerian discretisation, which had a volume of 15 mm × 15 mm × 8 mm, solid elements of the element type hexahedron with an element edge length of 0.1 mm were used. The friction drill was discretised with solid tetrahedral elements with an element edge length of 0.5 mm and with a refinement of the mesh at the tip of the tool of 0.3 mm. A direct solver and explicit scheme were used to solve the model.

3. Results

3.1. Material Data

In Table 1, the chemical compositions of the investigated HX220 steel are given, as measured by spark spectroscopy. The determined chemical compositions were within the normed range specified by DIN EN 10268 [26].

Table 1. Chemical composition of HX220 in mass %.

C	Si	Mn	P	S
0.0357	0.0305	0.4350	0.0366	0.0069

The experimental flow curves of HX220 in 0° to the rolling direction are shown for plastic strain up to 0.2 in Figure 5a. In each case, it was the middle curve of the five recorded curves. As expected, the flow stress decreased with increasing forming temperature. With respect to strain rate dependence, no strong influence could be observed for the particular temperatures. However, the yield stress increased slightly with a higher strain rate. Temperature-dependent anisotropy coefficients with the standard deviation of the HX220 are depicted in Figure 5b. For all three investigated forming temperatures, the anisotropy values at 0°, 45°, and 90° to the rolling direction, as well as the mean perpendicular anisotropy coefficient, showed values significantly higher than 1. Therefore, anisotropic plastic behaviour was present, in which more material flowed from the width to the length than from the thickness under tensile load in the longitudinal direction. In this case, the sheet metal exhibited greater resistance to a reduction in sheet thickness.

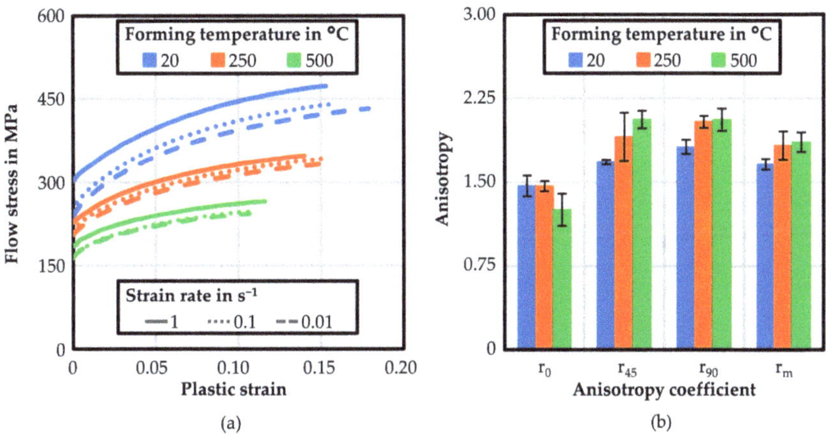

Figure 5. Experimental temperature and strain-ate- dependent flow curves at 0° to the rolling direction (**a**); and temperature-dependent anisotropy coefficients (**b**) for HX220 at a 1.2 mm sheet thickness.

The experimentally recorded flow curves were approximated and extrapolated using the Johnson–Cook hardening model. The resulting flow curves are shown in Figure 6a for a plastic strain up to two. The identified parameters of the Johnson–Cook hardening model are listed in Table 2. For a comparison of the material modelling, the tensile tests were simulated using the Johnson–Cook hardening model and Hill'48 yield criterion. Figure 6b shows a comparison of the forming force-displacement function of five experiments, and the numerical simulation exemplarily for a forming temperature of 20 °C and a strain rate of 1 s^{-1}. A very good agreement between the experimental results and the simulation was achieved. After the maximum forming force was reached, the simulated forming force decreased less strongly than in the experiments. This was because the calculation of the forming force in the simulation software did not take the damage evolution into account, which took place after the maximum forming force was reached and the necking occurred. Therefore, it was reasonable for the simulation to overestimate the experimental forces after the onset of necking.

The Johnson–Cook failure model was parameterised based on the simulations of the tensile tests. A three-dimensional representation of the parameterised Johnson–Cook failure model is shown in Figure 7a for a forming temperature of 20 °C varying the strain rate, and for a strain rate of 0.01 s^{-1} varying the forming temperature.

The failure strain decreased marginally with an increase of the strain rate from 0.01 s^{-1} to 1 s^{-1}. Increasing the forming temperature resulted in a large reduction of the failure strain. The calculated parameters are listed in Table 3 for HX220. For another comparison of the material modelling, the tensile tests were then simulated using the Johnson–Cook hardening and failure model. A comparison of the forming force-displacement function of five experiments and the numerical simulation exemplarily for a forming temperature of 20 °C and a strain rate of 1 s^{-1} is depicted in Figure 7b. The drop of the forming force in the simulation due to the deletion of the elements was in the range of the experiments. Overall, the simulation showed a good agreement with the failure model. Since the modelling of the material behaviour with the Johnson–Cook hardening and failure models matched well, the modelling approaches were used in the further simulations of the friction drilling process.

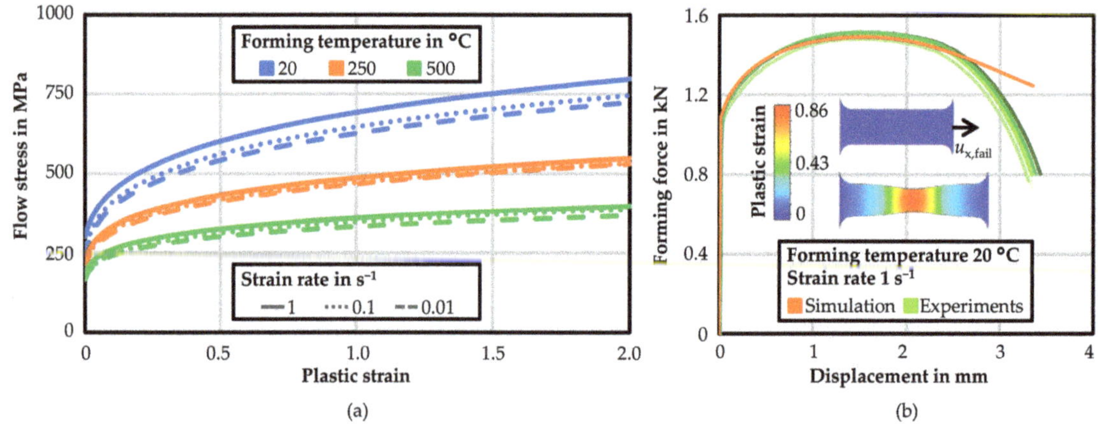

Figure 6. Extrapolated flow curves of HX220 for a plastic strain up to two (**a**); and comparison of the forming force–displacement curves from the experiments and simulation with average displacement at failure $u_{x,fail}$ (**b**).

Table 2. Parameters of the Johnson–Cook hardening model for HX220.

A in MPa	B in MPa	C	m	n	$\dot{\varepsilon}_{pl,0}$ in s^{-1}	T_{room} in °C	T_{melt} in °C
212.2	409.2	0	0.7521	0.3524	0.01	20	1500

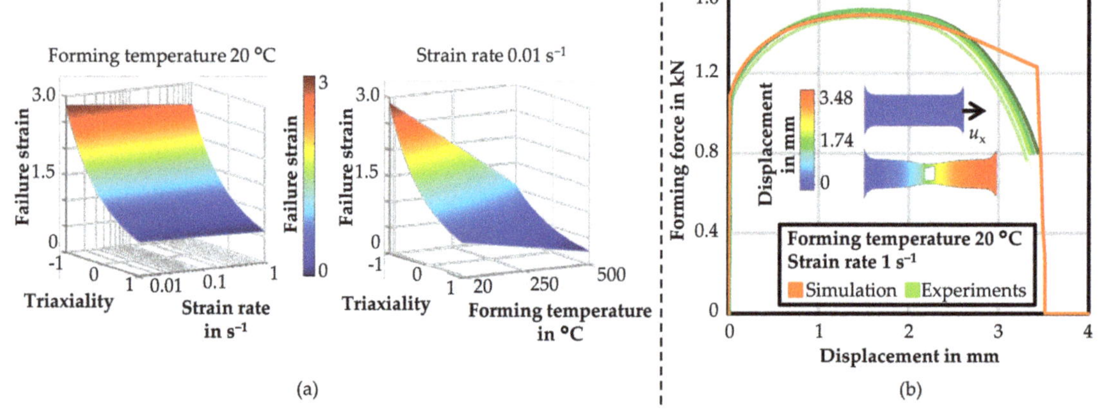

Figure 7. A 3D representation of the Johnson–Cook failure model for HX220 (**a**); and a comparison of the forming force–displacement curves from the experiments and simulation with displacement in x-direction u_x (**b**).

Table 3. Parameters of the Johnson–Cook failure model for HX220.

D_1	D_2	D_3	D_4	D_5
0.3846	0.8279	−1.0827	−0.0189	−1.9109

3.2. Experimental Results of Friction Drilling

Figure 8 illustrates the resulting bushings for HX220 and HX420 as optically measured 3D profiles. The upper bushing against the feed direction had a maximum length averaging of 0.72 mm for HX220 and 0.97 mm for HX420. The higher length of HX420 can be related to the higher sheet thickness of 1.5 mm compared to the thickness of 1.2 mm for HX220. A further cause of the higher length of HX420 was the higher tensile strength of 443 MPa in comparison to the tensile strength of 365 MPa for HX220. The gearlike structure of the upper bushings was due to the fact that the image was taken from above, and the small cracks in the upper bushing were projected onto the entire geometry. Lower bushings in the feed direction showed an inverse correlation. The maximum length for HX220 was on average 2.42 mm, and for HX420 slightly less than 2.38 mm. Cracks that appeared in the lower bushing of HX220 and not in HX420 can explain the cause of the almost equal lengths. Cracks indicated that the thickness of the lower bushing of HX220 was most likely reduced compared to HX420, and therefore increased the length of the lower bushing of HX220.

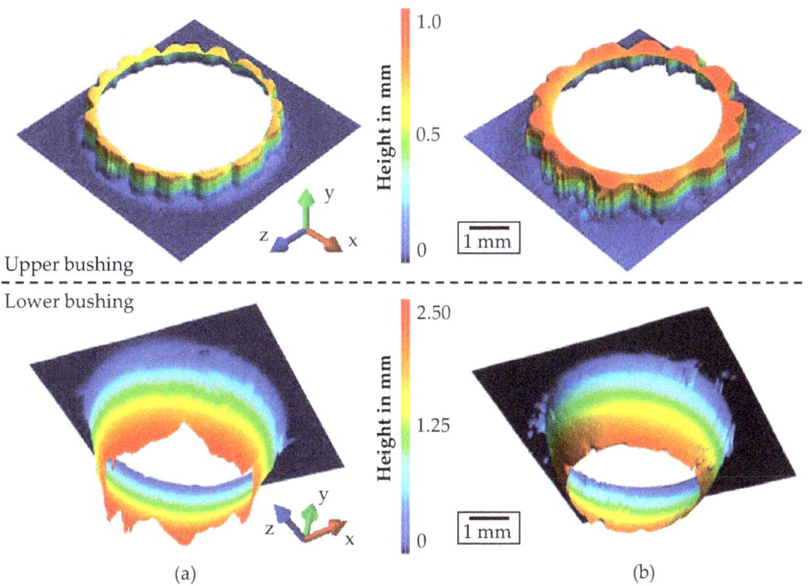

Figure 8. Optically measured 3D profiles of the upper and lower bushings for HX220 at a 1.2 mm sheet thickness (**a**) and HX420 at 1.5 mm (**b**).

Experimental force–time curves are shown in Figure 9a for HX220 and HX420, for three measurements each. The maximum force of HX220 was on average 0.71 kN, and for HX420 was 1.12 kN. Higher forces while friction drilling HX420 compared to HX220 were due to the higher sheet thickness and the higher tensile strength of the HX420. In addition to the higher maximum, the force–time curves of both materials showed the same progression. The force rose fast after contact of the drill with the specimen, then rose until the maximum force during forming and decreased quickly after the material was perforated. Figure 9b depicts the experimental temperature–time curves for HX220 and HX420 placed at 2 mm from the resulting bushing for three measurements. The diagram clearly shows that the temperature measurements were reproducible. A similar progression for both materials was visible. The maximum temperature was higher for HX420 than for HX220 due to the higher material strength of HX420, and due to the higher plastic work. In addition, the higher sheet thickness of HX420 resulted in a higher maximum temperature due to the increased contact area, since more material was in contact with the drill. The average maximum temperature for HX220 was 444.2 °C, and for HX420 was 519.2 °C. It was also noticeable that the temperature rose almost as fast for HX220, but fell faster than for HX420 due to the smaller sheet thickness.

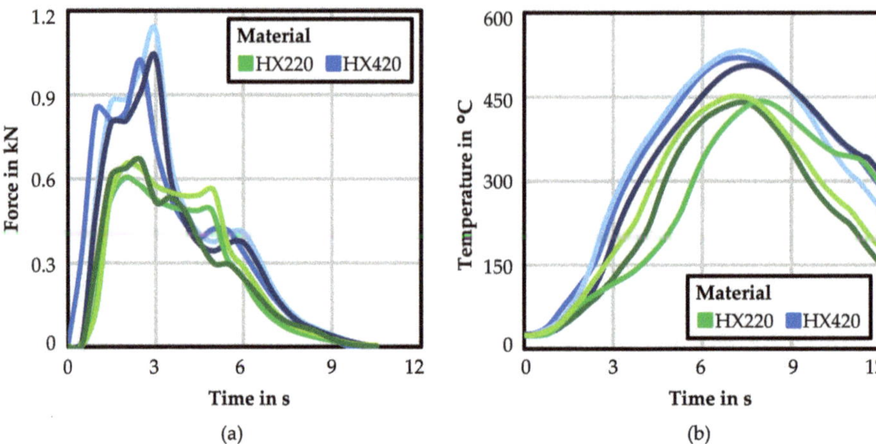

Figure 9. Experimental measured force–time curves (**a**) and temperature–time curves (**b**) for friction drilling HX220 at a 1.2 mm sheet thickness, as well as HX420 at 1.5 mm, for three repetitions each.

3.3. Numerical Results and Comparison to the Experiments

Figure 10 shows the results of the 3D FE simulation of a one-eighth geometry compared to the micrographs of the mounted separations for different strokes. Overall, the agreement between experiment and simulation was good for the different strokes. However, it was problematic that with increasing stroke, the plastic strain reached high values, the damage was accumulated, and thus many elements were removed. An even finer mesh in the area of the bushing did not improve the outcome and was computationally almost not practicable. For applications in which the upper and lower bushing were negligible, the simulation could be performed in 3D with FE. However, since the form of the bushing represented an important result variable in this study, further investigations in 2D with FE and in 3D with FV were carried out.

The results of all performed simulations and of the corresponding experiment are shown in Figure 11 as a comparison of the resulting geometry shapes of the bushing. Section views of the bushings were derived and measured digitally. The length of the bushing was measured from the upper bushing or from the upper side of the sheet to the lower bushing. While the diameter of the bushing was very similar for all simulations, the length of the upper and lower bushing differed greatly for the different simulation methods. Figure 11a,b depict the geometry shape from the experiment and the 3D FE simulation, as shown in Figure 10. The length of the bushing was 4.34 mm for the experiment and 3.35 mm for the 3D FE simulation. The rotationally symmetrical 2D FE model using the friction welding module from Simufact Forming 16.0 is shown in Figure 11c. It is clearly visible that no upper bushing was created while friction drilling, and the specimen was bent strongly, resulting in a shorter bushing length of 2.77 mm. For the numerical modelling of friction drilling, both the rotational and translational motion of the friction drill had to be modelled, and therefore a 3D FE simulation was necessary. Compared to rotational friction welding, the rotation also was required to be considered, and could not be simplified in a 2D FE simulation via an analytical approach for heat calculation. The 3D FV simulation in Figure 11d had a bushing with a 4.12 mm length, and was therefore the closest to the experiment. In addition, the thickness of the bushing was much more realistic than in the FE simulations. To prove the theory that both the rotational and translational motion of the friction drill had to be modelled, the 3D FV simulation was performed without the rotation boundary condition of the friction drill. The resulting bushing is illustrated in Figure 11e, and revealed similarities to the 2D FE simulation. The bushing for the 3D FV simulation without rotation was 2.05 mm.

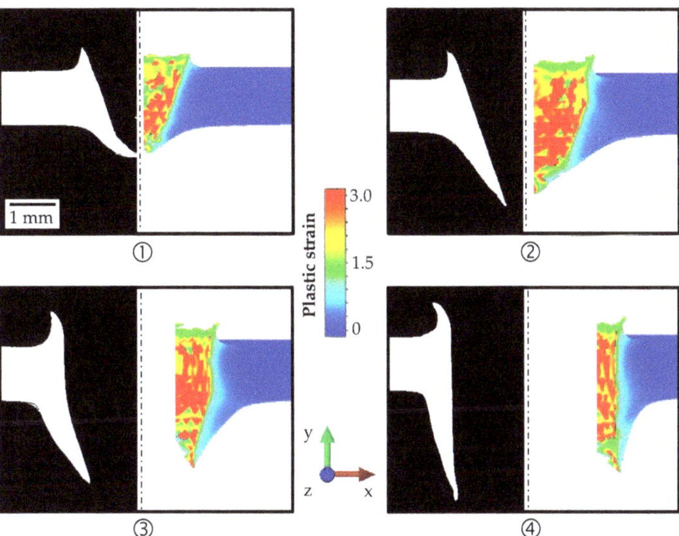

Figure 10. Comparison of the bushings for the experimental micrographs of the mounted separations and the 3D FE simulation at the states: entering (**1**) and breakthrough (**2**) of the conical part of the friction drill; and entering (**3**) and breakthrough (**4**) of the cylindrical part for HX220 at a 1.2 mm sheet thickness.

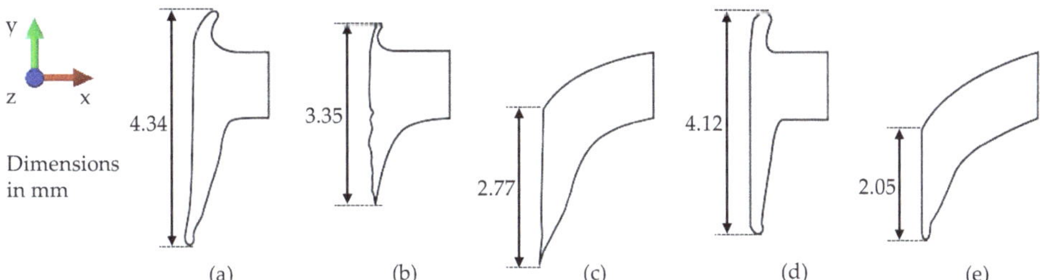

Figure 11. Geometry shapes of the resulting bushing for HX220 at a 1.2 mm sheet thickness: experiment (**a**); 3D FE (**b**); 2D FE (**c**); 3D FV (**d**); and 3D FV without rotation (**e**).

Figure 12 illustrates the results of the 3D FV simulation compared to the experimental micrographs for the different strokes of 2 mm, 4 mm, 6 mm, and 8 mm. In Abaqus/CAE, it was not possible to generate a section view of the contour plot for the 3D FV simulation. Therefore, the plastic strain is shown as seen from the outside. In order to ensure a better comparison with the experiment, micrographs of the specimens from outside were taken in addition to the micrographs of the mounted separations. The numerical results fit very well to the experimental investigations regarding the form of the upper and lower bushing at each stroke. High plastic strains were also achieved in the simulation with FV, but significantly fewer elements were removed compared to the 3D FE simulation, resulting in a better geometric approximation of the bushing.

A comparison of the temperature measurements from the experiments and the FV simulation are shown in Figure 13. In Figure 13a, the position of the spot-welded thermocouple (MP) and three measuring points (MP1–3) are depicted. The thermocouple was placed about 2 mm from the bushing's outer edge (i.e., approximately 4 mm from the centre of the former). In the simulation, the measuring points were positioned about 1 mm, 2 mm, and 3 mm from the bushing's outer edge (i.e., approximately 3 mm, 4 mm, and 5 mm from the centre of the former). The measurements did not ensure a full validation of the simulation, but a good estimation and verification was possible. In Figure 13b, the corresponding temperature–stroke curves from three experiments and from the FV simulation at the three measuring points are shown. In addition to the previous simulation with a friction coefficient of 0.3, the simulation was carried out with friction coefficients of 0.1 and 0.5 to study the influence of the coefficient. For the three friction coefficients, the development of the temperature in the simulation was similar. As expected, the maximum temperature increased with an increasing friction coefficient. The MP2 in the simulations with friction coefficients 0.1 and 0.3 agreed well with the experiments, whereby the temperature was slightly overestimated at 0.3 and underestimated at 0.1.

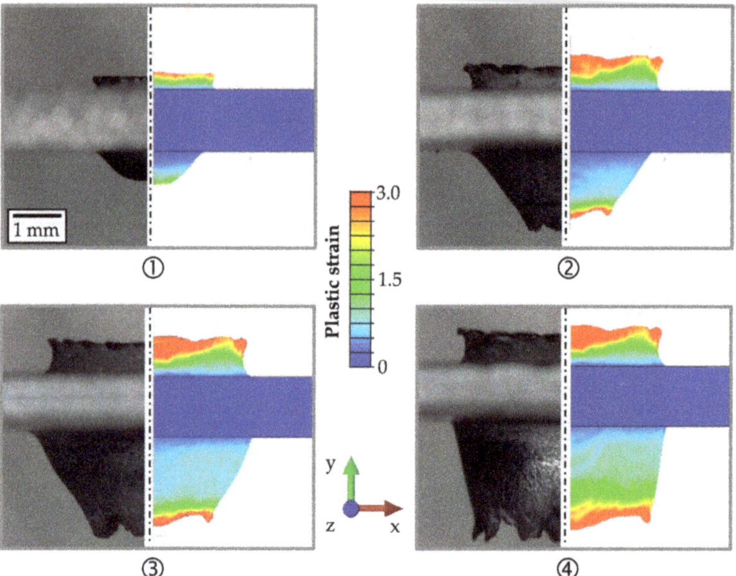

Figure 12. Comparison of the bushings for the experimental micrographs and the 3D FV simulation in the states: entering (**1**) and breakthrough (**2**) of the conical part of the friction drill; and entering (**3**) and breakthrough (**4**) of the cylindrical part for HX220 at a 1.2 mm sheet thickness.

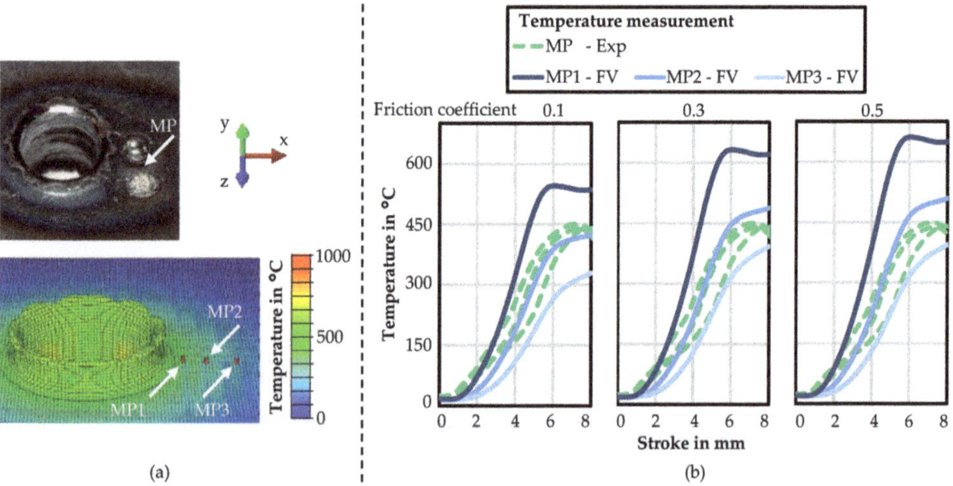

Figure 13. Temperature-measuring position in the experiments and in the FV simulation (**a**); and experimental measured and simulated temperature–stroke curves (**b**) for friction drilling HX220.

4. Discussion

Equivalent to the state of the art, the 3D FE simulation showed restriction in reproducing the experimental length of the bushing. In particular, the upper bushing was very short. For applications in which the geometry of the bushing is negligible, the simulation can be performed in 3D with FE. To improve the 3D FE simulation, a new model to split the mesh, such as node separation method, would also be helpful. Using 2D FE methods usually seems to be a time- and cost-saving option that still maintains good model quality. Nevertheless, it was not possible to model friction drilling in 2D with an adequate model quality compared to the other methods in this research. In addition to the analytical approach for generating the temperature, an analytical approach for calculating the rotation and its effects would be necessary for the 2D simulation. The bushing formation could be mapped well using the FV simulation in 3D and more realistic compared to the other methods. It also offers optimisation potential in terms of the bushing geometry. The development of cracks in the bushing could not be reproduced realistically with the current model.

5. Conclusions

This paper presented experimental und numerical investigations of friction drilling HSLA steel. Temperature, strain rate, and rolling direction dependent tensile tests of the HSLA HX220 were executed and used to parametrise the Johnson–Cook hardening and failure models. Friction drilling tests were performed and analysed for HX220 and for HX420. Further, the experiments of HX220 were numerically modelled using different methods. First, the process was modelled using FE in 3D, similar to most of the literature. Then, a 2D FE model was created using a welding module. Last, the FV method was used to reproduce and analyse the friction drilling process numerically.

Since the comparison of the simulations and the experiments showed good agreement, it could be assumed that the methods used for the material characterisation and modelling were appropriate. Based on the findings, we concluded that a more realistic material flow and geometry of the bushing was achieved by the FV simulation, rather than the 3D and 2D FE simulation. It is important to note that the integration of the rotation in the simulation was important for the correct mapping of the bushing geometry.

With the help of the improved modelling of the bushing geometry, it was possible to carry out numerical investigations regarding the bushing as an interlocking element for injection moulding in further work. An overmoulded bushing under load could be numerically mapped and compared with experiments so that further digital investigations can be carried out with verification. An investigation of the arrangement of bushings as a pattern and the effect on the bond strength would be of interest. Once a suitable bushing arrangement has been identified, the concept can be transferred to a simulated component and validated.

Author Contributions: Conceptualisation, B.-A.B., H.W. and E.S.; data curation, M.D. and E.S.; formal analysis, E.S.; funding acquisition, B.-A.B. and K.D.; methodology, E.S.; project administration, H.W. and A.H.; resources, H.W. and A.H.; software, H.W. and E.S.; comparison, M.D.; writing—original draft, E.S.; writing—review and editing, H.W., M.D., A.H., B.-A.B. and K.D. All authors have read and agreed to the published version of the manuscript.

Funding: This research was funded by the Federal Ministry for Economic Affairs and Energy on the basis of a decision of the German Bundestag. It was organised by the German Federation of Industrial Research Associations (Arbeitsgemeinschaft industrieller Forschungsvereinigungen, AiF) as part of the program for Industrial Collective Research (Industrielle Gemeinschaftsforschung, IGF) under grant number 20711N.

Institutional Review Board Statement: Not applicable.

Informed Consent Statement: Not applicable.

Data Availability Statement: The data presented in this study are available upon request from the corresponding author. The data are not publicly available due to industrial confidentiality.

Acknowledgments: The authors would like to express gratitude to the companies Flowdrill Fließformwerkzeuge GmbH and Volkswagen AG for providing the friction drills and the materials for the experimental investigations. Furthermore, the authors would like to thank the industrial partners in this research project for the scientific discussion. The Institute of Metals Science of the Leibniz University Hannover is thanked for performing the spark spectroscopy analysis.

Conflicts of Interest: The funders had no role in the design of the study; in the collection, analyses, or interpretation of data; in the writing of the manuscript; or in the decision to publish the results.

References

1. Dietrich, J. *Praxis in der Umformtechnik—Umform-und Zerteilverfahren, Werkzeuge, Maschinen*; Springer Vieweg: Wiesbaden, Germany, 2018.
2. Behrens, B.-A.; Jüttner, S.; Brunotte, K.; Özkaya, F.; Wohner, M.; Stockburger, E. Extension of the Conventional Press Hardening Process by Local Material Influence to Improve Joining Ability. *Procedia Manuf.* **2020**, *47*, 1345–1352. [CrossRef]
3. Stockburger, E.; Wester, H.; Uhe, J.; Brunotte, K.; Behrens, B.-A. Investigation of the forming limit behavior of martensitic chromium steels for hot sheet metal forming. In *Production at the Leading Edge of Technology*; Springer Vieweg: Berlin/Heidelberg, Germany, 2019; pp. 159–168. [CrossRef]
4. Miller, S.F.; Li, R.; Wang, H.; Shih, A.J. Experimental and Numerical Analysis of the Friction Drilling Process. *J. Manuf. Sci. Eng.* **2006**, *128*, 802–810. [CrossRef]
5. Miller, S.F.; Shih, A.J. Thermo-Mechanical Finite Element Modeling of the Friction Drilling Process. *J. Manuf. Sci. Eng.* **2007**, *129*, 531–538. [CrossRef]
6. Krasauskas, P.; Kilikevičius, S.; Česnavičius, R.; Pačenga, D. Experimental analysis and numerical simulation of the stainless AISI 304 steel friction drilling process. *Mechanika* **2014**, *20*, 590–595. [CrossRef]
7. Dehghan, S.; Ismail, M.I.S.; Ariffin, M.K.A.; Baharudin, B.T.H.T.; Sulaiman, S. Numerical simulation on friction drilling of aluminum alloy. *Mater. Sci. Eng. Technol.* **2017**, *48*, 241–248. [CrossRef]
8. Dehghan, S.; Ismail, M.I.S.; Souri, E. A thermo-mechanical finite element simulation model to analyze bushing formation and drilling tool for friction drilling of difficult-to-machine materials. *J. Manuf. Process.* **2020**, *57*, 1004–1018. [CrossRef]
9. Oezkaya, E.; Hannich, S.; Biermann, D. Development of a three-dimensional finite element method simulation model to predict modified flow drilling tool performance. *Int. J. Mater. Form.* **2019**, *12*, 477–490. [CrossRef]
10. Hynes, N.R.J.; Kumar, R. Simulation on friction drilling process of Cu_2C. *Mater. Today* **2018**, *5*, 27161–27165. [CrossRef]
11. Nimmagadda, S.; Balla, S.P.; Padmaja, A. 3D Finite Element Modeling and Simulation of Friction Drilling Process. *Int. J. Eng. Adv. Technol.* **2019**, *9*, 1070–1074. [CrossRef]
12. Bilgin, M.B.; Gök, K.; Gök, A. Three-dimensional finite element model of friction drilling process in hot forming processes. *J. Process Mech. Eng.* **2015**, *231*, 548–554. [CrossRef]
13. Wu, Y.; Wu, C.T.; Hu, W. Modeling of Ductile Failure in Destructive Manufacturing Processes Using the Smoothed Particle Galerkin Method. In Proceedings of the China LS-DYNA Users Conference, Shanghai, China, 23–25 October 2017; pp. 1–12.
14. Wu, C.T.; Wu, Y.; Hu, W.; Pan, Y. Modeling the Friction Drilling Process Using a Thermo-Mechanical Coupled Smoothed Particle Galerkin Method. In *Meshfree Methods for Partial Differential Equations IX*; Springer Nature Switzerland AG: Cham, Switzerland, 2019; Volume 129, pp. 149–166. [CrossRef]
15. Behrens, B.-A.; Brunotte, K.; Wester, H.; Stockburger, E. Investigation of the Process Window for Deformation induced Ferrite to improve the Joinability of Press-Hardened Components. In Proceedings of the 29th International Conference on Metallurgy and Materials, Brno, Czech Republic, 20–22 May 2020; pp. 579–584. [CrossRef]
16. DIN EN. *ISO 10275:2020: Metallic Materials—Sheet and Strip—Determination of Tensile Strain Hardening Exponent*; Beuth: Berlin, Germany, 2020.
17. DIN EN. *ISO 10113:2020: Metallic Materials—Sheet and Strip—Determination of Plastic Strain Ratio*; Beuth: Berlin, Germany, 2020.
18. Johnson, G.R.; Cook, W.H. Fracture Characteristics of three Metals subjected to various Strains, Strain Rates, Temperatures, and Pressures. *Eng.Fract. Mech.* **1985**, *21*, 31–48. [CrossRef]
19. Hill, R. A theory of the yielding and plastic flow of anisotropic metals. *Proc. R. Soc. A* **1948**, *193*, 281–297. [CrossRef]
20. Behrens, B.-A.; Uhe, J.; Wester, H.; Stockburger, E. Hot forming limit Curves for numerical Press Hardening Simulation of AISI 420C. In Proceedings of the 29th International Conference on Metallurgy and Materials, Brno, Czech Republic, 20–22 May 2020; pp. 350–355. [CrossRef]
21. Donea, J.; Huerta, A.; Ponthot, J.-P.; Rodriguez-Ferran, A. Arbitrary Lagrangian-Eulerian methods. In *The Encyclopedia of Computational Mechanics*; Wiley: Hoboken, NJ, USA, 2004; Chapter 14; Volume 1, pp. 413–437.
22. Behrens, B.-A.; Rosenbusch, D.; Wester, H.; Stockburger, E. Material Characterization and Modeling for Finite Element Simulation of Press Hardening with AISI 420C. *J. Mater. Eng. Perform.* **2021**, 1–8. [CrossRef]
23. Behrens, B.-A.; Brunotte, K.; Wester, H.; Rothgänger, M.; Müller, F. Multi-Layer Wear and Tool Life Calculation for Forging Applications Considering Dynamical Hardness Modeling and Nitrided Layer Degradation. *Materials* **2021**, *14*, 104. [CrossRef]
24. Behrens, B.-A.; Uhe, J.; Thürer, S.E.; Klose, C.; Heimes, N. Development of a Modified Tool System for Lateral Angular Co-Extrusion to Improve the Quality of Hybrid Profiles. *Procedia Manuf.* **2020**, *47*, 224–230. [CrossRef]

25. Furrer, D.; Semiatin, S.L. *ASM Handbook Volume 22A: Fundamentals of Modeling for Metals Processing*; ASM International: Novelty, OH, USA, 2009.
26. DIN EN. *10268:2013: Cold Rolled Steel Flat Products with High Yield Strength for Cold Forming—Technical Delivery Conditions*; Beuth: Berlin, Germany, 2006.

Article

Dynamic Softening and Hardening Behavior and the Micro-Mechanism of a TC31 High Temperature Titanium Alloy Sheet within Hot Deformation

Kexin Dang [1], Kehuan Wang [1,2,*] and Gang Liu [1,2]

1 National Key Laboratory for Precision Hot Processing of Metals, Harbin Institute of Technology, Harbin 150001, China; dangkexinhit@163.com (K.D.); gliu@hit.edu.cn (G.L.)
2 Institute of High Pressure Fluid Forming, Harbin Institute of Technology, Harbin 150001, China
* Correspondence: wangkehuan@hit.edu.cn; Tel.: +86-451-8641-8631

Abstract: TC31 is a new type of α+β dual phase high temperature titanium alloy, which has a high specific strength and creep resistance at temperatures from 650 °C to 700 °C. It has become one of the competitive candidates for the skin and air inlet components of hypersonic aircraft. However, it is very difficult to obtain the best forming windows for TC31 and to form the corresponding complex thin-walled components. In this paper, high temperature tensile tests were carried out at temperatures ranging from 850 °C to 1000 °C and strain rates ranging from 0.001 s^{-1} to 0.1 s^{-1}, and the microstructures before and after deformation were characterized by an optical microscope, scanning electron microscope, and electron back-scatter diffraction. The dynamic softening and hardening behaviors and the corresponding micro-mechanisms of a TC31 titanium alloy sheet within hot deformation were systematically studied. The effects of deformation temperature, strain rate, and strain on microstructure evolution were revealed. The results show that the dynamic softening and hardening of the material depended on the deformation temperature and strain rate, and changed dynamically with the strain. Obvious softening occurred during hot tensile deformation at a temperature of 850 °C and a strain rate of 0.001 s^{-1}~0.1 s^{-1}, which was mainly caused by void damage, deformation heat, and dynamic recrystallization. Quasi-steady flowing was observed when it was deformed at a temperature of 950 °C~1000 °C and a strain rate of 0.001 s^{-1}~0.01 s^{-1} due to the relative balance between the dynamic softening and hardening. Dynamic hardening occurred slightly with a strain rate of 0.001 s^{-1}. Mechanisms of dynamic recrystallization transformed from continuous dynamic recrystallization to discontinuous dynamic recrystallization with the increase in strain when it was deformed at a temperature of 950 °C and a strain rate of 0.01 s^{-1}. The grain size also decreased gradually due to the dynamic recrystallization, which provided an optimal forming condition for manufacturing thin-walled components with the desired microstructure and an excellent performance.

Keywords: titanium alloy; dynamic softening; dynamic hardening; recrystallization

Citation: Dang, K.; Wang, K.; Liu, G. Dynamic Softening and Hardening Behavior and the Micro-Mechanism of a TC31 High Temperature Titanium Alloy Sheet within Hot Deformation. *Materials* **2021**, *14*, 6515. https://doi.org/10.3390/ma14216515

Academic Editor: Sumantra Mandal

Received: 26 September 2021
Accepted: 25 October 2021
Published: 29 October 2021

1. Introduction

Titanium alloys are widely used in aerospace and other industrial fields due to their excellent properties, such as a high specific strength, corrosion resistance, and creep resistance [1–3]. In recent years, the demand for high-temperature titanium alloys has increased with the development of new generation aircraft aiming to be high-speed and lightweight [4,5]. The ultimate strength of the TC31 titanium alloy at 650 °C can reach more than 600 MPa [6,7]. Therefore, the TC31 titanium alloy meets the requirements of structural parts that are served short-time at 650 °C~700 °C such as the skins and air inlets of hypersonic aircrafts [2,5,6]. However, titanium alloys have high strength, poor ductility, and a low Young's modulus at room temperature, so it is usually necessary to heat titanium alloys to a certain temperature for hot forming, such as hot forging, hot pressing, superplastic forming, and hot gas pressure forming, etc. [8–11]. Hot gas pressure forming is a process that can form titanium alloy complex thin-walled components with a relatively high efficiency. However it is very challenging to control the deformation uniformity during the hot gas pressure forming process of thin-walled components, which may give rise to local thinning and cracking defects [12–15]. The deformation uniformity is affected by the coupling of dynamic hardening and dynamic softening during the hot deformation. The material flow behavior is complex and very sensitive to thermomechanical factors such as temperature, strain rate, and strain [16,17]; therefore, it is very challenging to accurately control the post-form microstructure and properties of high-temperature titanium alloy thin-walled parts.

Lots of studies have been performed on the hot deformation behavior and microstructure evolution of near-α titanium alloys. Hao et al. [18] studied the deformation mechanisms of the Ti-6Al-2Zr-1Mo-1V titanium alloy in a wide temperature range from −60 °C to 900 °C and found that the deformation mechanism was dislocation slip when the deformation temperature was below 600 °C, but the dynamic recrystallization and the spheroidization of lath α resulted in the flow softening and significantly decreased the flow stress when the deformation temperature was above 600 °C [19]. During the superplastic tensile deformation of the Ti55 titanium alloy at 925 °C and 0.00664 s^{-1} [20], the main mechanism was grain rotation accommodated by grain boundary sliding, and the softening during the hot tensile tests was mainly attributed to DDRX. During the isothermal multi-directional forging of the Ti-6.5Al-2.5Sn-9Zr-0.5Mo-1Nb-1W-0.25Si titanium alloy at the temperature range of 920 °C~1010 °C [21], the recrystallization mechanism changed from discontinuous dynamic recrystallization (DDRX) to continuous dynamic recrystallization (CDRX), and the grain size became smaller with the decrease in the deformation temperature. Wu et al. [22] pointed out that the hot tensile deformation of the TA32 high temperature titanium alloy rolled sheet at 700 °C~900 °C was affected by anisotropy, and the main dislocation slip types were cylindrical slip and conical slip along the rolling direction and transverse direction, respectively, but DDRX could effectively weaken the anisotropy at a higher deformation temperature. The flow softening mechanisms of the IMI834 high temperature titanium alloy during the isothermal compression deformation in the α+β two-phase region and the β single-phase region were dynamic recrystallization (DRX) and dynamic recovery (DRV), respectively [23]. When the deformation occurred at the β single-phase region, the cross slip and climb of dislocations easily occurred. The research on the hot deformation mechanisms of dual-phase titanium alloys has mainly concentrated on the TC4 titanium alloy [24]. At present, there are few reports on the hot deformation behavior and microstructure evolution of the new developed TC31 dual-phase high temperature titanium alloy.

In this paper, the hot deformation behaviors of the TC31 titanium alloy at 850 °C~1000 °C and 0.001 s^{-1}~0.1 s^{-1} were investigated firstly by uniaxial tensile tests, and the dynamic softening and hardening of the material were analyzed according to the obtained flow stress curves. Secondly, the microstructure evolution and mechanisms, especially the recrystallization behaviors, under different deformation temperatures, strain rates, and strains were systematically studied by scanning electron microscopy (SEM) and electron backscatter diffraction (EBSD). This work could provide some guidance for the forming of TC31 high temperature titanium alloy thin-walled components, and promote the industrial application of this newly developed titanium alloy.

2. Materials and Methods

2.1. Materials

The material used in this study was a TC31 high-temperature titanium alloy hot-rolled sheet with an average thickness of 1 mm, which was supplied by Baotai Group Co., Ltd. (Baoji, China). The chemical compositions obtained by PW4400 X-ray fluorescence spectrometer are listed in Table 1. The x-ray diffraction (XRD) analysis of the TC31 titanium alloy sheet was performed by an X'PERT PRO diffractometer (Malvern Panalytical Co. Ltd.,

Almelo, The Netherlands) and the results are shown in Figure 1a. The XRD patterns indicate that the TC31 titanium alloy consists of α phase and β phase. The initial microstructure of the hot-rolled sheet consists of a mostly equiaxed α phase, a few lamellar secondary α phase, and some strip and block shaped β phase. The β phase is distributed along the grain boundaries of the α phase, and the volume fraction of the β phase is about 20%, as shown in Figure 1b. The β transition temperature was confirmed to be 1025 °C by the metallographic method. The EDX elemental mapping of the TC31 titanium alloy is shown in Figure 1c. Regarding the TC31 titanium alloy, the α-stabilizing element includes Al, the β-stabilizing elements include Nb, Mo, W and Si, and the neutral elements include Sn and Zr [25]. It is clear that the distributions of neutral elements are uniform and the distributions of the remaining elements are non-uniform. The concentration of the Al element is higher in the α phase and the concentrations of Nb, Mo, W, Si elements are higher in the β phase, which is typical for titanium alloys. Nb and Mo are β isomorphous elements, so their concentration contrast in the α phase and the β phase is obvious; however, W and Si are β eutectoid elements and their concentration contrast in the α phase and the β phase are slight.

Table 1. Chemical composition of the TC31 titanium alloy.

Element	Al	Sn	Zr	Nb	Mo	W	Si	Ti
wt.%	5.21	3.45	3.47	1.26	1.24	0.39	0.20	Bal.

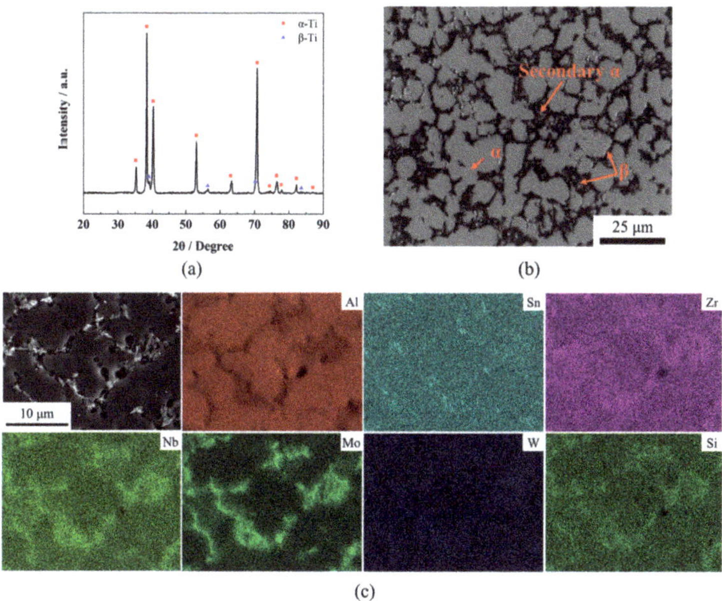

Figure 1. (**a**) The XRD of the TC31 titanium alloy; (**b**) microstructure of the as-received sheet; (**c**) the EDX elemental mapping of the TC31 titanium alloy.

2.2. *Uniaxial Tensile Tests and Microstructure Characterization*

The selection of the parameters was based on the hot gas pressure forming process for the TC31 titanium alloy. The forming parameters should ensure not only the formability but also the post-form properties. Regrading the TC31 titanium alloy, the peak stress at 850 °C is as high as 250 MPa at the strain rate of 0.01 s^{-1}, which is too high for the hot gas pressure forming process. Hence, the minimum tensile temperature was chosen as 850 °C. Titanium alloy thin-walled components are usually formed in the α+β two-phase zone in consideration of the post-form mechanical properties [2,9,13]. In this paper, the β transition temperature of the TC31 titanium alloy was confirmed to be 1025 °C by the metallographic method; therefore, the maximum tensile temperature was chosen as 1000 °C. The strain rate during the process of the hot gas pressure forming was about 0.01 s^{-1} [3,8,15]; therefore, the range of strain rate in this study was chosen from 0.001 s^{-1} to 0.1 s^{-1}.

Specimens for the high temperature tensile tests were machined by wire-cut electrical discharge machining from the as-received sheet with the tensile axis parallel to the rolling direction, and the dimensions are shown in Figure 2a. Firstly, the oxide layer of the TC31 titanium alloy was removed by acid pickling with a solution of 3% HF, 30% HNO_3, and 67% H_2O for 75 seconds before the tensile tests. Secondly, the surface of the sample was coated with an antioxidant of TO-12 glass protective lubricant supplied by Beijing Tianlichuang Company (Beijing, China) to prevent oxidation within hot deformation. Therefore, the oxidation state of the blank was neglected in this study. Uniaxial tensile tests were carried out on an INSTRON-5500R electronic universal material testing machine at 850 °C~1000 °C and 0.001 s^{-1}~0.1 s^{-1}. In order to obtain a uniform temperature distribution, the specimens were held at the preset temperature for 5 min before the tensile tests. In order to ensure the accuracy of the test results, each test was repeated three times. After the deformation, the sample was quenched with water immediately. Fixtures used in the tests to prevent the excessive deformation occurred at the clamping end of the thin plate samples which affect the accuracy of the test results during high temperature tensile tests at lower strain rates are shown in Figure 2b. Interrupted tensile tests at strains of 0.1, 0.3, 0.5, 0.7, and 0.9 were performed under different temperatures and strain rates to reveal the softening and hardening mechanisms of the TC31 titanium alloy within hot deformation.

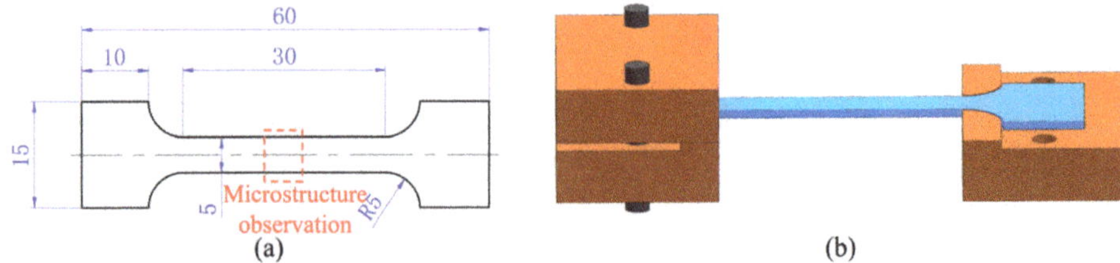

Figure 2. (**a**) Geometry dimensions of the tensile specimens (unit: mm); (**b**) schematic diagram for the clamping fixture.

The microstructures of the deformed specimens were characterized by Leica DMI3000M optical microscopy (OM) (Leica Microsystems Co. Ltd., Wetzlar, Germany), Quanta 200FEG field emission scanning electron microscopy (SEM) (FEI Systems Co. Ltd., Columbia, MD, USA), and electron backscatter diffraction (EBSD) (FEI Systems Co. Ltd.) respectively. The samples were taken from the center area (the red dotted box in Figure 2a). The samples were ground by 240#, 800#, 1200#, and 1500# metallographic sandpaper sequentially, and then electrically polished with a solution of 60% CH_3OH, 34% $C_4H_{10}O$, and 6% $HClO_4$ for 60 seconds. The samples used for OM and SEM also needed to be etched and the etching solution was composed of 13% HNO_3, 7% HF, and 80% H_2O. The EBSD characterization was performed at a step size of 0.3 μm and the test results were evaluated with OIM 6.1.4 software.

3. Results and Discussion

3.1. Dynamic Softening and Hardening Behavior

The true stress–true strain curves of the TC31 high-temperature titanium alloy under different deformation conditions are shown in Figure 3. It can be seen that the flow stress curves of the TC31 titanium alloy were similar when the deformation temperature was from 850 °C to 900 °C and the strain rate was from 0.001 s^{-1} to 0.1 s^{-1}. At the initial stage of deformation, the flow stress increased rapidly to a peak stress, and then decreased with the increasing strain. The deformation can be rapidly transferred and diffused when the strain was low; therefore, no localized necking occurred [26]. Regarding the curve of 850 °C in Figure 3c, it is well known that the ductility decreases with the increase in the strain rate and the decrease in the deformation temperature [16,17]. When the TC31 titanium alloy was deformed at a lower temperature of 850 °C and a higher strain rate of 0.1 s^{-1}, a fracture occurred before the true strain reached to 0.6; therefore, the curve is different to the rest of the curves. It should be noted that the other curves were interrupted at a strain of 0.6. When the deformation temperature increased from 850 °C to 900 °C, the flow stress decreased significantly with a maximum reduction of 54.5%.

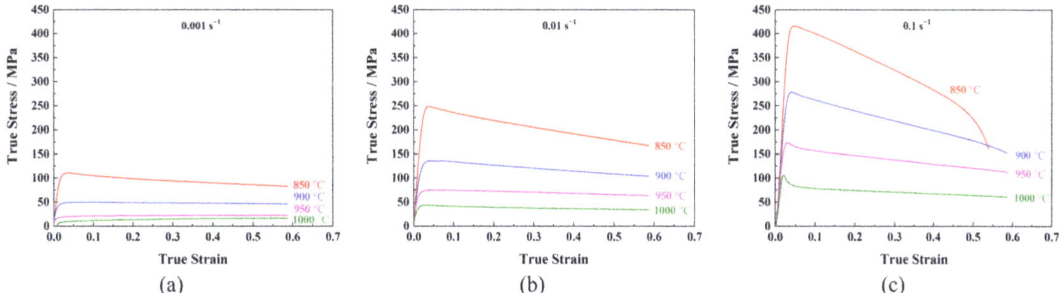

Figure 3. The flow behavior of the TC31 titanium alloy at different strain rates: (**a**) 0.001 s^{-1}; (**b**) 0.01 s^{-1}; (**c**) 0.1 s^{-1}.

The flow stress was almost steady with the increasing strain when the deformation temperature was from 950 °C to 1000 °C and the strain rate was from 0.001 s^{-1} to 0.01 s^{-1}, indicating that the plastic deformation was relatively uniform under such conditions [27]. Therefore, it is suggested that the hot forming temperature of this alloy should be above 900 °C. When the deformation temperature was close to the β transition temperature which is 1025 °C for the TC31 titanium alloy, more α phase with a hexagonal close-packed structure would transform into the β phase with a body-centered cubic structure under lower strain rates condition [28]. The β phase has more slip systems and a higher stacking fault energy than the α phase, which contributes to a more uniform deformation.

The flow stress curves of the TC31 titanium alloy showed a discontinuous yielding phenomenon [29,30]. When the strain rate was 0.1 s^{-1}, the flow stress reached a high peak value firstly at a lower strain condition, then the flow stress decreased rapidly. However, the flow stress decreased slowly with the increase in the strain after the lower yield point, showing the characteristics of continuous softening. The discontinuous yielding phenomenon became more and more obvious with the increase in the deformation temperature and strain rate [30,31], as shown in Figure 3c. One possible explanation for the stress drop was the stress-induced phase transformation (α→β) behavior which was observed during the hot deformation of the TC11 titanium alloy with α+β phase [30].

The peak stress of the TC31 titanium alloy increased with the increase in the strain rate and the decrease in the deformation temperature. When the strain rate was 0.1 s^{-1}, the peak stress decreased sharply with the increase in the temperature. However, when the strain rate was 0.001 s^{-1}, the values of the peak stress at 950 °C and 1000 °C were 23.2 MPa and 16.9 MPa, respectively. It can be inferred that the peak stress tended to be stable when the deformation temperature was higher than 950 °C at the strain rate of 0.001 s^{-1}, as shown in Figure 4.

To further analyze the evolution of the flow stress during the hot deformation of the TC31 titanium alloy, the values of work hardening rates at a constant deformation temperature and strain rate were calculated from the true stress–true strain curves by the following formula [32]:

$$\theta = \frac{\partial \sigma}{\partial \varepsilon} \quad (1)$$

where σ is the true stress and ε is the true strain. The variations between work hardening rate and hot deformation conditions are shown in Figure 5. The work hardening rate decreased rapidly with the increase in the strain at a lower strain condition (<0.05) as shown in Figure 5a,c, which can be attributed to the dynamic recovery of the TC31 titanium alloy [16,33]. Afterward, the variations in the work hardening rate were slight with the increase in the strain, as shown in Figure 5b,d. It can be seen from Figure 5a,b that the strain rate had a significant effect on the work hardening rate at 950 °C. The value of the work hardening rate was positive when the strain rate was 0.001 s^{-1} as shown in Figure 5b and the material was dynamically hardened. However, the work hardening rate changed from positive to negative when the strain rate increased from 0.001 s^{-1} to 0.1 s^{-1}, demonstrating that the dynamic softening of the flow stress occurred during the deformation with higher strain rates. The dynamic softening phenomenon was more obvious when the strain rate was higher [17,27]. Particularly, the work hardening rate decreased firstly and then increased with the strain ranging from 0.03 to 0.06 and a strain rate of 0.1 s^{-1}, which was attributed to the discontinuous yielding phenomenon [30,31]. The work hardening rate increased with the decrease in the deformation temperature at lower strain conditions as shown in Figure 5c, because much less recovery occurred at 850 °C than 1000 °C. However, with the increasing strain, the work hardening rate decreased with the decrease in the deformation temperature as shown in Figure 5d, which may be because of the grain growth with the increase in the deformation temperature [16,27].

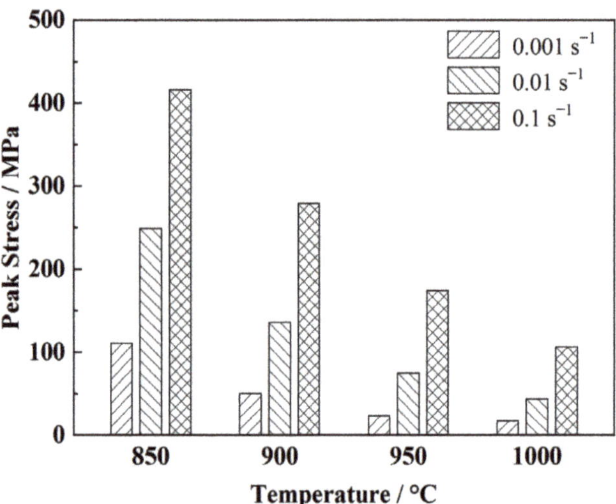

Figure 4. Peak stress at different temperatures and strain rates.

In order to quantitatively analyze the degree of dynamic softening during high temperature deformation, the flow softening coefficient ($M_{s-0.5}$) was introduced and can be calculated from the following formulas [34,35]:

$$M_{s-0.5} = \Delta\sigma/\sigma_p \tag{2}$$

$$\Delta\sigma = \sigma_p - \sigma_{0.5} \tag{3}$$

where σ_p is the peak stress and $\sigma_{0.5}$ is the flow stress at the strain of 0.5. The hot deformation conditions had a significant effect on the flow softening coefficient of the TC31 titanium alloy as shown in Figure 6. The dynamic softening occurred in the region of higher strain rates (0.01 s^{-1} to 0.1 s^{-1}) and the flow softening became more severe with the decrease in the deformation temperature. The values of the flow softening coefficients decreased with the decreasing strain rate and the increasing deformation temperature. Specifically, the values of the flow softening coefficients changed into negative values when the deformation temperature was 950 °C~1000 °C and the strain rate was 0.001 s^{-1}, which indicated the occurrence of dynamic hardening, but the minimum value was only −0.026 and the degree of dynamic hardening was relatively slight.

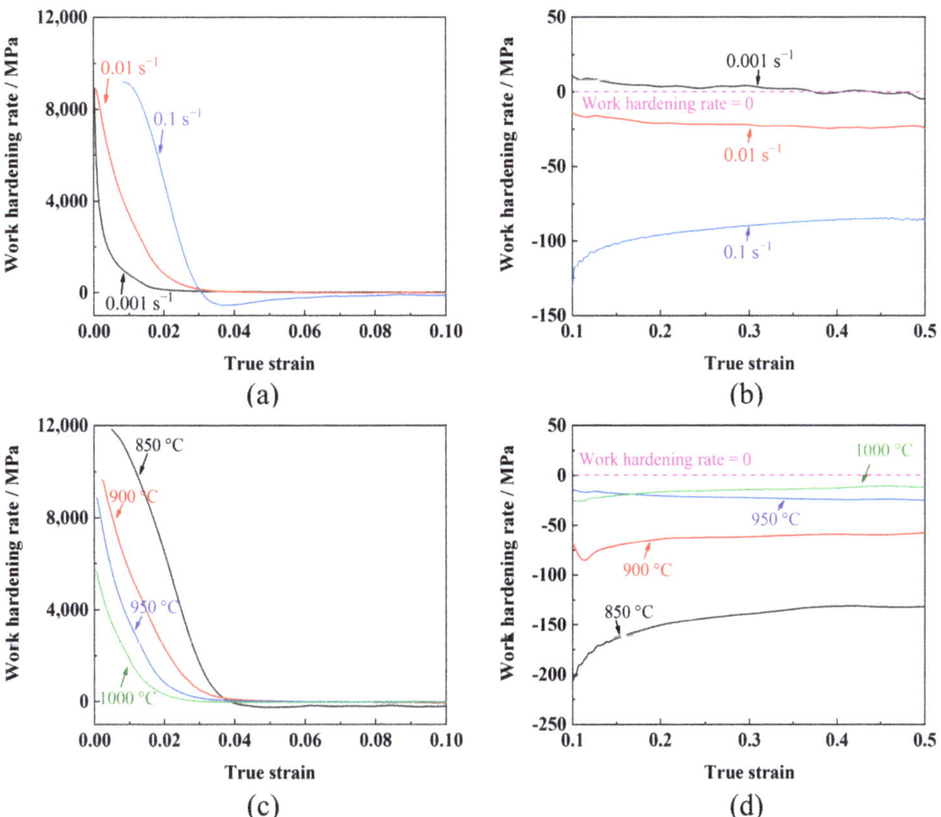

Figure 5. Work hardening rate at different deformation conditions: (**a**,**b**) 950 °C; (**c**,**d**) 0.01 s^{-1}; ((**a**,**c**) are lower strain conditions; (**b**,**d**) are relatively higher strain conditions).

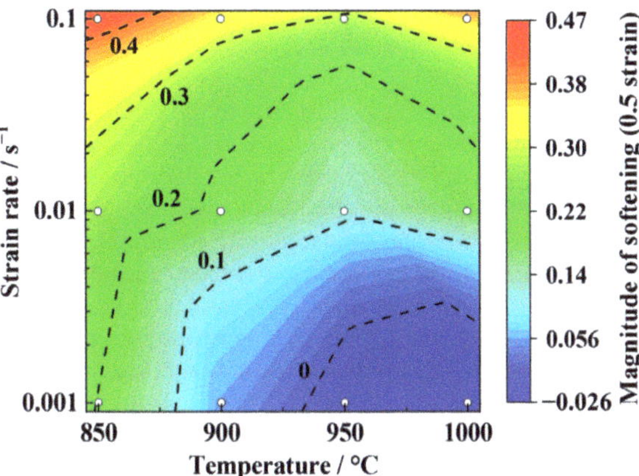

Figure 6. Flow softening coefficients at different temperatures and strain rates.

Compared with aluminum alloys, the thermal conductivity of titanium alloys is considerably lower and the rise in temperature caused by deformation heat cannot be neglected. It is very challenging to accurately measure the value of the rise in temperature during the process of high temperature tensile deformation. Therefore, the prediction for the rise in temperature during the hot tensile deformation is very important and helpful to understand the flow behavior of titanium alloys. The amount of the temperature increment (ΔT) can be estimated from the following formulas [36–38]:

$$\Delta T = \frac{p}{C}\int_0^{\varepsilon} \sigma d\varepsilon \tag{4}$$

$$p = 1 + \frac{\dot{\varepsilon}}{3} \tag{5}$$

where C is the heat capacity coefficient and the value of the TC31 titanium alloy is about 4 N·mm^{-2}·°C^{-1}, σ is the true stress, ε is the true strain, and p is the thermal transfer coefficient and is related to the strain rate. It can be concluded that the ΔT at 850 °C and 900 °C was much greater than that at 950 °C and 1000 °C because the flow stress decreased rapidly with the increase in the deformation temperature, particularly under the higher strain rate conditions. Therefore, the deformation heat will intensify the dynamic softening of the titanium alloy under the lower temperature conditions [35,38].

3.2. *Micro Mechanisms of Dynamic Softening and Hardening*

Micro variables such as grain size, dislocation density, grain boundary, recrystallization, phase volume fraction, phase morphology, and voids in the titanium alloy are closely related to the high temperature deformation conditions which include the deformation temperature, strain rate, and strain [16,32,38,39]. The evolution of those micro variables will determine the dynamic softening and hardening of the flow stress curves. To fully reveal the micro mechanisms of dynamic softening and hardening during the deformation, careful microstructure characterization and analysis are necessary.

3.2.1. The Effect of Deformation Temperature

The deformation temperature had a significant effect on the flow stress of the TC31 titanium alloy during high temperature deformation. The TC31 titanium alloy showed an obvious softening phenomenon when it was deformed at 850 °C, but the softening phenomenon gradually weakened with the increase in the deformation temperature. When the deformation temperature was from 950 °C to 1000 °C and the strain rate was 0.001 s^{-1}, slight dynamic hardening occurred during the deformation process. Figure 7 shows the evolution of the microstructure with increasing strain at 850 °C and 950 °C at a strain rate of 0.01 s^{-1}. It is well known that the β phase with a body-centered cubic structure has more slip systems than the α phase with a hexagonal close-packed structure; therefore, the deformation of the β phase is easier than the α phase and the β phase can withstand greater plastic deformation [20,40]. In the process of plastic deformation, the amount of the deformation for the α phase and the β phase is different, and the stress concentration will occur at the interface between the α phase and the β phase [1,18,28]. When the deformation temperature was 850 °C, the volume fraction and morphology of the β phase had fewer differences compared with the initial microstructure and the relatively large stress concentration that occurred at the phase boundaries with the increase in the strain. The volume fraction of the voids was about 11.2% when the true strain was 0.7. The appearance and growth of voids led to the rapid reduction in the flow stress and the softening of the TC31 titanium alloy, which was more obvious under lower temperature conditions [26].

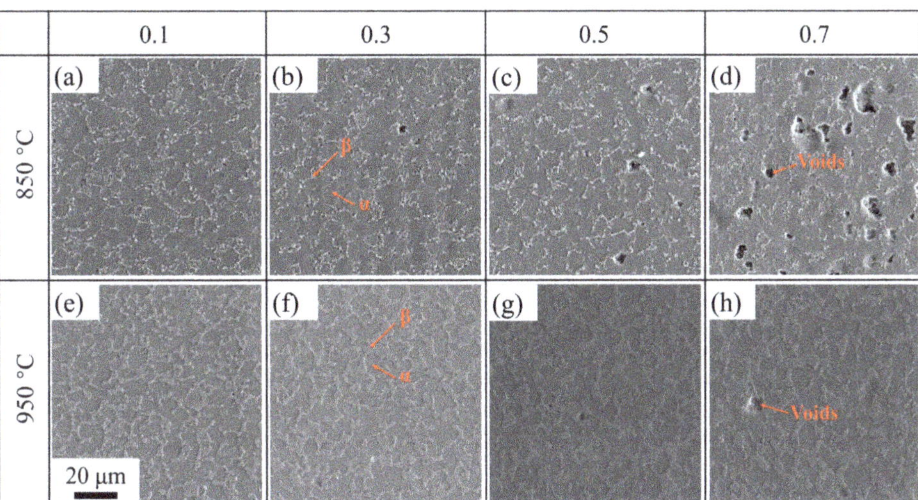

Figure 7. Microstructure at different temperatures and strains when deformed at 0.01 s^{-1}: (**a–d**) 850 °C; (**e–h**) 950 °C; (**a**,**e**) strain 0.1; (**b**,**f**) strain 0.3; (**c**,**g**) strain 0.5; (**d**,**h**) strain 0.7.

When the deformation temperature increased from 850 °C to 950 °C, the volume fraction of the β phase increased from 20% to 41% at the strain of 0.7 and the size of the β phase also had a significant coarsening. It is well known that the sliding resistance of the α/β phase interface is lower than that of the α/α interface and the β/β interface [40,41]. Therefore, the increase in the proportion of the α/β phase interface which is attributed to the rise of the volume fraction of β phase is beneficial to grain boundary sliding, and the deformation consistency is also improved simultaneously [22,24,41]. At the same time, the recovery and recrystallization at 950 °C was more efficient than those at 850 °C, which led to a lower dislocation density near the phase and grain boundaries [18,39,40]. Hence, the volume fraction of the voids was less than 1% when the deformation temperature was 950 °C and the strain was 0.7, and the flow stress was almost steady during the deformation, as shown in Figure 3b.

3.2.2. The Effect of the Strain Rate

When the TC31 titanium alloy was deformed at 950 °C, dynamic hardening occurred at a strain rate of 0.001 s^{-1} and dynamic softening occurred at the strain rates from 0.01 s^{-1} to 0.1 s^{-1}. The dynamic softening was more obvious with the higher strain rate, indicating that the strain rate had a significant effect on the evolution of the microstructure.

Figure 8 shows the distribution of the grain boundary at different strain rates, where the blue, green, and red lines represent high-angle grain boundaries (HAGBs), medium-angle grain boundaries (MAGBs), and low-angle grain boundaries (LAGBs), respectively. When it was deformed at the strain rate of 0.001 s^{-1}, the fraction of LAGBs decreased rapidly and the average grain size increased significantly, as shown in Figure 8a,d. Compared with the initial grain size of 2.80 μm, the grain sizes at the strain of 0.5 and 0.7 increased to 3.08 μm and 3.66 μm, respectively. When it was deformed with the strain rate of 0.1 s^{-1}, the fraction of LAGBs decreased from 61.8% to 49.3% at the strain of 0.5 and the grains were also refined compared with the initial microstructure, as shown in Figure 8c. The average grain sizes at the strain of 0.5 and 0.7 were 2.19 μm and 1.87 μm, respectively. When the strain rate was 0.01 s^{-1}, the refinement was slight and the average grain sizes were 2.52 μm and 2.39 μm at the strain of 0.5 and 0.7, respectively.

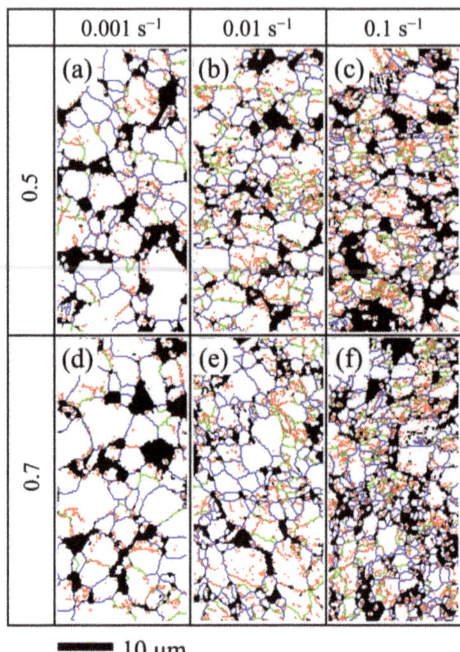

Figure 8. The distributions of the grain boundary at different strain and strain rates when deformed at 950 °C: (**a–c**) strain 0.5; (**d–f**) strain 0.7; (**a**,**d**) strain rate 0.001 s^{-1}; (**b**,**e**) strain rate 0.01 s^{-1}; (**c**,**f**) strain rate 0.1 s^{-1} (red lines represent LAGBs less than 5°, green lines represent MAGBs between 5°~15°, blue lines represent HAGBs more than 15°).

It can be summarized that during the deformation at 950 °C, the average grain size decreased with the increase in the strain rate at the strain of 0.5 or 0.7. At the lower strain rate of 0.001 s^{-1}, the grains appeared almost equiaxed with a few sub-grains and the sub-grains reduced rapidly compared with the initial microstructure [20]. The obvious reduction of the sub-grains could be attributed to DRX [42]. With the progress of DRX, the migration of HAGBs could consume the dislocation and reduce the sub-grains [36,39]. Fully DRX was completed before the strain reached 0.5 under 0.001 s^{-1} condition and then grain growth occurred with the increase in the strain. It should be noted that the recrystallized grains grew up significantly under the strain rate of 0.001 s^{-1}; therefore, the softening effect caused by DRX decreased with the decreasing strain rate and the flow stress exhibited dynamic hardening. Similar dynamic hardening was also reported during the superplastic deformation of the TC4 titanium alloy [24,34]. It can be seen from Figure 8a–f that the amount of the recrystallized grains without LAGBs inside increased and the average grain size decreased with the increasing strain rate. At a higher strain rate of 0.1 s^{-1}, the deformation inhomogeneity became more obvious. The deformation energy was higher around the grain boundaries than that at other sites [11,28]; therefore, most of the recrystallized grains nucleated near the grain boundaries. The deformation time of the sample deformed at a strain rate of 0.1 s^{-1} was too short for the growth of new recrystallized grains [24]. Therefore, the flow stress exhibited dynamic softening under the higher strain rate conditions.

The high-density dislocations around the grain boundaries could provide favorable sites for the nucleation of recrystallization [19,28]. In order to further analyze the effect of the strain rates on the deformation mechanism of the TC31 titanium alloy, the Kernel Average Misorientation (KAM) maps under different deformation conditions are shown in Figure 9. When the strain rate was 0.001 s^{-1}, the LAGBs had enough time to evolve into HAGBs leading to the decrease in the dislocation density. The mean value of KAM decreased rapidly from 1.069 at the initial stage to 0.735 at the strain of 0.5. When the strain was further increased to 0.7, the mean value of KAM was 0.716 and the reduction was not significant. It can be seen from Figure 9a,e that the dislocation density was at a lower level with a strain rate of 0.001 s^{-1} and the strain of 0.5 and 0.7, which would hinder the nucleation of dynamic recrystallization [24]. Therefore, the main deformation mechanism may have been grain boundary sliding and the grain size increased gradually [20,24]. The mean value of KAM decreased slightly with the increase in the strain when the strain rate was 0.01 s^{-1}. The mean values of KAM were 0.917 and 0.824 at the strain of 0.5 and 0.7, respectively. When the strain rate was 0.1 s^{-1}, only a few seconds were needed for the strain to reach to 0.5, and the mean values of KAM at the strain of 0.5 and 0.7 were 0.975 and 0.958, respectively. It can be seen from Figure 9c,d,g that the dislocation density remained almost steady with the increasing strain under the higher strain rate condition of 0.1 s^{-1}; therefore, the nucleation sites for dynamic recrystallization were abundant [28] and the dynamic softening of the TC31 titanium alloy was obvious.

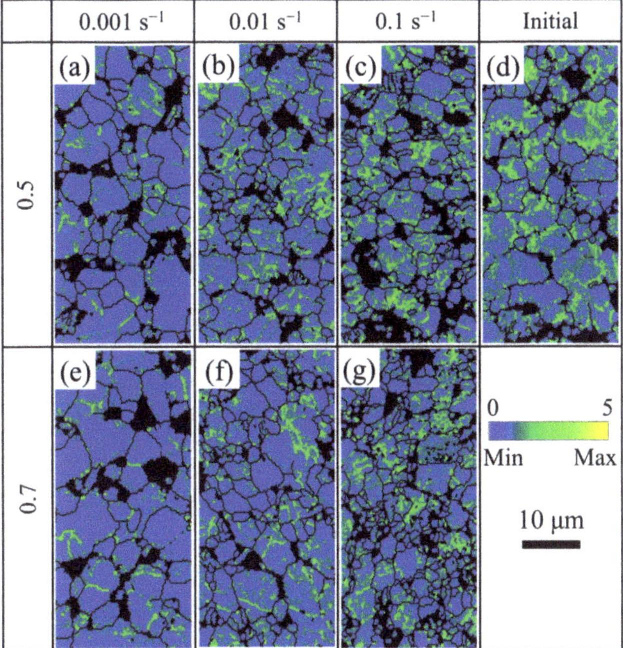

Figure 9. The distributions of KAM at different strains and strain rates when deformed at 950 °C: (**a–c**) strain 0.5; (**d**) initial; (**e–g**) strain 0.7; (**a**,**e**) strain rate 0.001 s^{-1}; (**b**,**f**) strain rate 0.01 s^{-1}; (**c**,**g**) strain rate 0.1 s^{-1}.

3.2.3. The Effect of Strain

The flow softening coefficient was only 0.12 when the deformation temperature was 950 °C and the strain rate was 0.01 s^{-1}, indicating that the dynamic softening phenomenon caused by recrystallization was very slight [8]. Figure 10 shows the evolution of recrystallization at different strains, where the blue area is the recrystallized grain. It can be seen from Figure 10a that the volume fraction of recrystallization of the as-received sheet was about 8.1%, indicating that the as-received sheet had not been annealed sufficiently. The volume fraction of recrystallization increased to 9.5% after the sample was soaked at 950 °C for 5 min as shown in Figure 10b. It can be concluded that the effect of static recrystallization was limited because of the short holding time [1]. The volume fractions of recrystallization were 12.3%, 22.8%, and 41.7% at the strain of 0.1, 0.5, and 0.9, respectively. The results show that the degree of recrystallization increased rapidly with the increase in strain when the strain rate is 0.01 s^{-1}. Because the strain rate was relatively higher, and the time during the deformation process at 950 °C was relatively shorter, the grain sizes decreased from 2.80 μm at the initial stage to 2.13 μm at the strain of 0.9 gradually. The relationships between the recrystallization fraction, the grain size, and the true strain are shown in Figure 11.

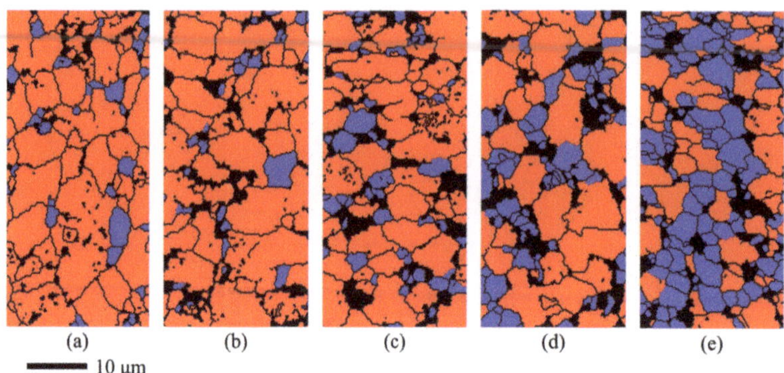

Figure 10. The distributions of recrystallization at different strains when deformed at 950 °C and 0.01 s^{-1}: (**a**) Initial; (**b**) 0; (**c**) 0.1; (**d**) 0.5; (**e**) 0.9 (blue represents the recrystallized grains, red represents the deformed grains).

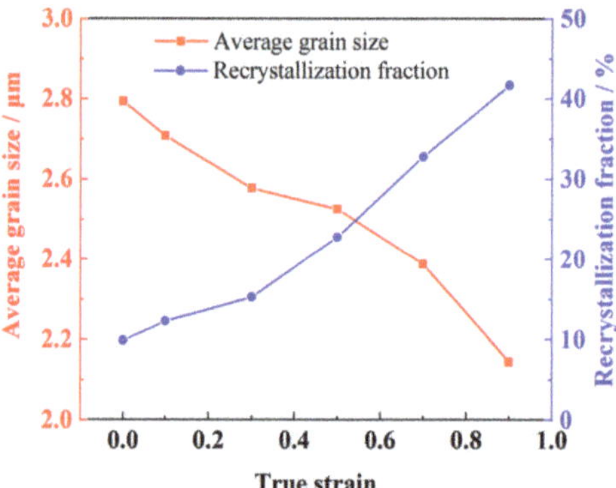

Figure 11. The relationships between the recrystallization fraction, the average grain size, and the true strain when deformed at 950 °C and 0.01 s^{-1}.

Figure 12 shows the distribution of the grain boundaries and the subgrain boundaries at different strains when the deformation temperature was 950 °C and the strain rate was 0.01 s^{-1}. A high density of subgrain boundaries including LAGBs and MAGBs with a total proportion of 75.6% was observed at the initial stage as shown in Figure 12a and the proportion decreased sharply to 44.9% at the strain of 0.9 as shown in Figure 12f. When the strain was lower, the distribution of the subgrain boundaries was not uniform and the subgrain boundaries migrated from the inside of the grain to the grain boundaries [20]. During the hot deformation process, the proportion of the subgrain boundaries decreased and the proportion of HAGBs increased gradually accompanied with some new fine grains formed around the grain boundaries [20,35]. The amount of new fine grains increased significantly when the strains were 0.7 and 0.9, indicating that the recrystallization was active.

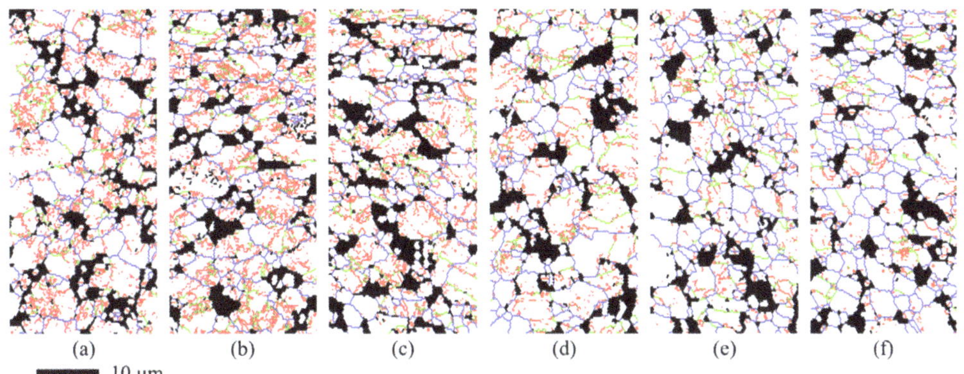

Figure 12. The distributions of the grain boundary at different strains when deformed at 950 °C and 0.01 s^{-1}: (**a**) Initial; (**b**) 0.1; (**c**) 0.3; (**d**) 0.5; (**e**) 0.7; (**f**) 0.9 (red lines represent LAGBs less than 5°, green lines represent MAGBs between 5°~15°, blue lines represent HAGBs more than 15°).

Figure 13 shows the distribution of misorientation at different strains. It can be seen from Figure 13a–c that the proportion of HAGBs was relatively lower and the distribution was uniform, but the proportion of LAGBs was very high [20,28]. The average misorientation increased approximately linearly with the increase in the strain, which was 14.87° and 29.45° at the strain of 0.1 and 0.9, respectively. The increase in the average misorientation was related to the dynamic recrystallization during hot deformation [39].

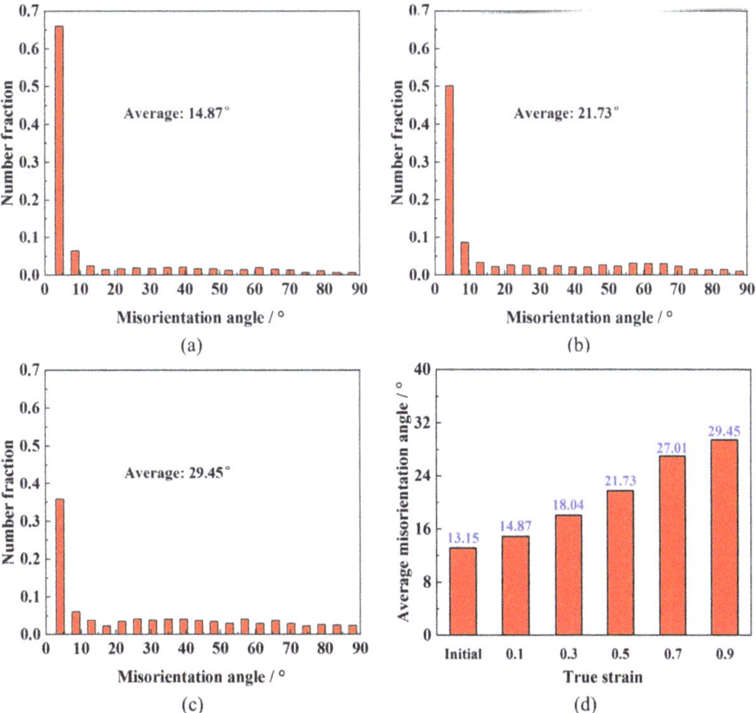

Figure 13. The distributions of the misorientation angle at different strains when deformed at 950 °C and 0.01 s^{-1}: (**a**) 0.1; (**b**) 0.5; (**c**) 0.9; (**d**) the evolution of the average misorientation angle with strain.

The variations in the fractions of LAGBs, MAGBs, and HAGBs during the hot tensile tests are summarized in Figure 14. The proportion of HAGBs increased slowly when the strain was less than 0.3 and increased rapidly when the strain was greater than 0.3. The proportion of LAGBs decreased gradually. However, the variation of MAGBs was not monotonous. The proportion of MAGBs increased gradually when the strain was less than 0.5 and decreased gradually when the strain was greater than 0.5. Dynamic recrystallization mechanisms include continuous dynamic recrystallization (CDRX) characterized by the rotation of the subgrain boundary and discontinuous dynamic recrystallization (DDRX) characterized by the nucleation and growth of grains with small sizes [40,43,44]. The MAGBs is the transitional stage of evolution from LAGBs to HAGBs; therefore, the variation of MAGBs is closely related to the types of recrystallization [20,39]. During the process of CDRX, the proportion of MAGBs increases with the increase in strain, but the proportion of MAGBs decreases with the increase in strain when DDRX occurred [39]. During the hot deformation of the Ti55 high-temperature titanium alloy, DDRX became the main mechanism of dynamic softening with the increase in the strain [20]. For the TC31 titanium alloy deformed at 950 °C and 0.01 s^{-1}, the CDRX was the dominant factor of dynamic softening when the strain was less than 0.5. With the increase in the strain, the effect of DDRX enhanced gradually and the DDRX became the dominant factor of dynamic softening when the strain was greater than 0.5.

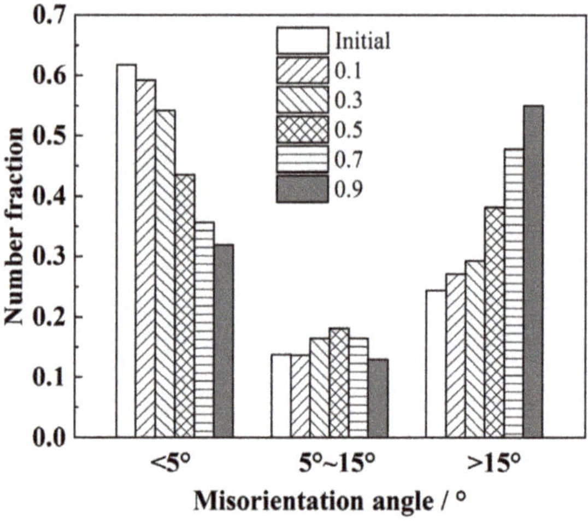

Figure 14. The evolution of the misorientation angle at different strains when deformed at 950 °C and 0.01 s^{-1}.

Figure 15 shows the IPF at different strains. When the strain was 0.3, it can be seen from Figure 15a that a few recrystallized grains with the grain size of about 1 μm appeared at the boundary of coarse grains, and their orientations were close to that of the adjacent coarse grains, as shown in Figure 15c. There were a large number of subgrain boundaries formed by the accumulation of dislocations inside the coarse grains. It was indicated that the CDRX occurred [21,39]. It can be seen from Figure 15b that the amount of the recrystallized grains increased significantly with the increase in the strain to 0.7, and the recrystallized grains were distributed in clusters. Their orientations were significantly different from that of the adjacent un-recrystallized coarse grains as shown in Figure 15d, which is considered to be a typical feature of DDRX [21,39]. The amount of the subgrain boundaries within the un-recrystallized grains reduced greatly and the grain boundaries of the un-recrystallized grains were serrated, demonstrating that the mechanism was DDRX [20,21].

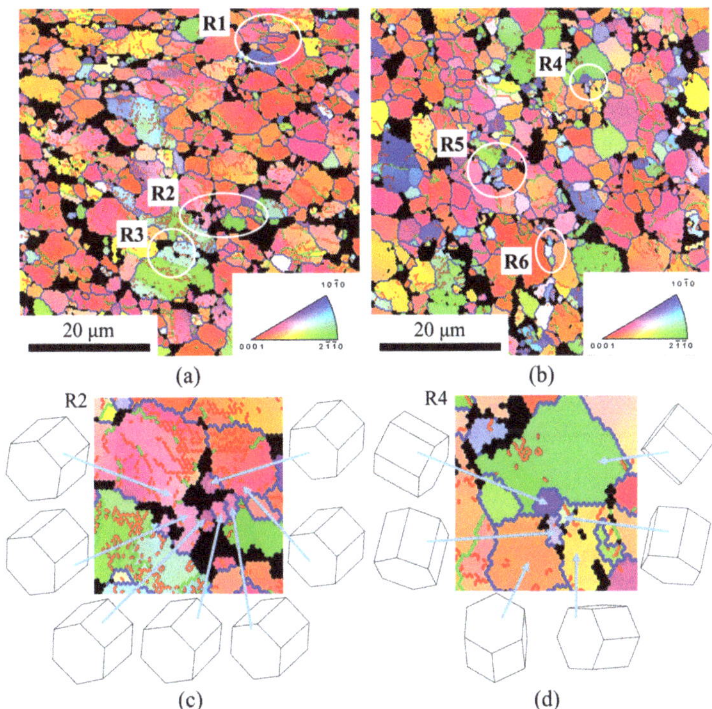

Figure 15. The IPF at different strains when deformed at 950 °C and 0.01 s^{-1}: (**a**,**c**) 0.3; (**b**,**d**) 0.7.

It is well known that the hot deformation parameters have a significant influence on the hot deformation behavior and the evolution of the microstructure [16,17,20]. Qiu et al. [27] pointed out that the flow softening of the SP700 titanium alloy was mainly attributed to DRX during the hot deformation with relatively low temperatures and relatively high strain rates, but the steady-state flow characteristics appeared with the increasing temperature and the decreasing strain rate. During the superplastic deformation of the TC4 titanium alloy, the flow stress increased with the increase in strain during the deformation with a strain rate of 0.0001 s^{-1} and a temperature ranging from 850 °C to 950 °C [24]. The dynamic hardening rate increased with the increasing deformation temperature due to the more obvious grain growth during the deformation [24]. The deformation behavior and microstructure evolution of the TC31 titanium alloy were similar with the aforementioned SP700 and the TC4 titanium alloy. In the future forming of the TC31 titanium alloy component, one should carefully choose the parameters such as temperature and strain rate to control the forming process and post-form properties according to the relationships between the parameters, the microstructure, and the flow stress. Based on the above analysis, the dynamic softening and hardening behaviors of the TC31 titanium alloy are summarized in Figure 16. In the upper left zone A, with a relatively lower temperature and a higher strain rate, the void damage and the deformation heat generation were the important reasons that caused the dynamic softening of the flow stress with the increase in the strain [37]. The softening effect of DRV and DRX increased with the increasing temperature and decreasing strain rate [18,23,27]. When the deformation temperature was higher than 950 °C and the strain rate was lower than 0.001 s^{-1}, such as in zone C in Figure 16, the dynamic hardening characteristics of the flow stress curves were mainly attributed to the grain growth [11,24]. In zone B, the effect of work hardening caused by dislocation accumulation and grain growth was approximately equal to that of the flow softening caused by DRV, DRX, or void damage [8,28]. In zone B, the recrystallization mechanism was mainly CDRX at the initial deformation stage, and transformed into DDRX with the increase in strain. Based on the above analysis, it is suggested that TC31 titanium alloy components are formed within zone B to obtain a relatively uniform deformation and good post-form properties.

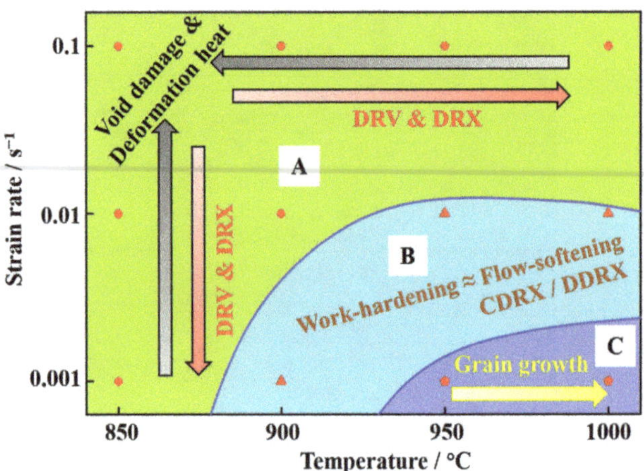

Figure 16. The hot deformation behaviors and the microstructure evolution mechanisms of the TC31 titanium alloy.

4. Conclusions

The dynamic softening and hardening behavior and microstructure evolution of the TC31 titanium alloy during high temperature tensile deformation have been systematically investigated in this paper. Based on the analysis of the experimental results, the main conclusions are summarized as follows:

1. The TC31 titanium alloy exhibited obvious softening behavior during hot tensile deformation at a temperature of 850 °C and a strain rate of 0.001 s^{-1}~0.1 s^{-1}; with an increase in the deformation temperature to 950 °C~1000 °C and an increase in the strain rate to 0.1 s^{-1}, discontinuous yielding occurred; quasi-steady flow appeared at a temperature of 950 °C~1000 °C and a strain rate of 0.01 s^{-1}; with a decease in the strain rate to 0.001 s^{-1}, slight dynamic hardening phenomenon occurred. Therefore, a careful selection of the forming temperature and the strain rate of the TC31 titanium alloy sheet is very important to control the dynamic softening or dynamic hardening during high temperature deformation.
2. When the deformation temperature increased from 850 °C to 950 °C, the volume fraction of the β phase increased from 20% to 41% after it deformed to a strain of 0.7 with a strain rate of 0.01 s^{-1}, whereas the volume fraction of voids was significantly reduced from 11.2% to less than 1%. The increased fraction of the β phase at higher temperatures improved the deformation compatibility and reduced the void damage. Therefore, a relatively high deformation temperature is recommended for the forming of complex TC31 titanium alloy components to avoid the void damage.
3. When the TC31 titanium alloy was deformed at 950 °C, the grains grew up at the strain rate of 0.001 s^{-1} and were refined at the strain rate from 0.01 s^{-1} to 0.1 s^{-1}, and the refinement was more significant under the higher strain rate conditions. The appropriate strain rate should be about 0.01 s^{-1} during the forming of the TC31 titanium alloy sheet considering both the grain coarsening and uniform deformation.
4. When the samples were deformed at a temperature of 950 °C and a strain rate of 0.01 s^{-1}, the proportion of MAGBs firstly increased when the strain was less than 0.5 and then decreased gradually when the strain was greater than 0.5. The main recrystallization mechanism transformed from CDRX to DDRX and the grain sizes decreased gradually with the increase in the strain.

Author Contributions: Investigation, formal analysis and writing—original draft preparation, K.D.; conceptualization, methodology, data curation and writing—review and editing, K.W.; funding acquisition, supervision and writing—review and editing, G.L.; All authors have read and agreed to the published version of the manuscript.

Funding: This research was funded by the Program of National Natural Science Foundation of China (No. U1937204 and No. 51905124), Fundamental Research Funds for the Central Universities (Grant Nos. HIT. NSRIF. 2020003), and the China Postdoctoral Science Foundation (2021T140153).

Institutional Review Board Statement: Not applicable.

Informed Consent Statement: Not applicable.

Data Availability Statement: Not applicable.

Conflicts of Interest: The authors declare no conflict of interest.

References

1. Wang, K.; Kopec, M.; Chang, S.; Qu, B.; Liu, J.; Politis, D.J.; Wang, L.; Liu, G. Enhanced formability and forming efficiency for two-phase titanium alloys by Fast light Alloys Stamping Technology (FAST). *Mater. Des.* **2020**, *194*, 108948. [CrossRef]
2. Liu, G.; Dang, K.; Wang, K.; Zhao, J. Progress on rapid hot gas forming of titanium alloys: Mechanism, modelling, innovations and applications. *Procedia Manuf.* **2020**, *50*, 265–270. [CrossRef]
3. Yasmeen, T.; Zhao, B.; Zheng, J.-H.; Tian, F.; Lin, J.; Jiang, J. The study of flow behavior and governing mechanisms of a titanium alloy during superplastic forming. *Mater. Sci. Eng. A* **2020**, *788*, 139482. [CrossRef]
4. Zhao, E.; Sun, S.; Zhang, Y. Recent advances in silicon containing high temperature titanium alloys. *J. Mater. Res. Technol.* **2021**, *14*, 3029–3042. [CrossRef]
5. Narayana, P.L.; Kim, S.-W.; Hong, J.-K.; Reddy, N.S.; Yeom, J.-T. Tensile properties of a newly developed high-temperature titanium alloy at room temperature and 650 °C. *Mater. Sci. Eng. A* **2018**, *718*, 287–291. [CrossRef]
6. Zhang, W.J.; Song, X.Y.; Hui, S.X.; Ye, W.J.; Wang, Y.L.; Wang, W.Q. Tensile behavior at 700 °C in Ti–Al–Sn–Zr–Mo–Nb–W–Si alloy with a bi-modal microstructure. *Mater. Sci. Eng. A* **2014**, *595*, 159–164. [CrossRef]
7. Zhang, W.-J., Song, X.-Y.; Hui, S.-X.; Ye, W.-J.; Wang, W.-Q. Phase precipitation behavior and tensile property of a Ti–Al–Sn–Zr–Mo–Nb–W–Si titanium alloy. *Rare Met.* **2018**, *37*, 1064–1069. [CrossRef]
8. Wang, K.; Wang, L.; Zheng, K.; He, Z.; Politis, D.J.; Liu, G.; Yuan, S. High-efficiency forming processes for complex thin-walled titanium alloys components: State-of-the-art and perspectives. *Int. J. Extrem. Manuf.* **2020**, *2*, 032001. [CrossRef]
9. Wang, K.; Shi, C.; Zhu, S.; Wang, Y.; Shi, J.; Liu, G. Hot Gas Pressure Forming of Ti-55 High Temperature Titanium Alloy Tubular Component. *Materials* **2020**, *13*, 4636. [CrossRef] [PubMed]
10. Wu, F.; Xu, W.; Yang, Z.; Guo, B.; Shan, D. Study on Hot Press Forming Process of Large Curvilinear Generatrix Workpiece of Ti55 High-Temperature Titanium Alloy. *Metals* **2018**, *8*, 827. [CrossRef]
11. Cheng, C.; Chen, Z.; Li, H.E.; Wang, X.; Zhu, S.; Wang, Q. Vacuum superplastic deformation behavior of a near-α titanium alloy TA32 sheet. *Mater. Sci. Eng. A* **2021**, *800*, 140362. [CrossRef]
12. Liu, G.; Wang, J.; Dang, K.; Yuan, S. Effects of flow stress behaviour, pressure loading path and temperature variation on high-pressure pneumatic forming of Ti-3Al-2.5V tubes. *Int. J. Adv. Manuf. Technol.* **2016**, *85*, 869–879. [CrossRef]
13. Liu, G.; Wang, J.; Dang, K.; Tang, Z. High Pressure Pneumatic Forming of Ti-3Al-2.5V Titanium Tubes in a Square Cross-Sectional Die. *Materials* **2014**, *7*, 5992–6009. [CrossRef]
14. Giuliano, G.; Polini, W. Influence of blank variable thickness on the material formability in hot gas sheet metal forming process. *Manuf. Lett.* **2020**, *24*, 72–76. [CrossRef]
15. Wang, K.; Jiao, Y.; Wu, X.; Qu, B.; Wang, X.; Liu, G. A novel composited process of solution treatment-hot gas forming and stress relaxation aging for titanium alloys. *J. Mater. Process. Technol.* **2021**, *288*, 116904. [CrossRef]
16. Guo, J.; Zhan, M.; Wang, Y.Y.; Gao, P.F. Unified modeling of work hardening and flow softening in two-phase titanium alloys considering microstructure evolution in thermomechanical processes. *J. Alloy. Compd.* **2018**, *767*, 34–45. [CrossRef]
17. Nan, Y.; Ning, Y.; Liang, H.; Guo, H.; Yao, Z.; Fu, M.W. Work-hardening effect and strain-rate sensitivity behavior during hot deformation of Ti–5Al–5Mo–5V–1Cr–1Fe alloy. *Mater. Des.* **2015**, *82*, 84–90. [CrossRef]
18. Hao, F.; Xiao, J.; Feng, Y.; Wang, Y.; Ju, J.; Du, Y.; Wang, K.; Xue, L.n.; Nie, Z.; Tan, C. Tensile deformation behavior of a near-α titanium alloy Ti-6Al-2Zr-1Mo-1V under a wide temperature range. *J. Mater. Res. Technol.* **2020**, *9*, 2818–2831. [CrossRef]
19. Gao, P.; Fu, M.; Zhan, M.; Lei, Z.; Li, Y. Deformation behavior and microstructure evolution of titanium alloys with lamellar microstructure in hot working process: A review. *J. Mater. Sci. Technol.* **2020**, *39*, 56–73. [CrossRef]
20. Liu, Z.; Li, P.; Xiong, L.; Liu, T.; He, L. High-temperature tensile deformation behavior and microstructure evolution of Ti55 titanium alloy. *Mater. Sci. Eng. A* **2017**, *680*, 259–269. [CrossRef]
21. Sun, Y.; Zhang, C.; Feng, H.; Zhang, S.; Han, J.; Zhang, W.; Zhao, E.; Wang, H. Dynamic recrystallization mechanism and improved mechanical properties of a near α high temperature titanium alloy processed by severe plastic deformation. *Mater. Charact.* **2020**, *163*, 110281. [CrossRef]

22. Wu, Y.; Fan, R.; Chen, M.; Wang, K.; Zhao, J.; Xiao, W. High-temperature anisotropic behaviors and microstructure evolution mechanisms of a near-α Ti-alloy sheet. *Mater. Sci. Eng. A* **2021**, *820*, 141560. [CrossRef]
23. Wanjara, P.; Jahazi, M.; Monajati, H.; Yue, S.; Immarigeon, J.P. Hot working behavior of near-α alloy IMI834. *Mater. Sci. Eng. A* **2005**, *396*, 50–60. [CrossRef]
24. Yang, J.; Wu, J. Grain Rotation Accommodated GBS Mechanism for the Ti-6Al-4V Alloy during Superplastic Deformation. *Crystals* **2021**, *11*, 991. [CrossRef]
25. Cao, S.; Zhang, S.-Z.; Liu, J.-R.; Li, S.-J.; Sun, T.; Li, J.-P.; Gao, Y.; Yang, R.; Hu, Q.-M. Interaction between Al and other alloying atoms in α-Ti for designing high temperature titanium alloy. *Comput. Mater. Sci.* **2021**, *197*, 110620. [CrossRef]
26. Lin, P.; He, Z.; Yuan, S.; Shen, J. Tensile deformation behavior of Ti–22Al–25Nb alloy at elevated temperatures. *Mater. Sci. Eng. A* **2012**, *556*, 617–624. [CrossRef]
27. Qiu, Q.; Wang, K.; Li, X.; Wang, J.; Gao, X.; Zhang, K. Hot deformation behavior and processing parameters optimization of SP700 titanium alloy. *J. Mater. Res. Technol.* **2021**, *15*, 3078–3087. [CrossRef]
28. Chuan, W.; Liang, H. Hot deformation and dynamic recrystallization of a near-beta titanium alloy in the β single phase region. *Vacuum* **2018**, *156*, 384–401. [CrossRef]
29. Fan, X.G.; Zhang, Y.; Gao, P.F.; Lei, Z.N.; Zhan, M. Deformation behavior and microstructure evolution during hot working of a coarse-grained Ti-5Al-5Mo-5V-3Cr-1Zr titanium alloy in beta phase field. *Mater. Sci. Eng. A* **2017**, *694*, 24–32. [CrossRef]
30. Jing, L.; Fu, R.D.; Wang, Y.P.; Qiu, L.; Yan, B. Discontinuous yielding behavior and microstructure evolution during hot deformation of TC11 alloy. *Mater. Sci. Eng. A* **2017**, *704*, 434–439. [CrossRef]
31. Wei, Z.; Peng, G.; Yongqing, Z.; Qian, L.; Shewei, X.; Jun, C. Discontinuous Yielding in High Temperature Deformation of Ti-5553 Alloy. *Rare Met. Mater. Eng.* **2015**, *44*, 2415–2418. [CrossRef]
32. Ji, X.; Emura, S.; Min, X.; Tsuchiya, K. Strain-rate effect on work-hardening behavior in β-type Ti-10Mo-1Fe alloy with TWIP effect. *Mater. Sci. Eng. A* **2017**, *707*, 701–707. [CrossRef]
33. Chen, G.; Ren, C.; Qin, X.; Li, J. Temperature dependent work hardening in Ti–6Al–4V alloy over large temperature and strain rate ranges: Experiments and constitutive modeling. *Mater. Des.* **2015**, *83*, 598–610. [CrossRef]
34. Souza, P.M.; Beladi, H.; Singh, R.; Rolfe, B.; Hodgson, P.D. Constitutive analysis of hot deformation behavior of a Ti6Al4V alloy using physical based model. *Mater. Sci. Eng. A* **2015**, *648*, 265–273. [CrossRef]
35. Souza, P.M.; Mendiguren, J.; Chao, Q.; Beladi, H.; Hodgson, P.D.; Rolfe, B. A microstructural based constitutive approach for simulating hot deformation of Ti6Al4V alloy in the α + β phase region. *Mater. Sci. Eng. A* **2019**, *748*, 30–37. [CrossRef]
36. Xu, J.; Zeng, W.; Zhou, D.; Ma, H.; He, S.; Chen, W. Analysis of flow softening during hot deformation of Ti-17 alloy with the lamellar structure. *J. Alloy. Compd.* **2018**, *767*, 285–292. [CrossRef]
37. Zhang, J.; Di, H. Deformation heating and flow localization in Ti–15–3 metastable β titanium alloy subjected to high Z deformation. *Mater. Sci. Eng. A* **2016**, *676*, 506–509. [CrossRef]
38. Gao, Y.; Ma, G.; Zhang, X.; Xu, J. Microstructure evolution and hot deformation behavior of Ti-6.5Al–2Zr–1Mo–1V alloy with starting lamellar structure. *J. Alloy. Compd.* **2019**, *809*, 151852. [CrossRef]
39. Huang, K.; Logé, R.E. A review of dynamic recrystallization phenomena in metallic materials. *Mater. Des.* **2016**, *111*, 548–574. [CrossRef]
40. Wang, X.X.; Zhan, M.; Gao, P.F.; Ma, P.Y.; Yang, K.; Lei, Y.D.; Li, Z.X. Deformation mode dependent mechanism and kinetics of dynamic recrystallization in hot working of titanium alloy. *Mater. Sci. Eng. A* **2020**, *772*, 138804. [CrossRef]
41. Roy, S.; Suwas, S. Deformation mechanisms during superplastic testing of Ti–6Al–4V–0.1B alloy. *Mater. Sci. Eng. A* **2013**, *574*, 205–217. [CrossRef]
42. Rupert, T.J.; Gianola, D.S.; Gan, Y.; Hemker, K.J. Experimental observations of stress-driven grain boundary migration. *Science* **2009**, *326*, 1686–1690. [CrossRef] [PubMed]
43. Li, Z.; Qu, H.; Chen, F.; Wang, Y.; Tan, Z.; Kopec, M.; Wang, K.; Zheng, K. Deformation Behavior and Microstructural Evolution during Hot Stamping of TA15 Sheets: Experimentation and Modelling. *Materials* **2019**, *12*, 223. [CrossRef] [PubMed]
44. Zhao, D.; Fan, J.; Zhang, Z.; Liu, X.; Wang, Q.; Chen, Z.; Tang, B.; Kou, H.; Jia, S.; Li, J. Microstructure and Texture Variations in High Temperature Titanium Alloy Ti65 Sheets with Different Rolling Modes and Heat Treatments. *Materials* **2020**, *13*, 2466. [CrossRef] [PubMed]

 *materials*

Article

Effect of Heating on Hot Deformation and Microstructural Evolution of Ti-6Al-4V Titanium Alloy

Dechong Li [1], Haihui Zhu [2], Shuguang Qu [2], Jiatian Lin [2], Ming Ming [3], Guoqing Chen [1], Kailun Zheng [2] and Xiaochuan Liu [3,*]

1 School of Materials Science and Engineering, Dalian University of Technology, Dalian 116024, China
2 School of Mechanical Engineering, Dalian University of Technology, Dalian 116024, China
3 School of Mechanical Engineering, Xi'an Jiaotong University, Xi'an 710049, China
* Correspondence: liuxiaochuan2020@xjtu.edu.cn

Abstract: This paper presents a systematic study of heating effects on the hot deformation and microstructure of dual-phase titanium alloy Ti-6Al-4V (TC4) under hot forming conditions. Firstly, hot flow behaviors of TC4 were characterized by conducting tensile tests at different heating temperatures ranging from 850 °C to 950 °C and heating rates ranging from 1 to 100 °C/s. Microstructure analysis, including phase and grain size, was carried out under the different heating conditions using SEM and EBSD. The results showed that when the heating temperature was lower than 900 °C, a lower heating rate could promote a larger degree of phase transformation from α to β, thus reducing the flow stress and improving the ductility. When the temperature reached 950 °C, a large heating rate effectively inhibited the grain growth and enhanced the formability. Subsequently, according to the mechanism of phase transformation during heating, a phenomenological phase model was established to predict the evolution of the phase volume fraction at different heating parameters with an error of 5.17%. Finally, a specific resistance heating device incorporated with an air-cooling set-up was designed and manufactured to deform TC4 at different heating parameters to determine its post-form strength. Particularly, the yield strength at the temperature range from 800 °C to 900 °C and the heating rate range from 30 to 100 °C/s were obtained. The results showed that the yield strength generally increased with the increase of heating temperature and the decrease of heating rate, which was believed to be dominated by the phase transformation.

Keywords: dual-phase titanium alloy; heating rate; microstructure; phase model; hot forming

Citation: Li, D.; Zhu, H.; Qu, S.; Lin, J.; Ming, M.; Chen, G.; Zheng, K.; Liu, X. Effect of Heating on Hot Deformation and Microstructural Evolution of Ti-6Al-4V Titanium Alloy. *Materials* **2023**, *16*, 810. https://doi.org/10.3390/ma16020810

Academic Editor: Antonino Squillace

Received: 12 December 2022
Revised: 9 January 2023
Accepted: 11 January 2023
Published: 13 January 2023

1. Introduction

Dual-phase titanium alloys have been widely used in the aerospace industry for manufacturing complex-shaped thin-walled components [1,2] due to their high strength-to-weight ratio, excellent corrosion resistance and high heat resistance [3]. Titanium alloys have strong deformation resistance and severe spring-back drawback that forms at room temperature, which may damage forming tools and reduce the accuracy of formed components [4]. In order to overcome the low formability at room temperature, superplastic forming (SPF) [5], hot stamping [6] and hot metal gas forming (HMGF) [7] were developed for forming complex-shaped thin-walled parts in recent years, featuring low deformation resistance and high forming precision [8,9]. Heating to elevated temperatures is the first and key step in hot forming processes. The microstructure, such as phase and grain size, varies with the forming temperature [10,11], which determines the subsequent hot flow behavior and post-form mechanical properties [12]. To facilitate the design of reasonable heating variables, particularly heating rate and temperature, a thorough understanding of the correlation between micro and macro-properties, i.e., microstructure evolution and hot flow, at various heating parameters is of vital importance to the hot forming of TC4 sheets.

The relationship between the mechanical properties, phase transformation and grain size evolution of titanium alloys at different heating temperatures has been studied [13,14].

Compared to the extensive research on the forming temperature, studies on the heating rate effect on the hot forming of titanium alloy sheets are limited to date. Xiao et al. [15] conducted uniaxial tensile tests on TC4 sheets in the temperature range of 923–1023 K. The real stress-strain curves showed that the flow stress decreased with the increase of temperature and the decrease of strain rate, but the effect of heating rate on the mechanical properties before and after forming was not considered. Wang et al. [16] studied both β phase transformation during rapid heating and martensitic phase transformation during cooling through high-speed machining of TC4. A stress-temperature-induced phase transformation model based on Avrami and Clausius-Clapeyron equations was proposed to predict the phase transformation from α to β during the rapid heating process. However, the mechanism of phase transformation at specific heating rates was not discussed in this study. Elmer et al. [17] conducted in-situ X-ray diffraction experiments to observe the phase transformation of TC4 alloys during heating using synchrotron radiation. It was observed that the diffusion of the β stable element V was inhibited as the heating rate increased, which led to a decrease in the β phase transformation temperature. However, the study was mainly limited to the phase transformation kinetics without investigating the phase transformation effect on the high-temperature deformation and the post-form properties. In addition, Surya et al. [18,19] studied the effects of input factors such as speed, feed and depth of cut on the material removal rate and the surface roughness of TC4 titanium alloy, and established a high-precision mathematical prediction model to determine the optimal parameter combination for cutting performance. This modeling idea is also applicable to the high-temperature deformation of TC4. It is, therefore, necessary to build a material model based on temperature and microstructure and to formulate a reasonable forming process by predicting the phase distribution of materials at high temperatures, in order to comprehensively understand the subsequent heat flow and strength variations after forming [20].

To quantify the heating rate, the conventional environmental heating furnace has some limitations, such as uncontrollable heating rate and low thermal efficiency. In order to accurately control the temperature change rate during the heating process, various heating methods were studied. Mori et al. [21] and Maeno et al. [22] applied a resistance heating method to the hot stamping of high-strength steel and dual-phase titanium alloy parts. Due to a high heating rate being used, negligible oxidation occurred on the sheet surface, and the current passed directly through the stamping parts, thus improving the heating efficiency. Therefore, the resistance heating technology is suitable for adjusting the current output and controlling the temperature change during the heating process [23]. Wang et al. [24] used resistance heating to heat the workpiece to 850–950 °C at heating rates of 4 °C/s and 100 °C/s, respectively, then cooled it to 700 °C to simulate the workpiece heating to the target temperature at different heating rates, and finally transferred it to the die for the hot stamping process, in order to investigate the effect of the heating rate on material ductility and post-form strength. However, the results of this study were not applicable to isothermal forming processes at specific temperatures without transfer processes, and little research has been performed in this field.

In this study, the heating effect on the microstructure evolution and hot flow of TC4 was systematically studied using the resistance heating method. The hot flow and phase evolution at different heating parameters were characterized by using a Gleeble thermal-mechanical simulator (Gleeble, New York, NY, USA), and the post-form strength was compared using the specimens that were heat treated using a self-developed resistance heating device. Additionally, a phenomenological phase transformation model was established based on the phase transformation kinetics during the heating process. The performed research is able to provide important guidance for the design of heating parameters, i.e., heating rate and temperature, for the various hot forming processes of TC4 sheets.

2. Experimentation

2.1. Material

Dual-phase titanium alloy TC4 sheets with a thickness of 1.5 mm were used in this study. The chemical composition is shown in Table 1, and the microstructure is shown in Figure 1, in which the white region represents the β phase and the dark region represents the α phase. The initial microstructure consisted of an equiaxed α phase with a volume fraction of 87% and a fine β phase with a volume fraction of 13%. The initial heat treatment temper and annealing condition had a yield strength and ultimate tensile strength of 1056 MPa and 1071 MPa, respectively. In addition, the elongation of TC4 was 18.5%.

Table 1. Chemical composition of the as-received TC4 in weight percentage.

Ele.	Al	V	Fe	C	N	H	O	Ti
wt%	6.1	4.2	0.15	<0.01	<0.01	0.007	0.13	Bal.

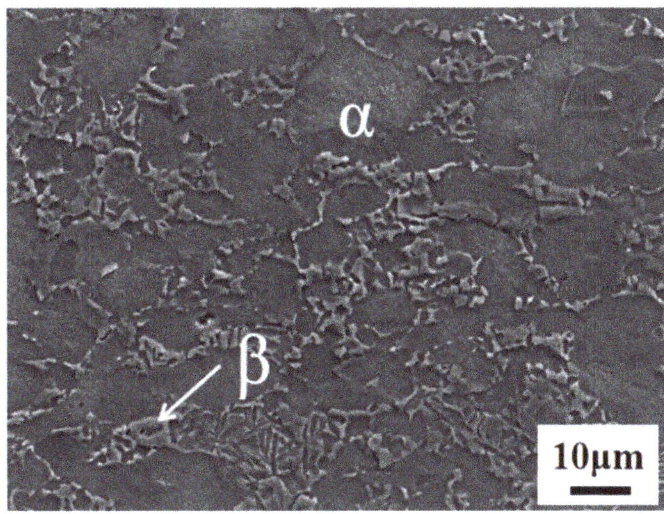

Figure 1. Microstructure of the as-received TC4 titanium alloy.

2.2. Resistance Heating Tests

2.2.1. Gleeble Hot Tensile Tests and Microstructure Characterization

In order to examine the effect of the heating process on the hot flow behaviors and microstructure evolution of TC4, Gleeble hot tensile tests were performed at various heating temperatures and heating rates using a Gleeble 3800 thermal-mechanical simulator, which enabled the maximum heating rate to reach 10,000 °C/s and allowed for precise control of temperature within ±1 °C.

Figure 2a shows the experimental flow chart of the high-temperature thermodynamic testing. Regarding the Gleeble hot tensile tests, conditions of different heating rates, temperatures and strain rates were performed as follows. The heating rates of 1 °C/s and 10 °C/s were selected, enabling the rate magnitudes of practical processes to be covered. Three temperatures of 850 °C, 900 °C and 950 °C were used at each heating rate. Once the specimen reached the target temperature, hot tensile tests at a strain rate of 0.1/s were conducted. The high-temperature tensile properties of TC4 at different heating rates and heating temperatures were then measured. The temperature was continuously monitored by the thermocouple welded at the center of the specimen to precisely feedback-control the heating rates and heating temperatures during testing. In the experimental group of heat treatments used to characterize the microstructure of TC4 during heating, the same heating history as that of the tensile tests at high temperatures was used. The specimen was immediately water-quenched after a soaking time of 1 s to maintain the microstructure. Microstructure observations, i.e., SEM and EBSD, were used to characterize the phase transformation and grain evolution of the specimens, and the quantitative microstructure results with statistical significance were subsequently obtained.

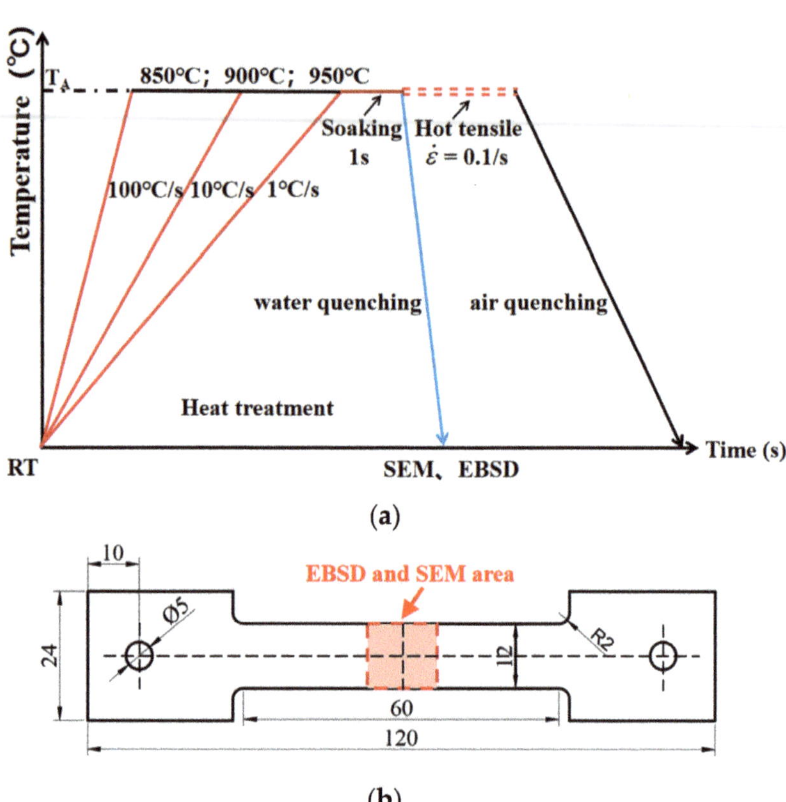

Figure 2. (**a**) Temperature profiles of the Gleeble hot tensile and heat treatment tests; (**b**) Geometries of Gleeble hot tensile tests and microstructure specimens.

Figure 2b shows the specimen size and shape for the tensile and heat treatment tests. The dog-bone-shaped specimen was used with a length of 60 mm, a width of 12 mm and a thickness of 1.5 mm, which was machined from the as-received sheet using Electrical Discharge Machining along the rolling direction. The length of the parallel zone of the specimen could guarantee a homogeneous temperature zone of 10 mm in length in the middle. Therefore, secondary sampling for EBSD and SEM was carried out in the center of the sample to ensure the uniformity of its temperature history.

2.2.2. Post-Form Strength Characterization

In order to investigate the rapid heating of TC4 sheets as well as the in-line cooling to characterize the post-form strength at different heating parameters, a resistance heating experimental device for titanium alloy sheets was designed and manufactured as schematically shown in Figure 3.

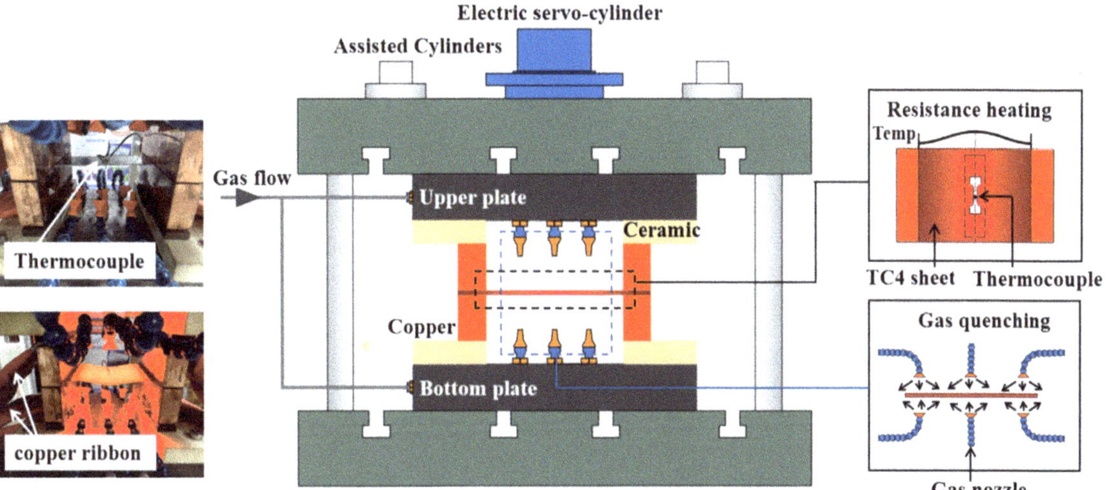

Figure 3. Schematic diagram of the resistance heat treatment device.

The current in the device was output by the high-frequency switching power supply, flowing directly from the copper ribbon to the copper electrode and passing through the TC4 sheet. Al_2O_3 ceramics were bolted between the copper electrode and the upper/lower plates to facilitate the insulation, ensuring that the current flew to the TC4 sheet completely through the copper electrode.

A plurality of gas channels was positioned inside the upper and lower plates, and connected with an adjustable gas source to provide pressure up to 0.8 MPa. Meanwhile, the outlet of each gas channel was connected with gas nozzles that were evenly distributed on both sides of the TC4 sheet, thus achieving uniform cooling of the sheet. The flowmeter was used to monitor the gas flow in real-time so as to control the cooling rate of the TC4 sheet. The k-type thermocouple was welded on the surface of the sheet, and the other end was inserted into the temperature recorder to facilitate the real-time monitoring and recording of the temperature.

Therefore, the device was able to perform the static treatment experiments at different heating rates, heating temperatures and cooling conditions to simulate the industrial forming process of titanium alloys.

Figure 4 illustrates the temperature histories of the resistance heating experiments. Three heating rates of 30 °C/s, 60 °C/s and 100 °C/s were selected. At each heating rate, three temperatures of 800 °C, 850 °C and 900 °C were used. Once the sheet reached the target temperature, the sheet immediately cooled to room temperature at a cooling rate of 200 °C/s using the gas flow, thus restoring the microstructure after heating. The temperature was continuously monitored by the thermocouple welded at the center of the sheet to precisely monitor the heating rates, cooling rates and heating temperatures. Sub-sized dog-bone-shaped specimens were designed and machined (shown in Figure 4) from the sheets to measure the strength and investigate the effect of temperature on the mechanical properties of TC4.

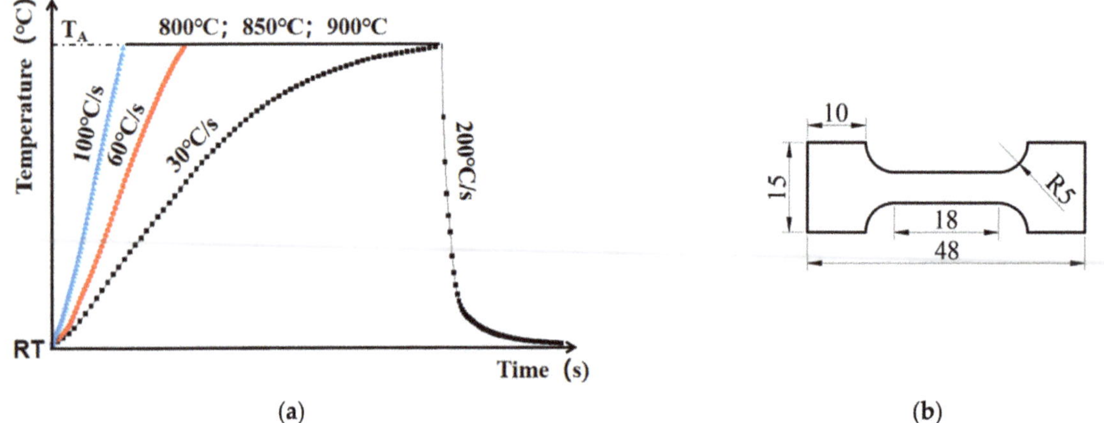

(a) (b)

Figure 4. (**a**) Temperature histories of resistance heating experiments; (**b**) Geometry of the Gleeble tensile test specimens.

3. Results and Discussion

3.1. Flow Behaviors of TC4 at Different Forming Temperatures and Heating Rates

The effects of forming temperature and heating rate on the flow behavior of TC4 were investigated based on the Gleeble tensile tests and microstructure results. The true stress-strain curves shown in Figure 5 indicated that the maximum flow stress and elongation of TC4 decreased with the increase of the forming temperature. Specifically, when the heating rate was 1 °C/s, the maximum flow stress was 230 MPa, 149 MPa and 96 MPa at the forming temperatures of 850 °C, 900 °C and 950 °C, respectively, suggesting a 58% reduction with an increase in temperature from 850 °C to 950 °C, as shown in Figure 6a. Similarly, the maximum flow stress of TC4 decreased by 53% from 242 MPa to 114 MPa with the increase in forming temperature from 850 °C to 950 °C at a heating rate of 10 °C/s. In addition, the maximum flow stress at the heating rate of 10 °C/s was higher than that at the heating rate of 1 °C/s. Notably, when the heating temperature was 950 °C, the maximum flow stress increased by 19% from 96 MPa to 114 MPa with the increase in heating rate from 1 °C/s to 10 °C/s.

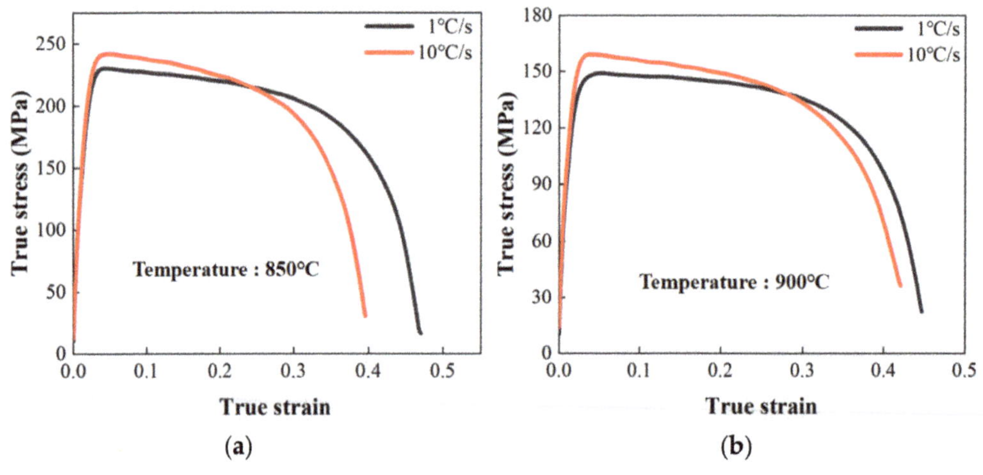

(a) (b)

Figure 5. *Cont.*

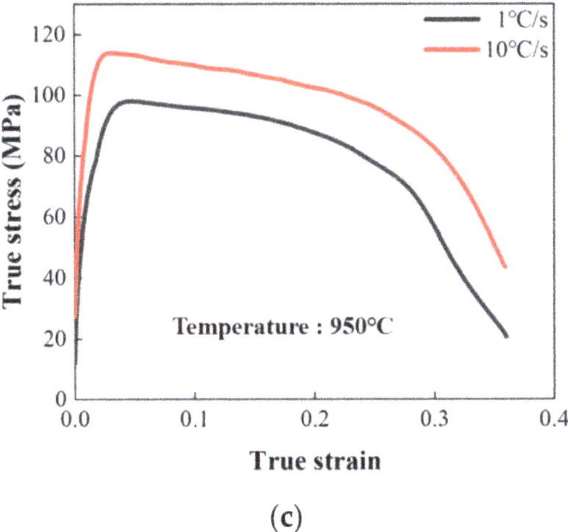

Figure 5. Effect of heating rate on the hot stress strain behaviors of TC4 at heating temperatures of (**a**) 850 °C; (**b**) 900 °C; (**c**) 950 °C.

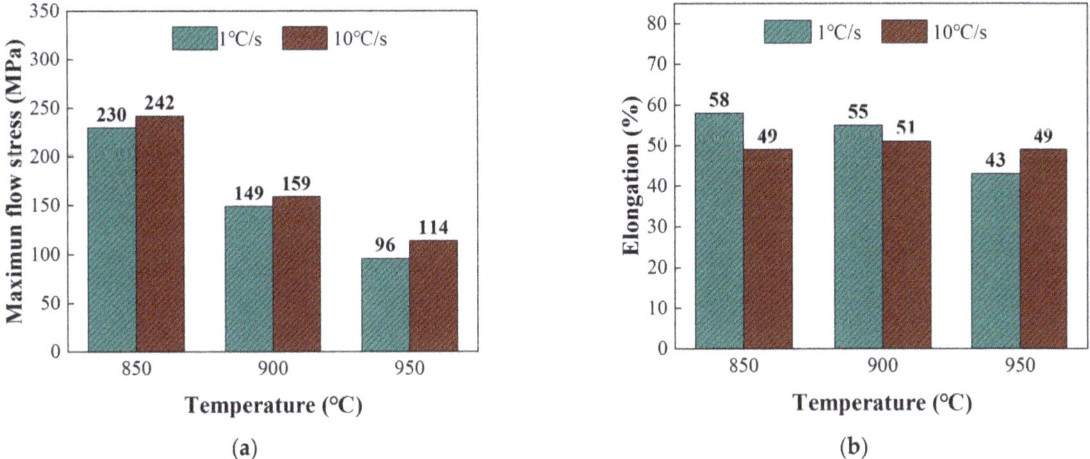

Figure 6. Effects of forming temperature and heating rate on (**a**) the maximum flow stress and (**b**) elongation.

It was found that the elongation of TC4 decreased with an increase in forming temperature as well, as shown in Figure 6b. Specifically, the elongation decreased from 58% at a forming temperature of 850 °C to 43% at a forming temperature of 950 °C when a heating rate of 1 °C/s was applied. However, the effect of forming temperature on the elongation was negligible when the heating rate increased to 10 °C/s. When the heating temperature was 950 °C, the elongation was increased by 14% at a heating rate of 10 °C/s, compared with that at a heating rate of 1 °C/s.

3.2. *Effects of Forming Temperature and Heating Rate on the Phase Transformation*

The microstructural phase distributions of TC4 were obtained while investigating the effects of different forming temperatures and heating rates on the maximum flow stress and elongation of TC4, as shown in Figures 7 and 8, in which the white region represents the β phase and the dark region represents the α phase. The β phase is a high-temperature stable phase, which gradually changes from the α phase during the heating process and mainly depends on the temperature and heating rate of the process. It could be seen that the volume fraction of the β phase increased with the increase in heating temperature when the heating rate was constant and decreased with the increase in heating rate when the heating temperature was constant. Moreover, the β phase microstructure was found to be finer at a lower heating temperature and a higher heating rate.

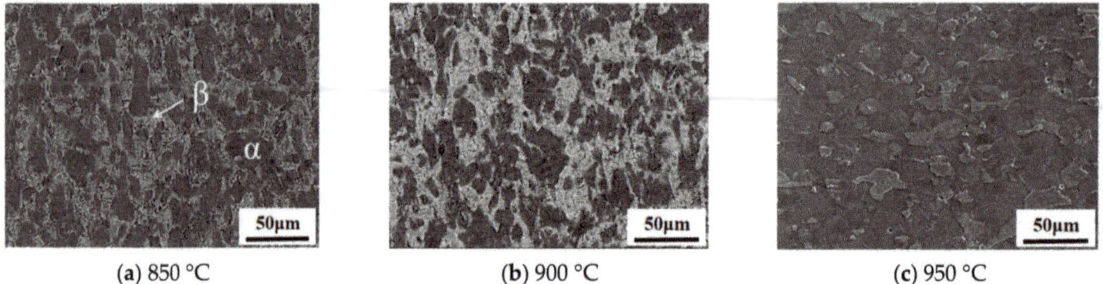

(**a**) 850 °C (**b**) 900 °C (**c**) 950 °C

Figure 7. SEM observations of the α and β phases at different forming temperatures and a constant heating rate of 1 °C/s.

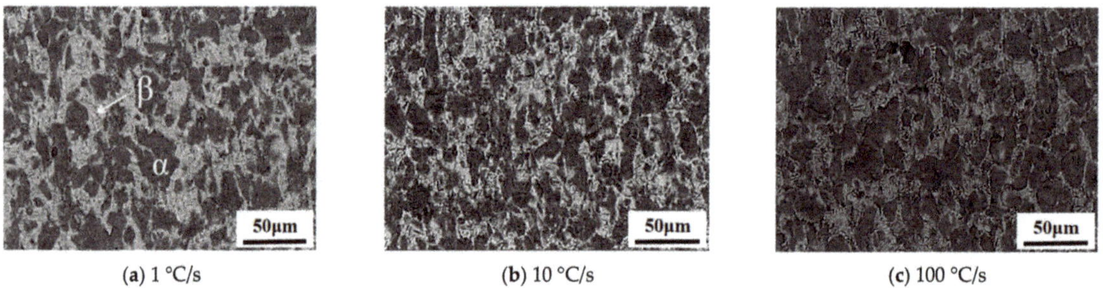

(**a**) 1 °C/s (**b**) 10 °C/s (**c**) 100 °C/s

Figure 8. SEM observation of the α and β phases at different heating rates and a constant forming temperature of 900 °C.

In order to investigate the effect of the β phase volume fraction on the mechanical properties clearly, a statistical analysis of the β phase volume fraction was conducted, as shown in Figure 9a, which was found to be 34%, 47% and 72% at the forming temperatures of 850 °C, 900 °C and 950 °C, respectively, when a heating rate of 1 °C/s was applied. It had been reported [25] that the α phase with a hexagonal compact packing structure was approximately three times stronger than the β phase with a body-centered cubic structure at an elevated temperature. Meanwhile, the dynamic recrystallization of TC4 increased with the increase in forming temperature, thus softening the material and reducing its deformation resistance. As a combined result of the dynamic recrystallization and phase transformation, the maximum flow stress of TC4 decreased with the increase in forming temperature.

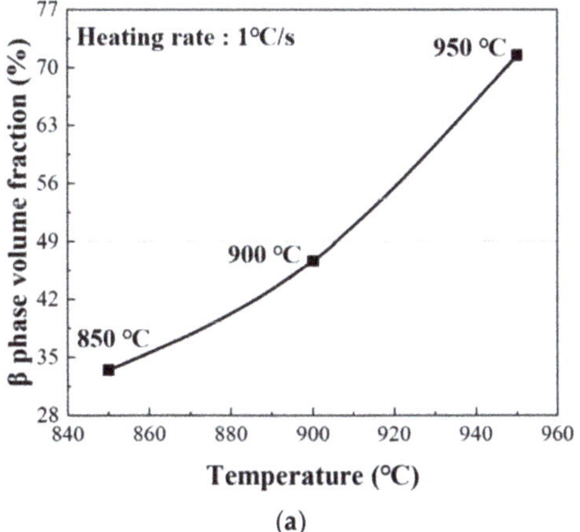

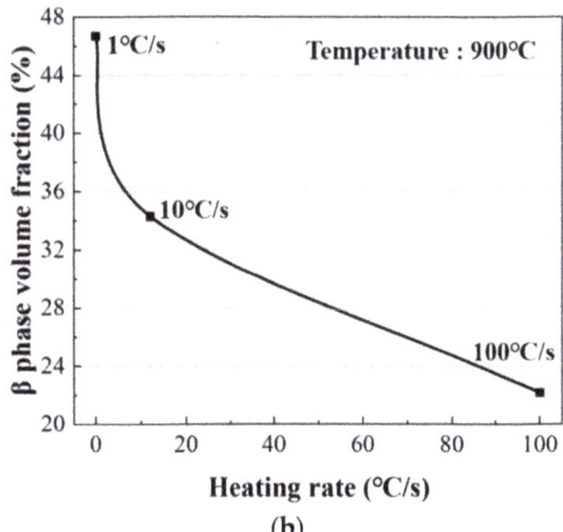

Figure 9. Volume fractions of the β phase at different (**a**) forming temperatures and (**b**) heating rates.

The volume fraction of the β phase decreased with the increase in heating rate, as shown in Figure 9b. Specifically, when using a forming temperature of 900 °C, it was 47% at a heating rate of 1 °C/s and increased to 34% and 20% with the increasing heating rate of 10 °C/s and 100 °C/s, respectively. This was due to the fact that the phase transformation was determined by the diffusion of alloying elements. As the heating rate increased, the β phase stable element V lacked enough time to diffuse, resulting in an insufficient transformation of the β phase at this temperature. Consequently, the volume fraction of the β phase for the high heating rate condition was less than that of the equilibrium phase transformation process obtained from the isothermal condition [26]. Therefore, a higher heating rate applied in the hot deformation contributed to a higher maximum flow stress with less phase transformation.

In theory, the increase in forming temperature enhances the volume fraction of the phase, thus improving the elongation. However, severe grain coarsening at high temperatures may overshadow the decrease in elongation. At the same time, the increase in heating rate might lead to a decrease in the volume fraction of the α to β phase transformation, the combined effect of the two leading to a lower elongation during rapid heating at the forming temperatures of 850 °C and 900 °C.

The grain microstructure and size of TC4 were also measured at different heating rates based on the EBSD observations, as shown in Figure 10. It was found that the average grain size decreased from 10.16 μm at 10 °C/s to 8.76 μm at 100 °C/s, indicating that the rapid heating suppressed grain coarsening. The rate of the phase transformation was significant at 950 °C, and thus a large number of primary α transformed into β, significantly deteriorating the heating rate effect. In contrast, a slower heating rate caused the material to remain at an elevated temperature for a longer period, reducing the volume fraction of the α phase as well as the grain boundary movement hindrance of the β phase. Therefore, it was concluded that the grain size played a dominant role in the mechanical properties of TC4. The finer grains caused by rapid heating allowed the deformation to be dispersed in more grains, thus reducing the dislocation plugging in each grain and effectively improving the elongation at elevated temperatures.

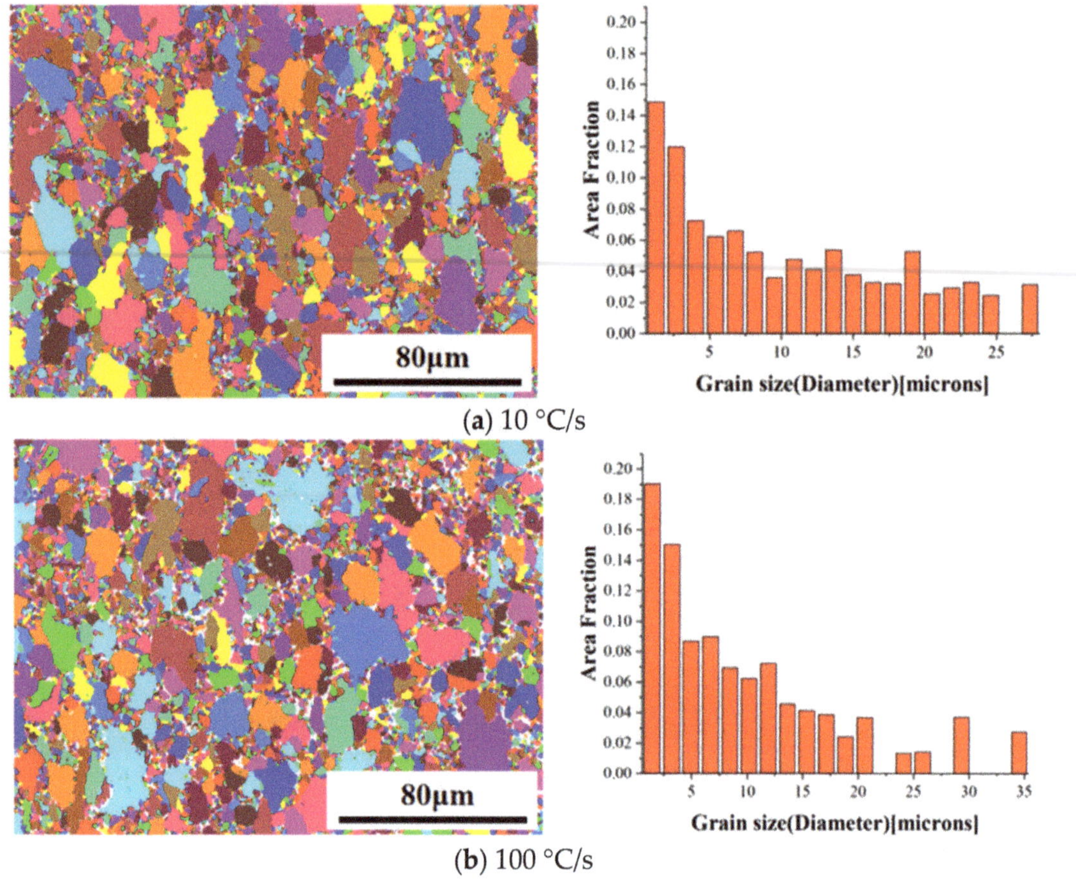

Figure 10. EBSD of the grain microstructure and size of TC4 at different heating rates and a constant forming temperature of 950 °C.

3.3. Model of Phase Transformation at Different Heating Temperatures and Rates

For the dual-phase titanium alloys, the β phase with a bcc crystal structure is softer than the α phase with an hcp crystal structure. Hence, the volume fraction of the β phase has a large influence on the mechanical properties of the material. In addition, this parameter was found to be temperature-dependent based on the experimental observations. Consequently, the Johnson-Mehl-Avrami (JMA) equation describing the phase transformation from α to β under the isothermal conditions was used in the present research to model the relationship between the volume fraction of the β phase and the deformation temperature, as shown in Equation (1) [27]:

$$f_\beta = \left(\frac{T}{1270}\right)^{10} \quad (1)$$

where f_β was the volume fraction of the β phase, and T was the deformation temperature.

Considering the non-isothermal conditions, especially at a high heating rate, the phase transformation process is not equilibrated any longer. Therefore, the effect of the heating rate on the non-equilibrium phase transformation process was defined as Equation (2) [28]:

$$f_T = 1 - \exp\{-[K(\mathrm{T} - \mathrm{T_0})/\mathrm{H}]^{\mathrm{n}}\} \tag{2}$$

where f_T was the temperature-dependent transformation volume fraction, n was the exponent of the transformation from α to β, $\mathrm{T_0}$ was the activation transformation temperature at the current heating rate, H was the heating rate, and K was a temperature-dependent constant, which was modelled by using the Arrhenius equation as below.

$$K(\mathrm{T}) = K_0 exp(-\frac{Q}{RT}) \tag{3}$$

where K_0 was a constant and Q was the diffusion activation energy of the phase transformation from α to β.

The initial temperature of the phase transformation was affected by the heating rate. As a result, the phase transformation temperature demonstrated hysteretic behavior with the increase in the heating rate, as shown in Equation (4).

$$T_0 = T_1 * [1 + H/(a * H_{\max})] \tag{4}$$

where T_1 was the activation temperature of the β phase equilibrium transformation, and $H_{\max}$ is the ultimate heating rate of Gleeble and is fixed at 1000 °C/s.

Consequently, the phase variation volume fraction of TC4 during continuous heating could be predicted by Equation (5), enabling the model to predict the volume fraction of the β phase at different temperatures and heating rates.

$$f_\beta = f_0 + f_T \tag{5}$$

where, f_0 is the β phase volume fraction of TC4 titanium alloy at room temperature.

The experimental conditions of 850 °C–1 °C/s, 950 °C–1 °C/s, 900 °C–1 °C/s, 900 °C–10 °C/s and 900 °C–100 °C/s were selected as the orthogonal experiments, and each parameter in the equation was then calibrated using the Matlab fitting method as shown in Table 2. As shown in Figure 11, the experimental results of the volume fraction of the β phase were consistent with the modeling results with an average error of 5.17%, indicating a good accuracy of the proposed model. Based on the present research, the effect of cooling on the post-form strength and microstructure of TC4 will be used in the future to develop a theoretical model to predict the mechanical strength of formed components.

Table 2. Material constants of a phase transformation volume fraction prediction model.

Parameter	Value	Parameter	Value	Parameter	Value
f_0	0.13	K_0 (s^{-1})	4.52×10^{16}	Q (kJ/mol)	445
T_1 (K)	923	a	0.86	$H_{\max}$ (°C/s)	1000
n	0.35	R (J/mol/K)	8.314		

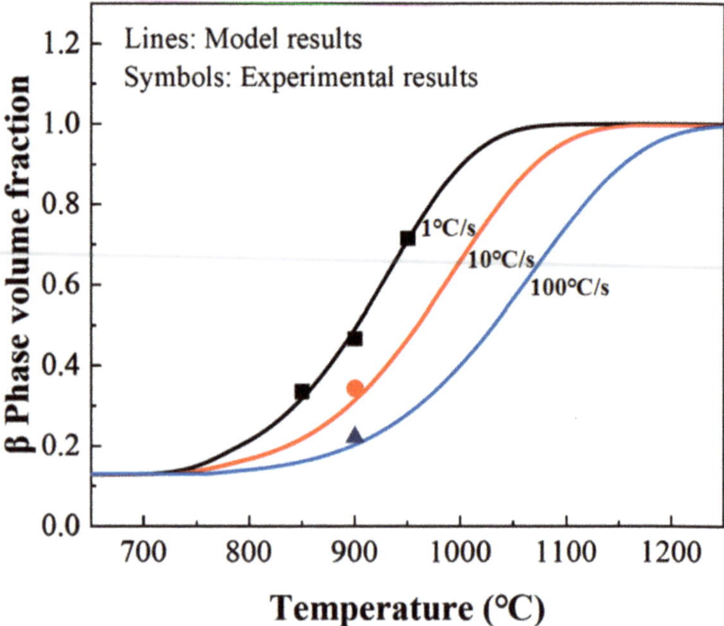

Figure 11. Comparisons of the experimental and modelling results on the volume fraction of the β phase at different heating rates and heating temperatures.

3.4. Effects of the Forming Temperature and Heating Rate on the Post-Form Strength of TC4

This study investigated the effects of forming temperature and heating rate on the strength of TC4 sheets that were heat treated using a self-developed resistance heating device. The strength properties at room temperature were characterized using the sub-size specimen shown in Figure 4b. For an easier comparison of the yield and ultimate tensile strength, engineering stress-strain curves were plotted in Figure 12, and the distribution of determined yield strength at various heating rates is shown in Figure 13.

It was clear that the heating rate affected the yield strength significantly, which decreased gradually with the increase in heating rate at the forming temperatures of 850 °C and 900 °C, but first increased and then decreased at the forming temperature of 800 °C. This trend was more obvious at a higher forming temperature. The difference in the yield strength between the heating rates of 30 °C/s and 100 °C/s was 16% at 800 °C, which was lower than that at 900 °C (24%). This was because when the heating rate was lower, the diffusion of alloying element V was more adequate, leading to a higher volume fraction of the β phase. It was known that the metastable β phase transformed to the fine and dislocated irregular acicular α′ phase, instead of the α phase, after cooling at a rate that was higher than the critical value of the martensitic phase transformation of TC4 [29]. Due to the α′ phase being stronger than the α phase, the more the β phase generated at a lower heating rate, the more it transformed into the α′ phase, resulting in a higher yield strength.

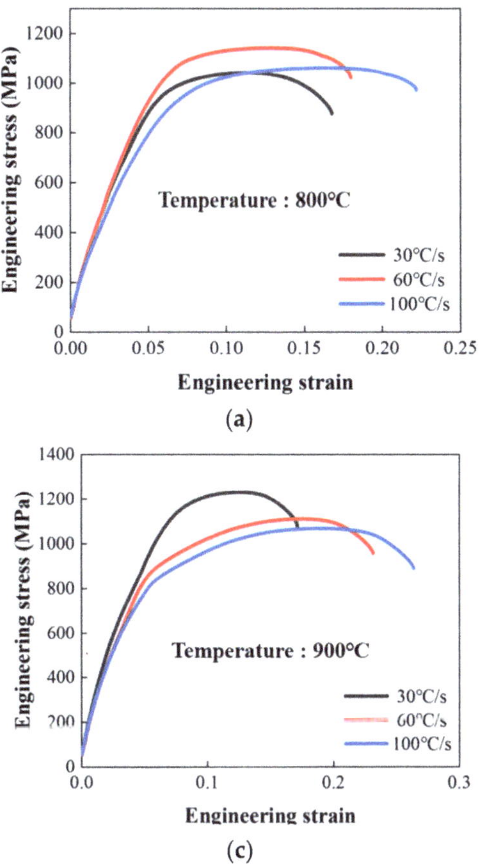

Figure 12. Effect of heating rate on the engineering stress-strain curves at the heating temperatures of (**a**) 800 °C (**b**) 850 °C; (**c**) 900 °C.

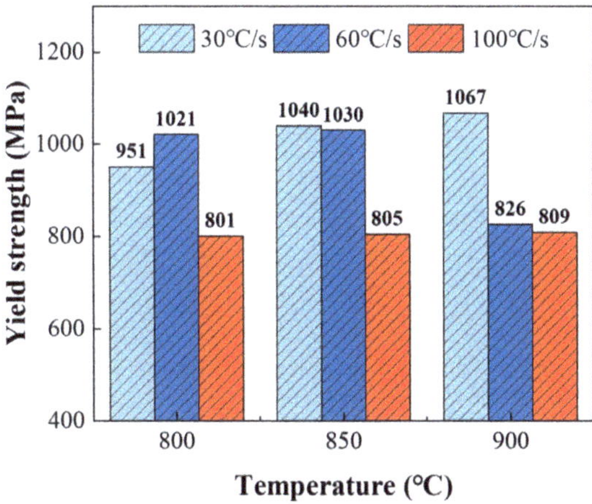

Figure 13. Yield strength after cooling at a cooling rate of 200 °C/s at different heating temperatures and rates.

4. Conclusions

In this paper, the effects of heating rate and forming temperature on the plastic deformation, microstructure and post-forming properties of a dual-phase titanium alloy TC4 were systematically studied. The results of which would facilitate the precise control of the shape and mechanical properties of the formed titanium components under hot forming conditions. Therefore, the present research would contribute to the manufacture of complex-shaped thin-walled aerospace components made from dual-phase titanium alloys. The effects of the forming temperature and heating rate on the maximum flow stress and elongation of TC4 were investigated by conducting hot tensile tests and microstructural analysis. Subsequently, the yield strength at different heating rates was compared using the specimens that were heat treated using a resistance heating device. The main conclusions were drawn as below:

1. In the forming temperature range between 850 °C and 950 °C, both the maximum flow stress and elongation decreased with an increase in forming temperature. When the heating rate was 10 °C/s, the flow stress was larger than that at the heating rate of 1 °C/s, while the elongation remained constant.
2. For the microstructure evolution under various heating conditions, the volume fraction of the β phase increased with an increase in heating temperature and a decrease in heating rate. The average grain size decreased with an increased heating rate. A higher volume fraction of the β phase and finer grains improved the material ductility.
3. Based on the microstructure observation results, a model was established to predict the volume fraction of the β phase under different heat treatment conditions. The prediction error of the model was 5.17%, which would contribute to a qualitative analysis of the mechanical properties of TC4 titanium alloy under high-temperature deformation conditions.
4. More martensite transformation was involved in the metastable β phase at a higher heating temperature and a lower heating rate during the rapid cooling process, leading to a higher yield strength.

Author Contributions: Conceptualization, D.L., H.Z. and K.Z.; Formal analysis, D.L. and S.Q.; Funding acquisition, K.Z.; Investigation, J.L.; Methodology, D.L., H.Z. and S.Q.; Project administration, K.Z.; Supervision, X.L.; Visualization, M.M.; Writing—original draft, D.L., H.Z., S.Q., M.M., K.Z. and X.L.; Writing—review & editing, G.C., K.Z. and X.L. All authors have read and agreed to the published version of the manuscript.

Funding: The authors would like to thank the funding from the National Natural Science Foundation of China (Grant Nos: 52105359, U1908229, 52005076, 52205412).

Institutional Review Board Statement: Not applicable.

Informed Consent Statement: Not applicable.

Data Availability Statement: Not applicable.

Conflicts of Interest: The authors declare no conflict of interest.

References

1. Wang, K.; Wang, L.; Zheng, K.; He, Z.; Politis, D.J.; Liu, G.; Yuan, S. High-Efficiency Forming Processes for Complex Thin-Walled Titanium Alloys Components: State-of-the-Art and Perspectives. *Int. J. Extrem. Manuf.* **2020**, *2*, 032001. [CrossRef]
2. Li, F.; Mo, J. Springback Analysis of TC4 Titanium Alloy Complex Part with Double Curvature under Single Point Incremental Forming Process. *Adv. Mater. Res.* **2012**, *468–471*, 1094–1098. [CrossRef]
3. Seshacharyulu, T.; Dutta, B. Influence of Prior Deformation Rate on the Mechanism of B→α+β Transformation in Ti–6Al–4V. *Scr. Mater.* **2002**, *46*, 673–678. [CrossRef]
4. Sirvin, Q.; Velay, V.; Bonnaire, R.; Luc, P. Mechanical Behavior and Modelisation of Ti-6Al-4V Titanium Sheet under Hot Stamping Conditions. *AIP Conf. Proc.* **2017**, *1896*, 20016.
5. Giuliano, G. Constitutive Equation for Superplastic Ti–6Al–4V Alloy. *Mater. Des.* **2008**, *29*, 1330–1333. [CrossRef]
6. Zhou, J.; Yang, X.; Wang, B.; Mu, Y. Investigation of Interfacial Heat Transfer Characterization for TC4 Alloy in Triple-Layer Sheet Hot Stamping Process. *Int. J. Adv. Manuf. Technol.* **2022**, *121*, 1–14. [CrossRef]

7. Talebi Anaraki, A.; Chougan, M.; Loh-Mousavi, M.; Maeno, T. Hot Gas Forming of Aluminum Alloy Tubes Using Flame Heating. *J. Manuf. Mater. Process.* **2020**, *4*, 56. [CrossRef]
8. He, Z.; Teng, B.; Che, C.; Wang, Z.; Zheng, K.; Yuan, S. Mechanical Properties and Formability of TA2 Extruded Tube for Hot Metal Gas Forming at Elevated Temperature. *Trans. Nonferr. Met. Soc. China* **2012**, *22*, s479–s484. [CrossRef]
9. Zheng, K.; Zheng, J.; He, Z.; Liu, G.; Politis, D.J.; Wang, L. Fundamentals, Processes and Equipment for Hot Medium Pressure Forming of Light Material Tubular Components. *Int. J. Lightweight Mater. Manuf.* **2019**, *3*, 1–19. [CrossRef]
10. Meng, M.; Fan, X.G.; Guo, L.G.; Zhan, M. Achieving Fine-Grained Equiaxed Alpha via Thermo-Mechanical Loading under off-Equilibrium State in Two-Phase Ti-Alloys. *J. Mater. Process. Technol.* **2018**, *259*, 397–408. [CrossRef]
11. Picu, C.; Majorell, A. Mechanical Behavior of Ti-6Al-4V at High and Moderate Temperatures-Part II: Constitutive Modeling. *Mater. Sci. Eng. A* **2002**, *326*, 306–316. [CrossRef]
12. Zhang, J.; Ju, H.; Xu, H.; Yang, L.; Meng, Z.; Liu, C.; Sun, P.; Qiu, J.; Bai, C.; Xu, D.; et al. Effects of Heating Rate on the Alloy Element Partitioning and Mechanical Properties in Equiaxed A+β Ti-6Al-4V Alloy. *J. Mater. Sci. Technol.* **2021**, *94*, 1–9. [CrossRef]
13. Gao, S.; Sang, Y.; Li, Q.; Sun, Y.; Wu, Y.; Wang, H. Constitutive Modeling and Microstructure Research on the Deformation Mechanism of Ti-6Al-4V Alloy under Hot Forming Condition. *J. Alloys Compd.* **2022**, *892*, 162128. [CrossRef]
14. Gao, S.; He, T.; Li, Q.; Sun, Y.; Sang, Y.; Wu, Y.; Ying, L. Anisotropic Behavior and Mechanical Properties of Ti-6Al-4V Alloy in High Temperature Deformation. *J. Mater. Sci.* **2022**, *57*, 651–670. [CrossRef]
15. Xiao, J.; Li, D.; Li, X.; Jin, C.; Zhang, C. Constitutive Modeling of the Rate-Dependent Behavior of Ti-6Al-4V Using an Arrhenius-Type Law. *Adv. Mater. Res.* **2012**, *591–593*, 949–954. [CrossRef]
16. Wang, Q.; Zhanqiang, L.; Yang, D.; Mohsan, A.U.H. Metallurgical-Based Prediction of Stress-Temperature Induced Rapid Heating and Cooling Phase Transformations for High Speed Machining Ti-6Al-4V Alloy. *Mater. Des.* **2017**, *119*, 208–218. [CrossRef]
17. Elmer, J.; Palmer, T.; Babu, S.; Specht, E. In Situ Observations of Lattice Expansion and Transformation Rates of α and β Phases in Ti–6Al–4V. *Mater. Sci. Eng. A* **2005**, *391*, 104–113. [CrossRef]
18. Siva Surya, M. Optimization of Turning Parameters While Turning Ti-6Al-4V Titanium Alloy for Surface Roughness and Material Removal Rate Using Response Surface Methodology. *Mater. Today Proc.* **2022**, *62*, 3479–3484. [CrossRef]
19. Surya, M.S.; Prasanthi, G.; Kumar, A.K.; Sridhar, V.K.; Gugulothu, S.K. Optimization of Cutting Parameters While Turning Ti-6Al-4 V Using Response Surface Methodology and Machine Learning Technique. *Int. J. Interact. Des. Manuf.* **2021**, *15*, 453–462. [CrossRef]
20. Zheng, K.; Yong, L.; Yang, S.; Fu, K.; Zheng, J.; He, Z.; Yuan, S. Investigation and Modeling of the Preheating Effects on Precipitation and Hot Flow Behavior for Forming High Strength AA7075 at Elevated Temperatures. *J. Manuf. Mater. Process.* **2020**, *4*, 76. [CrossRef]
21. Mori, K.; Okuda, Y. Tailor Die Quenching in Hot Stamping for Producing Ultra-High Strength Steel Formed Parts Having Strength Distribution. *Cirp Ann. Technol.—CIRP Ann.-Manuf. Technol.* **2010**, *59*, 291–294. [CrossRef]
22. Maeno, T.; Yamashita, Y.; Mori, K.-I. Hot Stamping of Titanium Alloy Sheets into U Shape with Concave Bottom and Joggle Using Resistance Heating. *Key Eng. Mater.* **2016**, *716*, 915–922. [CrossRef]
23. Mori, K.; Maki, S.; Tanaka, Y. Warm and Hot Stamping of Ultra High Tensile Strength Steel Sheets Using Resistance Heating. *CIRP Ann.—Manuf. Technol.* **2005**, *54*, 209–212. [CrossRef]
24. Wang, K.; Kopec, M.; Chang, S.; Qu, B.; Liu, J.; Politis, D.J.; Wang, L.; Liu, G. Enhanced Formability and Forming Efficiency for Two-Phase Titanium Alloys by Fast Light Alloys Stamping Technology (FAST). *Mater. Des.* **2020**, *194*, 108948. [CrossRef]
25. Kim, J.; Semiatin, S.; Lee, Y.-H.; Lee, C.S. A Self-Consistent Approach for Modeling the Flow Behavior of the Alpha and Beta Phases in Ti-6Al-4V. *Metall. Mater. Trans. A* **2011**, *42*, 1805–1814. [CrossRef]
26. Haozhang, Z.; Qian, M.; Hou, W.; Xinyu, Z.; Gu, J. The β Phase Evolution in Ti-6Al-4V Additively Manufactured by Laser Metal Deposition Due to Cyclic Phase Transformations. *Mater. Lett.* **2017**, *216*, 50–53. [CrossRef]
27. Majorell, A.; Srivatsa, S.; Picu, R.C. Mechanical Behavior of Ti-6Al-4V at High and Moderate Temperatures-Part I: Experimental Results. *Mater. Sci. Eng. A* **2002**, *326*, 297–305. [CrossRef]
28. Blázquez, J.; Conde, C.F.; Conde, A. Non-Isothermal Approach to Isokinetic Crystallization Processes: Application to the Nanocrystallization of HITPERM Alloys. *Acta Mater.* **2005**, *53*, 2305–2311. [CrossRef]
29. Ahmed, T.; Rack, H. Phase Transformations during Cooling in A+β Titanium Alloys. *Mater. Sci. Eng. A* **1998**, *243*, 206–211. [CrossRef]

Article

Tailoring Titanium Sheet Metal Using Laser Metal Deposition to Improve Room Temperature Single-Point Incremental Forming

Michael McPhillimy [1,*], Evgenia Yakushina [2] and Paul Blackwell [1,2]

[1] Department of Design, Manufacturing & Engineering Management, University of Strathclyde, Glasgow G1 1XJ, Scotland, UK

[2] Advanced Forming Research Centre, Inchinnan PA4 9LJ, Scotland, UK

* Correspondence: michael.t.mcphillimy@strath.ac.uk

Abstract: Typically, due to their limited formability, elevated temperatures are required in order to achieve complex shapes in titanium alloys. However, there are opportunities for forming such alloys at room temperature using incremental forming processes such as single-point incremental forming (SPIF). SPIF is an innovative metal forming technology which uses a single tool to form sheet parts in place of dedicated dies. SPIFs ability to increase the forming limits of difficult-to-form materials offers an alternative to high temperature processing of titanium. However, sheet thinning during SPIF may encourage the early onset of fracture, compromising in-service performance. An additive step prior to SPIF has been examined to tailor the initial sheet thickness to achieve a homogeneous thickness distribution in the final part. In the present research, laser metal deposition (LMD) was used to locally thicken a commercially pure titanium grade 2 (CP-Ti50A) sheet. Tensile testing was used to examine the mechanical behaviour of the tailored material. In addition, in-situ digital image correlation was used to measure the strain distribution across the surface of the tailored material. The work found that following deposition, isotropic mechanical properties were obtained within the sheet plane in contrast to the anisotropic properties of the as-received material and build height appeared to have little influence on strength. Microstructural analysis showed a change to the material in response to the LMD added thickness, with a heat affected zone (HAZ) at the interface between the added LMD layer and non-transformed substrate material. Grain growth and intragranular misorientation in the added LMD material was observed. SPIF of a LMD tailored preform resulted in improved thickness homogeneity across the formed part, with the downside of early fracture in a high wall angle section of the sheet.

Keywords: Titanium; sheet metal; incremental forming; additive manufacture

Citation: McPhillimy, M.; Yakushina, E.; Blackwell, P. Tailoring Titanium Sheet Metal Using Laser Metal Deposition to Improve Room Temperature Single-Point Incremental Forming. *Materials* **2022**, *15*, 5985. https://doi.org/10.3390/ma15175985

Academic Editors: Mateusz Kopec and Denis Politis

Received: 24 June 2022
Accepted: 24 August 2022
Published: 30 August 2022

1. Introduction

Single-point incremental forming (SPIF) is a method of producing sheet metal parts without the need for expensive tooling. In place of the conventional fixed forming die setup, a computer numerical control (CNC) machine controls a tool as it performs incremental localised deformation on the sheet [1]. It is thanks to the incremental form of deformation that SPIF is capable of increasing the forming limits of difficult to form materials such as titanium [2], with Jackson et al. [3] highlighting plain strain stretching and Allwood et al. [1] through thickness shear as key deformation modes. Lightweight materials such as titanium are widely used in aerospace and automotive industries for their high strength to weight ratio [4]. However, their alpha-grained hexagonally close-packed (hcp) microstructure limits their formability at room temperature due to a restricted number of slip systems. As such, there are increasing demands for advanced forming technologies to improve the low temperature formability of this lightweight material [5]. The increased forming limits achievable with SPIF makes room temperature forming of titanium possible, offering a cost-effective alternative to forming titanium at elevated temperatures [6].

A disadvantage of SPIF is the localised sheet thinning during SPIF which limits the potential geometric capabilities of the process [7]. A combination of the compressive load induced by the tool and stretching of the sheet was found to be responsible for the thinning [8]. Jeswiet et al. [9] attributed high forming angles to increased levels of thinning in the sheet cross-section which can result in inhomogeneous thickness distributions across SPIF parts. This is especially true in sheet parts containing wall angles at 70° to the sheet plane or higher [10]. Reducing the severity of sheet thinning in room temperature SPIF is integral to the successful forming of titanium sheet parts. Different approaches from other researchers have been taken to achieve this. Mohsen et al. [11] combined SPIF of AA6082-T6 sheet with hydraulic bulging, increasing the formability by 26% and reducing sheet thinning by 45%. Ambrogio et al. [12] machined pockets into specific zones of a AA1050-O aluminium sheet to weaken these areas to better control the thinning distribution during ISF. In another study, Ambrogio et al. [13] used additive manufacturing (AM) to avoid excessive sheet thinning and improve accuracy via two routes, an AM generated backing plate and sintered material to reinforce the area of pronounced deformation in aluminium sheet. Kumar et al. [14] found decreasing the wall angle, step depth and spindle speed and increasing tool diameter reduced thinning, so improving uniform thickness distribution of SPIF parts.

Tailored blanks offer a potential route for generating preforms for optimising forming. Raut et al. [15] designed aluminium and steel blanks with areas of thicker and stronger material in zones of expected higher stress and loading to reduce part weight without compromising strength. AM reinforcements have been seen to offer better final part properties than other types of tailored blanks due to the full metallurgical bonding between deposit and substrate [16]. Laser metal deposition (LMD) is particularly suited for tailoring sheet metal due to its good heat dissipation, its capability of printing thin layers and fast deposition speeds which limit sheet deformation [17]. Such advantages led researchers to propose the use of LMD to increase the thickness and stiffness of aluminium sheet with locally applied patches [16]. The material properties of LMD parts can be advantageous to specific applications as found in the following studies. Yu et al. [18] discovered LMD generated Ti6Al4V parts resulted in superior yield and ultimate tensile strengths than cast and annealed wrought material, with consistent microstructure between layers and tracks and a low dilution depth and heat-affected zone (HAZ). Barro et al. [19] found AM generated titanium grade 4 parts for dental components performed better than milled parts. LMD is capable of processing a wide range of metals making it the most versatile AM process [20]. With LMD the geometry is printed on a substrate surface by melting the surface with a laser whilst simultaneously applying metal powder. The melt pool is protected from oxidation by an inert gas, usually argon or helium, and the metal powder is fed into the melt pool by a nozzle. Typically, the part is kept stationary as the deposition head follows a tool path to achieve the final part geometry.

The purpose of this investigation is to tailor a commercially pure titanium (CP-Ti) SPIF preform with local thickening via LMD to improve thickness homogeneity in the final SPIF sheet part. The development of this novel SPIF + LMD hybrid process may potentially provide a route to improve geometric accuracy at high forming angles in cold formed CP-Ti sheet parts. A commercially pure titanium grade 2 (CP-Ti50A) sheet will be locally thickened with CP-Ti powder using LMD. The two variables of interest are LMD build height (in the build direction) and LMD track direction (with respect to the sheet rolling direction, RD). The impact which LMD build height and direction have on the tailored materials mechanical and directional properties is to be examined. Tensile testing will be performed to examine the mechanical behaviour of the tailored material with in-situ digital image correlation (DIC) to measure surface strain distribution. Material characterisation techniques will be used to examine the microstructure of the tailored material. Following this, a forming trial will be performed to test the capability of the LMD tailored preform for room temperature SPIF. LMD variables are to be kept constant to avoid confounding factors from influencing the results of the study. All the details are reported in the following sections.

2. Materials and Methods

The sheet used for the LMD substrate material was an uncoated cold rolled sheet of CP-Ti50A of dimensions 500 mm × 500 mm × 1.6 mm (W × L × T), selected for its good room temperature ductility relative to other titanium alloys. The chemical composition of the material is listed in Table 1 [21] and its mechanical properties in Table 2. Examination of the as-received (A-R) sheet substrate showed a microstructure of fine equiaxed grains with average grain size of 5.4 ± 0.2 μm, Figure 1a. The true stress–strain curve of the material in three directions with respect to the sheet rolling direction (RD) showed anisotropy of the material, Figure 1b. The powder metal used for laser metal deposition (LMD) was a CP-Ti grade 2 powder with particle size distribution of 45–150 μm. It was gas atomised with a spherical particle shape, apparent density 2.62 g/cm^3 and oxygen content 0.16% [22].

Table 1. Chemical composition of CP-Ti50A.

Constituent Elements	Ti	N	C	H	Fe	O
Composition, wt%	Balance	0.03	0.08	0.015	0.20	0.18

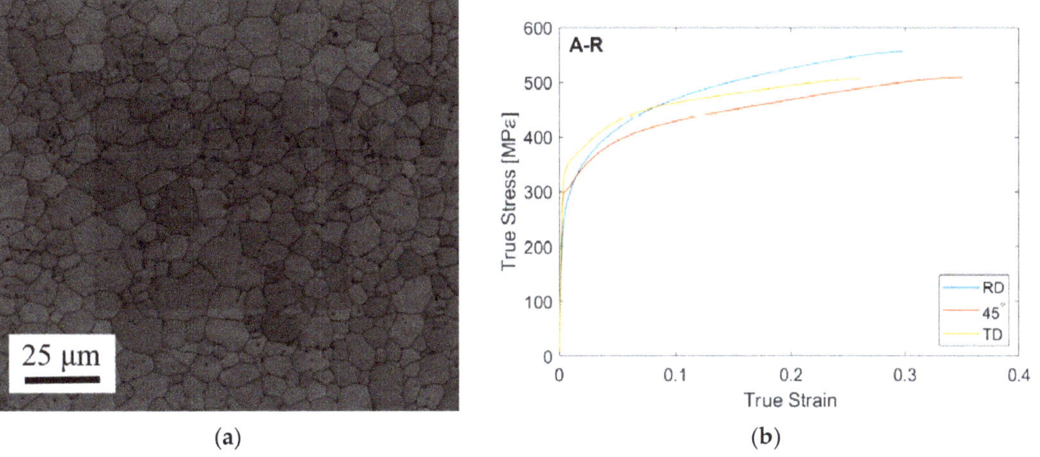

Figure 1. Substrate material properties: (**a**) Microstructure of A-R (CP-Ti50A); (**b**) True stress–strain plot of A-R CP-Ti50A in three directions at room temperature.

Table 2. Mechanical properties of A-R CP-Ti50A.

Sheet Orientation	Yield Strength (MPa)	Tensile Strength (MPa)	Poisson's Ratio
RD	353 ± 1.25	432 ± 0.78	0.33
45° to RD	343 ± 1.62	389 ± 1.21	0.75
TD	378 ± 0.83	420 ± 0.95	0.9

A non-contact blue light 3D scanning technique was used to scan a 1.6 mm thick CP-Ti50A sheet part previously formed by SPIF (SPIF part). The part was designed with wall angles (θ) at 20°, 40°, 60° and 80° to the sheet plane. GOM ATOS® software was used to generate a three-dimensional representation of a section of the part, Figure 2. Through-thickness thinning was measured to an accuracy of ±20–40 μm across each angled section.

Added thickness from a spray, used to minimise light reflection, was responsible for an additional 0.05 mm thickness. Five measurements were taken from each section and the mean calculated, Table 3. The thinning measurements determined the quantity of material to be added by LMD. The same scanning technique was used to generate a digital model of the SPIF + LMD hybrid sheet part created as part of this study, to measure its thickness profile.

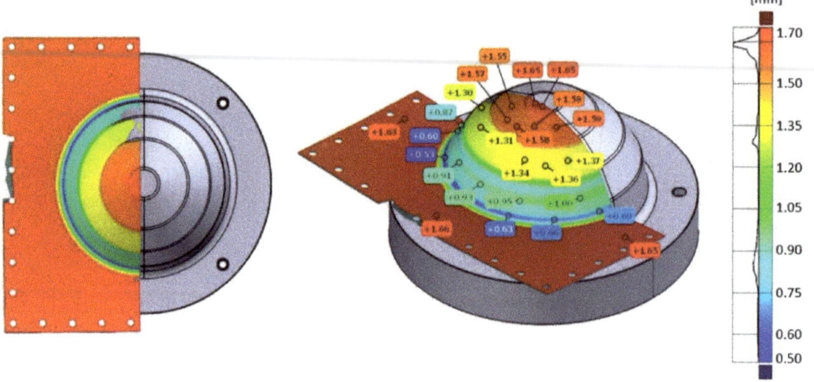

Figure 2. GOM scan to measure thickness reduction across SPIF formed part.

Table 3. Thickness measurements from the scanned SPIF part to inform the necessary LMD thickening.

Region	Thickness (mm)	Thinning (mm)
$\theta = 0°$	1.65 ± 0.01	N/A
$\theta = 20°$	1.57 ± 0.02	0.07
$\theta = 40°$	1.34 ± 0.03	0.31
$\theta = 60°$	0.93 ± 0.05	0.72
$\theta = 80°$	0.62 ± 0.06	1.03

Two LMD tailored sheets were designed for this investigation. The first for conducting material analysis on the LMD tailored material, Figure 3, and the second a LMD tailored preform for SPIF, Figure 4c. Designing the first sheet with nine LMD deposit (pad) locations made it possible to investigate the influence of the LMD build height and orientation on the materials mechanical and microstructural properties. Each pad measured 111 mm $\times$ 71 mm (L $\times$ W). Three pads were orientated with their length parallel to the sheet rolling direction (RD), three at 45° to the RD (45° to RD) and three at 90°, or transverse, to the RD (TD). From each pad three tensile specimens were extracted. The added thickness via LMD (A) in the build direction (Z) was achieved by stacking 0.15 mm layers in a crosshatch pattern (ex. x4 layers achieves 0.6 mm added thickness). The tensile specimen gauge lengths were made in-line with the LMD pad length (Y) which determined the tensile specimen orientation in relation to the sheet rolling direction (RD). The resulting material variable configurations are provided in Table 4. Additional tensile specimens were designed to be extracted directly from the A-R sheet for testing (sets a, e and i). For the SPIF preform a uniform thickness of 0.3 mm was deposited using the same LMD parameter across a second CP-Ti50A sheet. The deposited material aligned with the areas of thinning in the previously formed SPIF geometry.

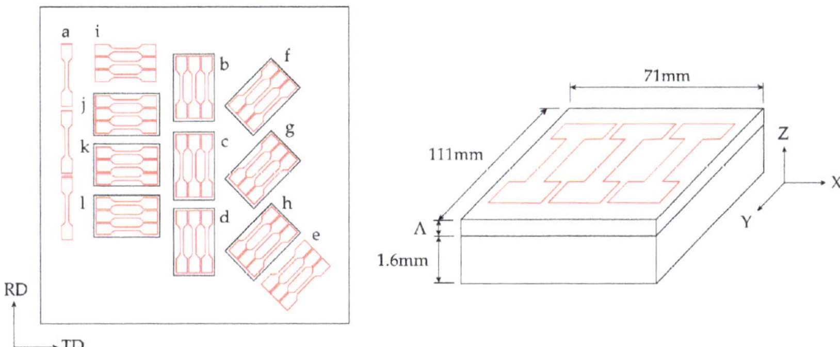

Figure 3. LMD pad locations (internal black boxes) on sheet with the tensile test specimen extraction locations (red) indicated, and tensile gauge lengths in-line with pad length (Y).

Table 4. Material variables.

Material Location	Tensile Gauge Orientation (Y direction)	LMD Thickness (A) (mm)
Set a	RD	0
Set b	RD	0.3
Set c	RD	0.6
Set d	RD	0.9
Set e	45° to RD	0
Set f	45° to RD	0.3
Set g	45° to RD	0.6
Set h	45° to RD	0.9
Set i	TD	0
Set j	TD	0.3
Set k	TD	0.6
Set j	TD	0.9

A modular fixture consisting of two sub-modules was designed to support the hybrid LMD + SPIF process. One sub-module (LMD fixture) secured the sheets during LMD and a subsequent stress relief heat treatment. The LMD fixture comprised of three parallel plates made from hardened tool steel with a central cavity of 442 mm diameter in the top plate to allow the die to pass through during SPIF. The LMD fixture, with sheet clamped, fits into the larger sub-module (SPIF fixture). The SPIF fixture positions and lowers the LMD fixture over a die as the sheet is formed with a 25.4 mm diameter hemispherical tool. Physical alignment grooves and guiding pillar/bushing assemblies in the SPIF fixture ensure accuracy and repeatability.

LMD was performed by a commercial supplier using a TRUMPF 2 kW TruDisk Laser and Oerlikon Metco Twin-10-C powder feeder. The laser and powder feed movement were provided with a 5-axis REIS RL80 Gantry Manipulation System with a 3 m × 2.25 m × 1.5 m work envelope, Figure 4a. The following LMD machine parameters were developed experimentally on-site: laser power (P) of 675.0 W, scanning speed (v) of 15.0 mm/s, track separation of 1.4 mm, and carrier and nozzle shielding gas flow rates of 4.0 L/min and 10.0 L/min, respectively. These parameters achieved a low linear laser energy density (LED) of 45 J/mm ensuring a low depth of heat penetration, a narrow melt region and minimal HAZ [23]. The shielding (argon) gas carried the powder into the molten pool where it was melted and, on solidification, layered with a step size of 0.15 mm. These machine parameters were used for performing LMD on both sheets, Figure 4b,c.

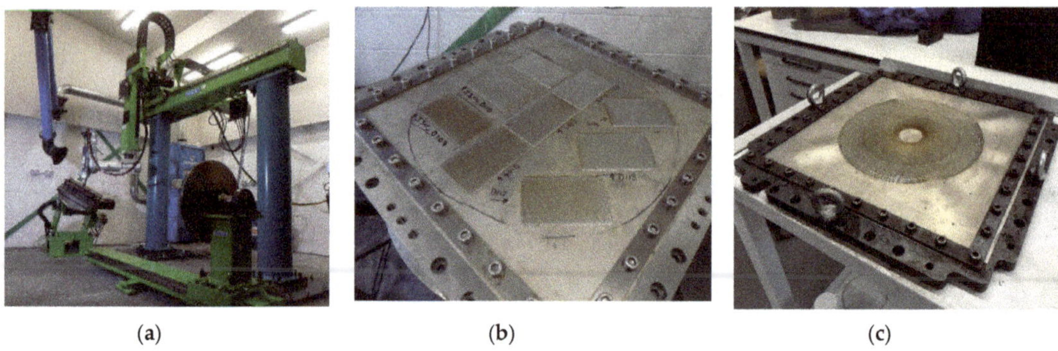

Figure 4. LMD fixture and sheet for LMD: (**a**) LMD processing cell. (**b**) Fixture with CP-Ti50A sheet mounted on T-slot table; (**c**) Completed LMD deposits on sheet.

X-ray diffraction (XRD) was used to measure the residual stress remaining after LMD. The results of XRD showed some variation in the stress measurements acting in the RD and TD directions, Table 5. The maximum stress was made to be 168 ± 6.69 MPa, below the typical yield strength (YS) of CP-Ti50A which is 345 MPa [21].

Table 5. XRD measurement results.

Material Location	Measurement Direction	Measured Stress (MPa)
Set b	TD	137 ± 6.07
	RD	27 ± 3.97
Set c	TD	93 ± 5.25
	RD	75 ± 5.78
Set d	TD	133 ± 5.45
	RD	55 ± 4.51
Set j	TD	−32 ± 6.21
	RD	0 ± 5.13
Set k	TD	51 ± 5.78
	RD	77 ± 5.47
Set l	TD	168 ± 6.69
	RD	10 ± 4.55

A stress relieving heat treatment was performed using a Carbolite LCF furnace to ensure no sheet movement affected the LMD build quality. Following this, tensile specimens were extracted by wire electrical discharge machining (EDM) from the LMD part using a AgieCharmilles CUT 400 Sp machine. The geometry of the tensile specimens is provided in Figure 5. A speckle pattern was spray painted on the surface of each tensile specimen for digital image correlation (DIC) analysis, used to track the sample surface displacement and generate an accurate strain map. Selected specimens had the speckle pattern applied to the LMD deposit side and others on the substrate side.

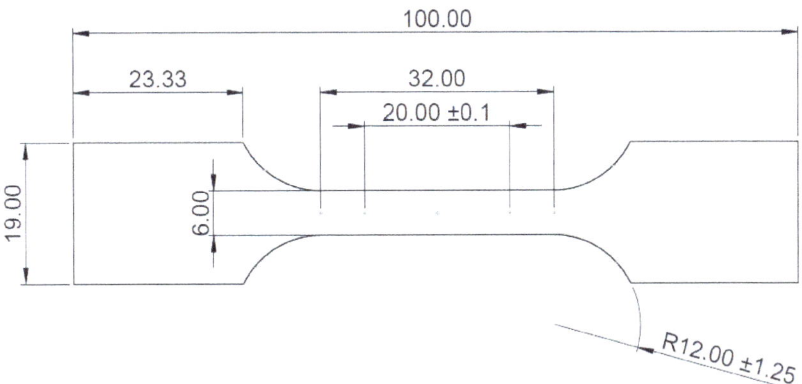

Figure 5. Geometry of tensile specimen based on ASTM-E8/EM (dimensions in mm).

A 150 kN Zwick/Roell Z150 testing machine was used to perform room temperature (23 °C) uniaxial tensile tests. All tests were carried out at a constant strain rate of 0.001 s^{-1}, controlled by testXpert II software and in accordance with ISO 6892-1:2019. An extensometer measured the elongation of the specimen. The test ended at tensile specimen failure. Twelve sets of tensile specimens were extracted from the LMD tailored sheet with different configurations of the LMD build height and build orientation, see Table 4. For each configuration, three repeat tests were performed to validate results. In total, 36 tensile tests were performed. In-situ DIC analysis measured local surface strain during tensile deformation. A two-camera setup captured strain data with a fixed frequency of 1 Hz and strain maps were calculated by measuring the displacement of the surface speckle pattern using DaVis 8.0 software supplied by LaVision.

Following mechanical testing, material characterisation of the LMD tailored material was conducted. The LMD surface morphology was analysed using an Alicona Infinite Focus microscope. The tensile samples were mounted, cleaned with acetone and the fracture face examined using a Quanta 250 FEG SEM. Following this, grinding and polishing steps were performed to prepare the sample surfaces for further characterisation. Material hardness was measured using a Vickers Micro Hardness Tester by applying a force of 1 kg for 15 s using a diamond indenter. Samples were polished and the surface etched to expose the grain boundaries for optical microscopy. An electropolishing step was used to prepare material samples for microstructural examination. A polishing current of 0.14 A and voltage 35 V was used to polish a 1 cm^2 area with A3 electrolyte. Electron backscatter diffraction (EBSD) was performed with step size 0.2 µm to observe the material microstructure using an SEM operated with accelerating voltage of 20 kV.

Room temperature SPIF was performed on the LMD + SPIF preform, Figure 6. A DMU 125 FD Duo Block 5-axis CNC machine was used to perform SPIF. The SPIF fixture was secured to the machine bed, Figure 6a. Scissor lifts lifted the top plate of the SPIF fixture into position above the die, the LMD fixture with the LMD + SPIF preform attached was installed into the SPIF fixture, and the scissor lifts removed. The toolpath was generated using Autodesk Fusion 360. Vericut was used to test for collisions and the programme uploaded to CIMCO and made available to the CNC machine operator. SPIF was stopped after every 10 mm of forming and the CNC machine door unlocked to check the part for signs of failure, measure the temperature of the tool, and lubricate the sheet and tool. SPIF was completed and the formed part (LMD + SPIF part) taken for analysis, Figure 6b.

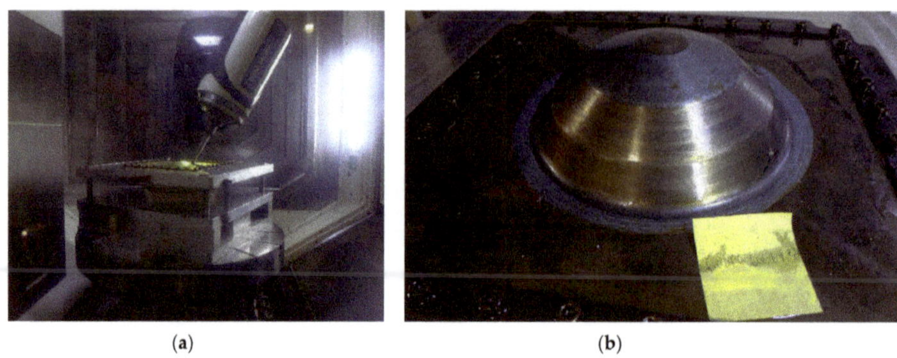

Figure 6. Performing SPIF on LMD tailored part: (a) Modular fixture with sheet attached during SPIF. (b) Final SPIF part with fracture.

3. Results

3.1. Microstructural Response to LMD

Optical microscopy was used to view the microstructural development across the cross-section of the A-R, Figure 7a, and LMD material samples, Figure 7b,c. Near-surface grain growth is observable in the LMD tailored samples to a depth equal to the added thickness (Zone 1), Figure 8a,d. Near the base of each sample there is a fine equiaxed microstructure which is indicative of the non-transformed substrate material (Zone 3), Figure 8c,f. The microstructure in Zone 3 is comparable to the A-R material, Figure 7a. At the interface between Zone 1 and 3 is the HAZ (Zone 2), Figure 8b,e. The microstructural change in the HAZ was likely a result of the thermal input from the laser during LMD. With the equipment available it was not possible to measure the temperature during deposition, however it is expected to have risen above the CP-Ti transus temperature of 882 °C [24]. This thermal input from the laser and subsequent cooling is likely the cause for the $\alpha > \beta > \alpha$ phase change. As such, Zone 2 contains a platelet alpha microstructure, typical for CP-Ti when cooled from the high temperature body centered cubic (bcc) β-phase field [25]. An increase in the HAZ thickness from 0.9 mm to 1 mm and a decrease in the un-transformed substrate from 0.7 mm to 0.5 mm is observed in response to the additional LMD added thickness, Figure 7b,c.

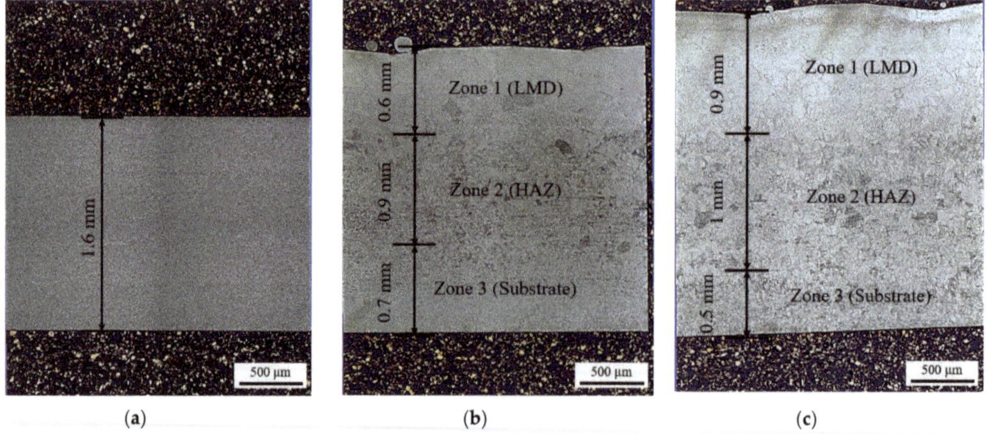

Figure 7. Optical images of material sample cross-sections at ×10 magnification with approximate LMD layer, heat affected zone (HAZ) and unchanged substrate defined: (a) A-R material; (b) 0.6 mm LMD sample; (c) 0.9 mm LMD sample.

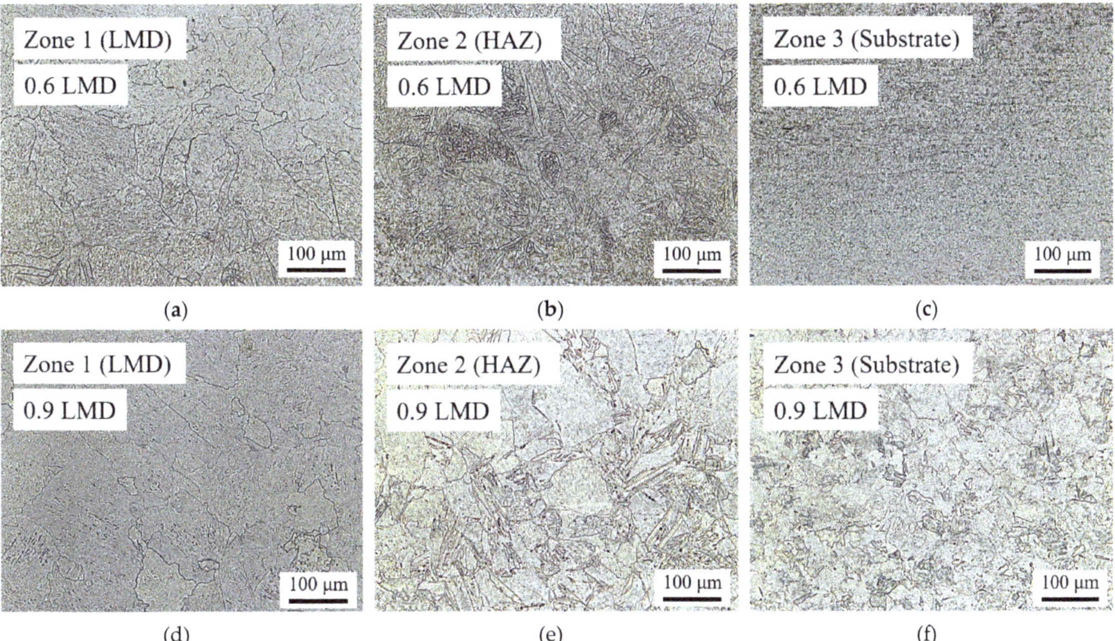

Figure 8. Optical images of 0.6 mm and 0.9 mm LMD samples across cross-section at ×20 magnification: (**a**) Zone 1 (LMD tailored material) of 0.6 mm LMD sample; (**b**) Zone 2 (Heat affected zone, HAZ) of 0.6 mm LMD sample; (**c**) Zone 3 (Non-transformed substrate) of 0.6 mm LMD sample; (**d**) Zone 1 (LMD tailored material) of 0.9 mm LMD sample; (**e**) Zone 2 (Heat affected zone, HAZ) of 0.9 mm LMD sample; (**f**) Zone 3 (Non-transformed substrate) of 0.9 mm LMD sample.

The difficult sample preparation process for CP-Ti meant the microstructure was not clearly visible by optical microscopy. As such, EBSD analysis was performed on samples extracted from the LMD thickness to view its microstructure in greater detail. Grinding to depths of 0.3–0.6 mm below the material surface exposed the microstructure within Zone 1. Inverse pole figure (IPF) maps were generated, Figure 9. The RGB colour code: red for (0001), green for $(12\bar{1}0)$ and blue for $(01\bar{1}0)$ as shown in the standard stereographic triangle corresponds to the crystallographic orientation of each grain. The microstructure of the LMD part remains randomly orientated, Figure 9b,c, similar to the A-R microstructure, Figure 9a. Low magnification maps were used to measure the average grain size of the LMD tailored material using the intercept method. For the 0.3 mm thickened material, 158 grains were measured with an average grain size of 54.5 ± 0.2 µm and size range of 8.1 µm–250 µm (SD = 56.9). For the 0.9 mm thickened material 158 grains were measured with an average grain size of 39 ± 0.2 µm and size range of 2.2 µm–326.2 µm (SD = 57.7). The A-R material has an average grain size of 5.4 ± 0.2 µm (SD = 3.3). A grain growth of approximately x10 is observed in response to LMD. The higher SD of the LMD tailored microstructure suggests significant spread in grain size, in contrast to the lower SD of the A-R material which suggests a normal distribution of grain size. Twinning was observed in the LMD material. Some twins are seen to have nucleated and terminated on grain boundaries, labelled T1 in Figure 9d, whereas other twins nucleated on a boundary and terminated within the grain interior, labelled T2. Very small amounts of leftover transformed beta (β) phase is present in the LMD tailored material, likely a leftover during a phase transition in response to the LMD heating cycle.

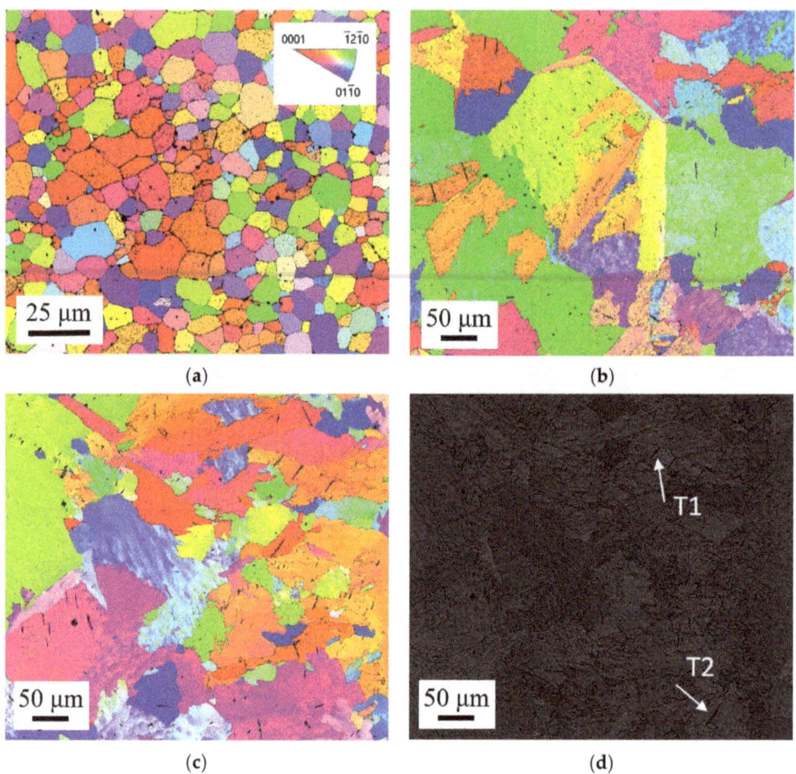

Figure 9. EBSD analysis of material specimens extracted from LMD tailored material: (**a**) IPF map for A-R specimen with no LMD thickening (**b**) IPF map of 0.3 mm LMD layer; (**c**) IPF map of 0.9 mm LMD layer; (**d**) Band contrast image of 0.9 mm LMD layer showing twinning.

A high magnification coloured IPF map was performed to examine the microstructure of the LMD part in greater detail, Figure 10a. The map contains several single grains and a fragmented large grain with intragranular misorientations. The single grains are presumed to be the microstructure of the melted deposit material. The fragmented grain has an internal substructure containing partially and fully enclosed grains with low angle grain boundaries (LAGB) represented by blue lines, and several fully enclosed grains with high angle grain boundaries (HAGB) with misorientation over 15°, represented by black lines. The substructure has a single crystal orientation as determined by the IPF colouring. A misorientation angle measurement across a fully enclosed HAB grain, labelled L1 in Figure 10b, reveals that the misorientation is just over 15° which confirms the internal HAGBs evolved from LAGBs with the accumulation of dislocations. An incomplete LAGB misorientation, labelled L2 in Figure 10c, has a misorientation of just under 8° which suggests further accumulation of dislocations could result in a fully enclosed LAB grain which could develop into a HAB grain. The size and shape of the LAB segments are close to the fine equiaxed microstructure of the A-R CP-Ti50A material, Figure 9a. The occurrence of incomplete grains is a strong indication that continuous dynamic crystallisation (CDRX) occurred [26].

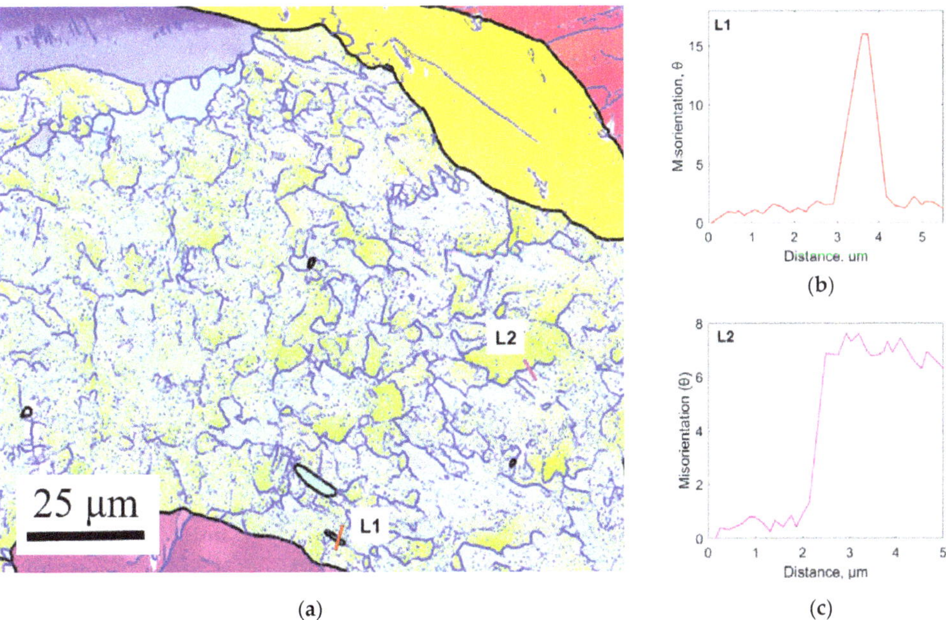

(a) (b) (c)

Figure 10. EBSD analysis of LMD tailored metal: (**a**) x2000 magnified coloured IPF map of 0.9 mm LMD layer; (**b**) misorientation across fully enclosed HAGB; (**c**) misorientation across segmented LAGB.

3.2. Mechanical Response to LMD Tailoring

Tensile test specimens from the tailored sheet were prepared along the RD, 45° to RD and TD directions for uniaxial tensile testing. The tensile test for each specimen was repeated three times and averaged. The tensile test data was plotted as shown in Figure 11a,c,e and the tensile specimens at fracture are seen in Figure 11b,d,f. The details of the tensile test are tabulated in Table 6. The CP-Ti50A sheet material in the A-R form exhibits anisotropic tensile behaviour, Figure 1b. In the RD, the A-R material has a higher ultimate tensile strength (UTS) and greater elongation before necking and fracture than at 45° to RD. This anisotropy is a direct response to the pronounced texture and limited number of slip systems in hexagonally close-packed (hcp) materials such as CP titanium. Despite the anisotropic tensile behaviour of the substrate material, the specimens extracted from the LMD part exhibited isotropic mechanical behaviour. This result suggests LMD tailoring of sheet preforms will allow for good design flexibility due to the omnidirectional mechanical behaviour across the thickened preform. In the RD, the LMD thickened material specimens showedreduced strength and a small reduction in ductility with more abrupt hardening suggesting increased brittleness compared to the A-R material, Figure 11a. The results show a 2.4–5.6% increase in the yield stress (YS), 4–6% reduction in elongation (E) and a 6.2–9.9% decrease in the ultimate tensile strength (UTS) compared to the A-R material. At 45° to the RD the LMD material exhibited a 5.2–9.9% increase in the YS, 11–14% reduction in E which indicates lower ductility, and similar UTS values of a range 0–4.6% higher than the A-R material, Figure 11c. In the TD there was little change in the mechanical behaviour of the LMD tailored material compared to the A-R material with a 1.1–2.7% decrease in the YP, 3–6% decrease in the E and a 3.5–5.3% decrease in UTS values, Figure 11e. In general, the results indicate that LMD thickening reduces elongation and formability. A larger decrease in ductility was seen in the RD and 45° to RD directions. A noteworthy observation is the lack of influence the number of stacked LMD layers had on the strength properties of the thickened material. This suggests the major change in the crystallographic microstructure occurred at the initial introduction of the LMD material and remained constant as more layers were deposited. The tensile results indicate there are unlikely to be significant defects in the added LMD material, such as pores. The material performs well and would suggest the introduction of an added thickness via LMD is a potential route for tailoring preforms for SPIF. In summary, the isotropic nature of the LMD tailored material and lack of influence of the LMD build height ensure that anisotropy and geometry are less relevant when designing LMD tailored SPIF preforms.

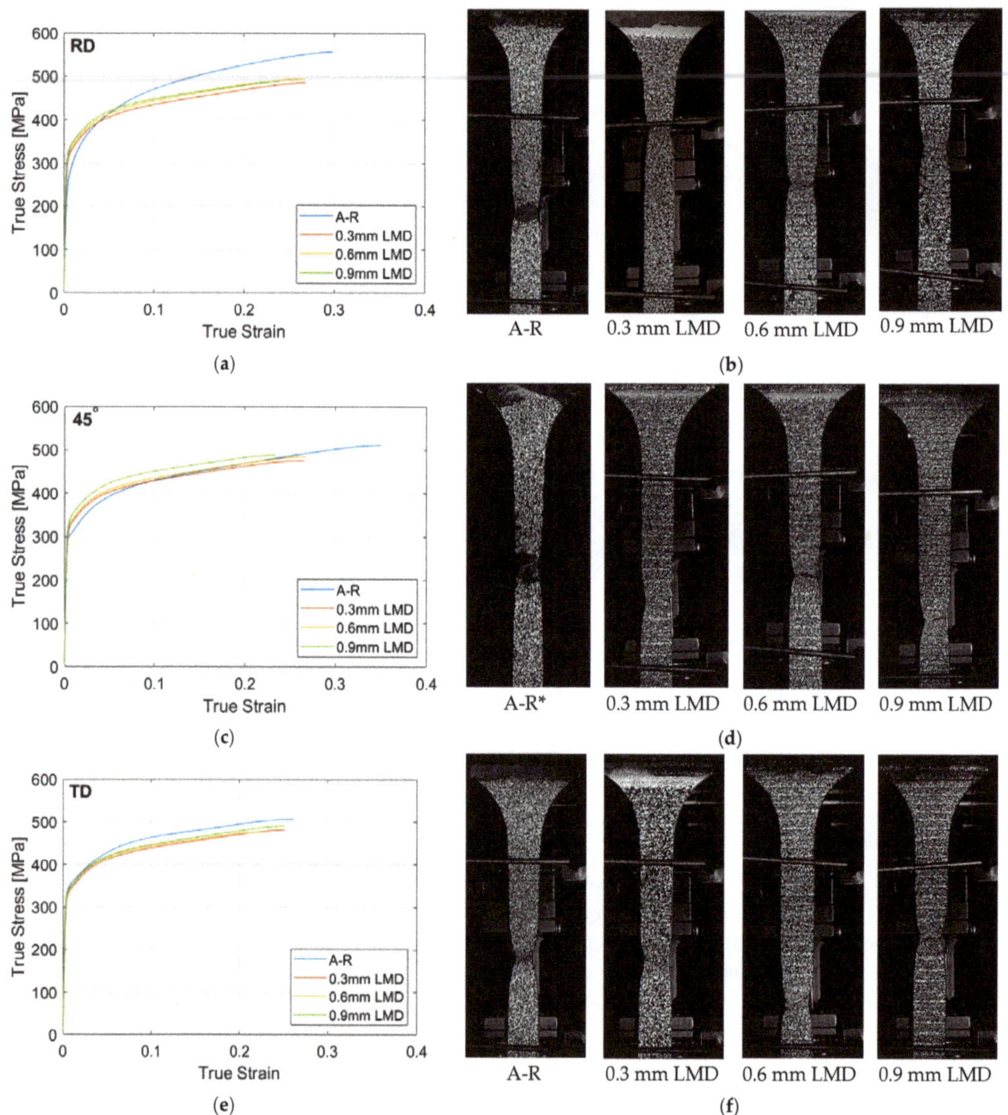

Figure 11. Results of uniaxial tensile testing: (**a**) True stress (α) vs. strain (ε) curves in the sheet RD; (**b**) RD tensile specimens at fracture; (**c**) True α vs. ε curves in the diagonal direction (45°); (**d**) 45° diagonal tensile specimens at fracture * Extensometer was removed due to amount of elongation and plot was calculated from raw strain data; (**f**) True α vs. ε curves in the TD direction; (**e**) TD tensile specimens at fracture.

Table 6. Material properties for CP-Ti50A sheet as-received and with LMD added thickness of variable levels.

Material Location (See Table 4 for Variables)	YS (0.2% Proof Stress) (MPa)	E (% Increase, 20 mm Gauge)	UTS (MPa)
Set a	353 ± 1.25	43 ± 0.75	432 ± 0.78
Set b	362 ± 1.02	38 ± 3.12	393 ± 2.34
Set c	367 ± 0.75	40 ± 3.51	401 ± 0.84
Set d	374 ± 2.36	38 ± 0.93	406 ± 0.39
Set e	343 ± 1.62	52 ± 0.29	389 ± 1.21
Set f	361 ± 0.32	40 ± 3.55	388 ± 1.09
Set g	368 ± 0.98	41 ± 2.85	394 ± 1.41
Set h	377 ± 1.73	38 ± 1.01	407 ± 1.67
Set i	378 ± 0.83	43 ± 0.84	420 ± 0.95
Set j	368 ± 2.49	38 ± 1.16	398 ± 3.63
Set k	373 ± 1.07	37 ± 2.05	402 ± 0.72
Set l	374 ± 1.44	40 ± 3.02	405 ± 1.50

3.3. Strain Analysis of LMD Material

Optical non-contact measurement techniques were used to show effective strain displacement in the plastic region of the tensile specimens during tensile deformation. At 90% extension in the A-R RD sample, strain was seen to concentrate near the gauge centre and localised necking was observed, followed by diffused necking and reduction in the gauge width before failure. High strain localisation was seen to initiate at the edge of each sample before failure, at which a crack nucleated and travelled through the specimen. The DIC results indicate the necking morphology changed in response to the introduction of an LMD layer. With the specimens from the LMD part, strain localised in a smaller area during necking and eventual fracture. The A-R specimen had a diffused neck of 10 mm and localised neck of 5 mm at 90% extension, Figure 12a. The LMD tailored specimens in the RD had diffused necks of 5 mm and localised necks of 1–1.5 mm at approximately 90% extension, Figure 12b–d. Strain was seen to localise within shallow grooves between LMD tracks compared to the homogeneous strain distribution across the surface of the AR specimen. Similar strain behaviour was observed for all samples oriented 45° to RD and TD, Figures 13 and 14. This suggests the surface morphology from the LMD tracks influenced strain distribution. The LMD tailored specimens exhibited lower elongation and earlier fracture nucleation and failure compared to the A-R specimens as indicated by the true stress–strain curves. This is likely a result of cracks nucleating from defects in the heterogeneous LMD surface, causing necking to initiate at earlier strain during tensile deformation in comparison to the more homogeneous A-R material specimens.

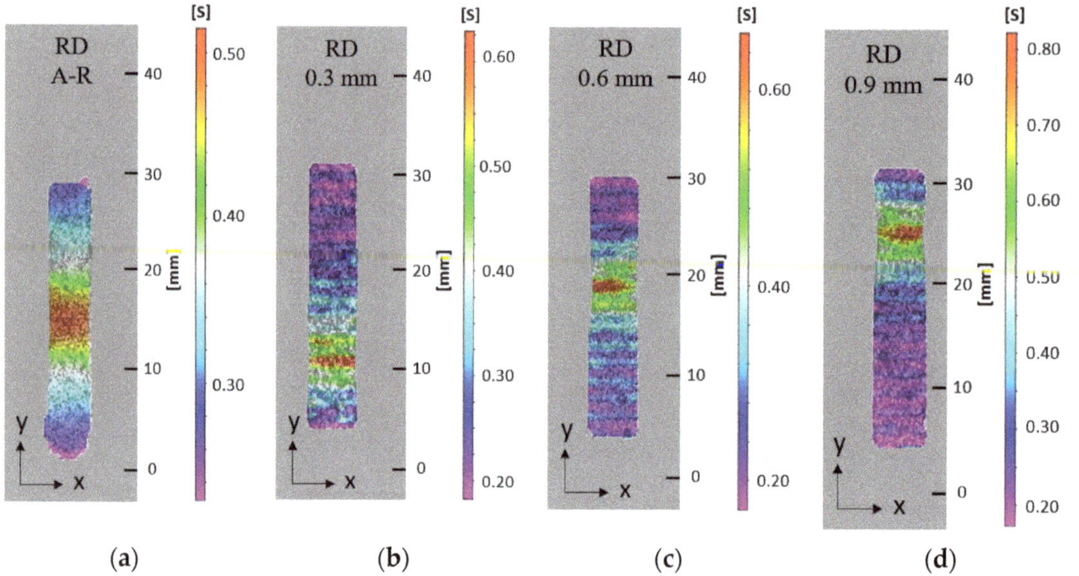

Figure 12. Effective surface strain at 90% extension on tensile samples orientated in substrate RD: (**a**) A-R material; (**b**) 0.3 mm LMD sample; (**c**) 0.6 mm LMD sample; (**d**) 0.9 mm LMD sample.

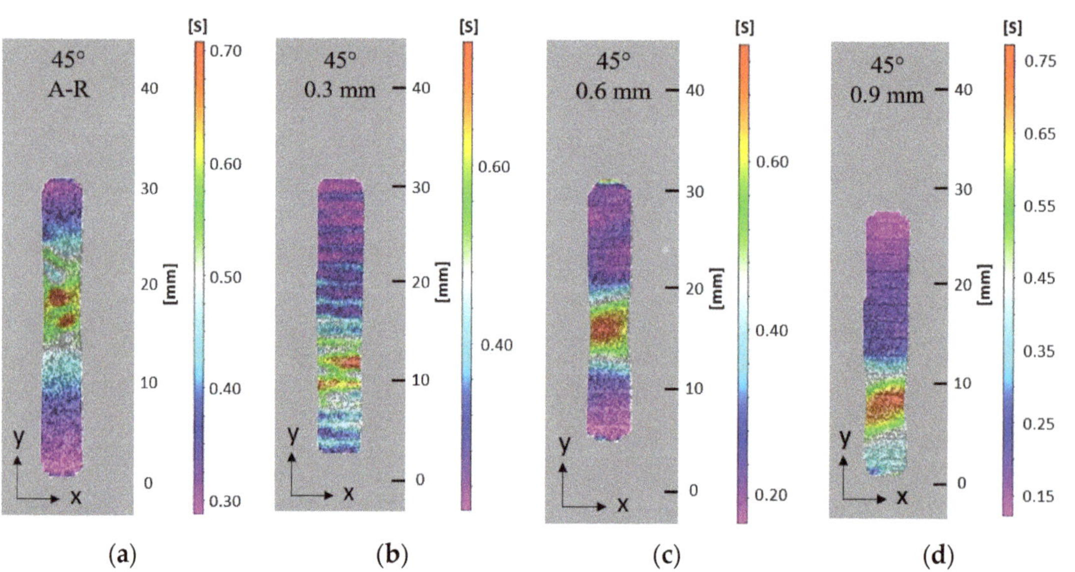

Figure 13. Effective surface strain at 90% extension on tensile samples orientated 45° to substrate RD: (**a**) A-R material; (**b**) 0.3 mm LMD sample; (**c**) 0.6 mm LMD sample; (**d**) 0.9 mm LMD sample.

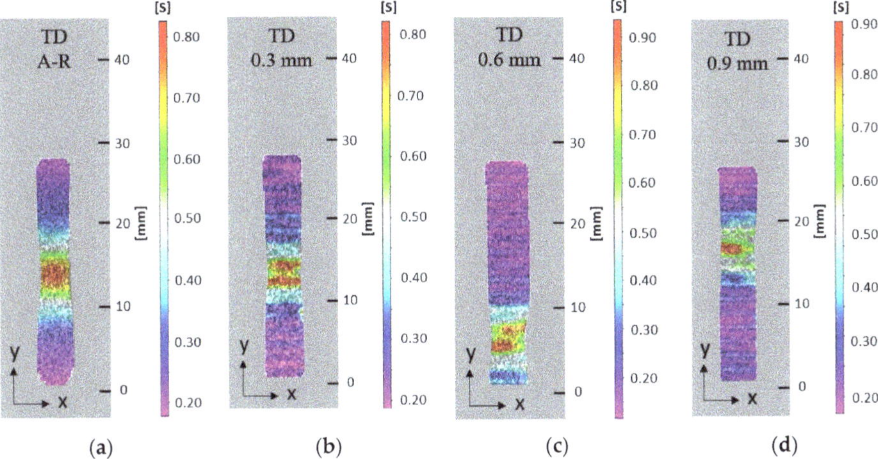

Figure 14. Effective surface strain at 90% extension on tensile samples orientated in substrate TD: (**a**) A-R material; (**b**) 0.3 mm LMD sample; (**c**) 0.6 mm LMD sample; (**d**) 0.9 mm LMD sample.

A speckle pattern was sprayed on the reverse side of selected specimens to compare the strain distribution on the LMD surface (front) and reverse side with no LMD (back), Figure 15. At 90% extension the diffused neck measured 5 mm on both front and back of the RD and TD specimens with 0.9 mm added LMD material. Despite the same diffused necking, the high strain which indicates localised necking was 1–2 mm on the front and 3–4 mm on the back for the RD and TD specimens. This indicates surface morphology likely impacted surface strain distribution. Higher surface strain at 90% extension occurred on the front compared to the back, with 0.80 and 0.90 strain values for front RD and TD specimens, respectively, compared to 0.72 and 0.68 strain values for back RD and TD specimens, respectively.

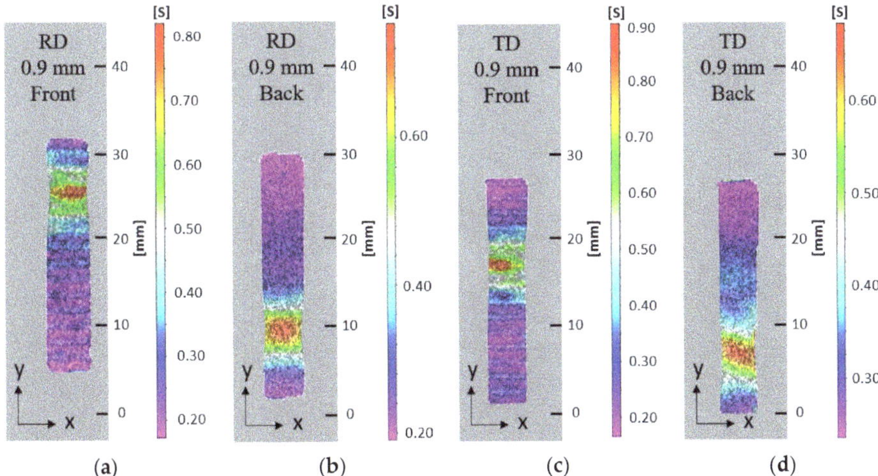

Figure 15. Effective surface strain at 90% extension on the LMD (front) surface and the reverse (back) side with no LMD: (**a**) Front of RD 0.9mm LMD sample; (**b**) Back of RD 0.9 mm LMD sample; (**c**) Front of TD 0.9 mm LMD sample; (**d**) Back of TD 0.9 mm LMD sample.

3.4. Material Study

Due to the influence of the LMD deposit morphology on the surface strain distribution, an Alicona microscope was used to measure the depth of grooves between each LMD track. A 3D representation of the surface was stitched together from layered images using OmniSurf 3D software, Figure 16a. The LMD surface morphology of the analysed areas contain parallel LMD tracks with relatively flat surfaces and shallow grooves. The average groove depth across all LMD sample surfaces was 73.5 ± 10.6 µm. Unconsolidated CP-Ti powder was seen to collect in the track valleys, Figure 16b. This is likely a result of this region being located at the cooler outer regions of the laser melt pool. The unconsolidated powder particles may have functioned as surface defects, acting as crack nucleation sites for early onset fracture. Machining is a potential route to improve the quality of the final LMD surface.

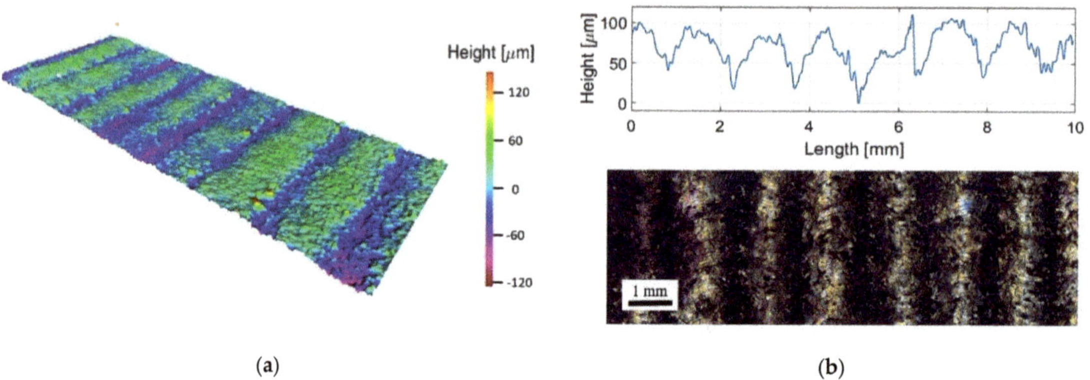

Figure 16. Surface morphology of RD 0.9 mm LMD sample: (**a**) LMD build surface morphology; (**b**) LMD track form with optical image showing unconsolidated powder particles in the grooves between tracks.

Fractography analysis was performed on the fracture face of the A-R tensile specimen and the sample with 0.9 mm added thickness by LMD, both with the gauge length orientated in the sheet RD. Fractography was performed using the FEI Quanta 250 FEG SEM at high magnification. The centre of the fracture face on the A-R sample (Zone 1) is characterised by a fine dimple texture accompanied by microscopic voids typical of ductile fracture, Figure 17b. The edge of the A-R sample has small dimples, micro-voids and evidence of shear, Figure 17c. The LMD surface (Zone 1) and non-transformed substrate material (Zone 2) have distinct fracture morphologies, Figure 18a. No delamination of the LMD material is evident with good bonding across the interface region of Zone 1 and Zone 2. Within Zone 1 are large equiaxed voids with groups of coalesced microscopic voids at their base, Figure 18b. Zone 2, Figure 18c, has similar fracture morphology to the edge of the A-R specimen, Figure 17c, with micro-voids propagating in the direction of the stress axis and visible surface shear, suggesting the microstructure in Zone 2 was not altered significantly by LMD.

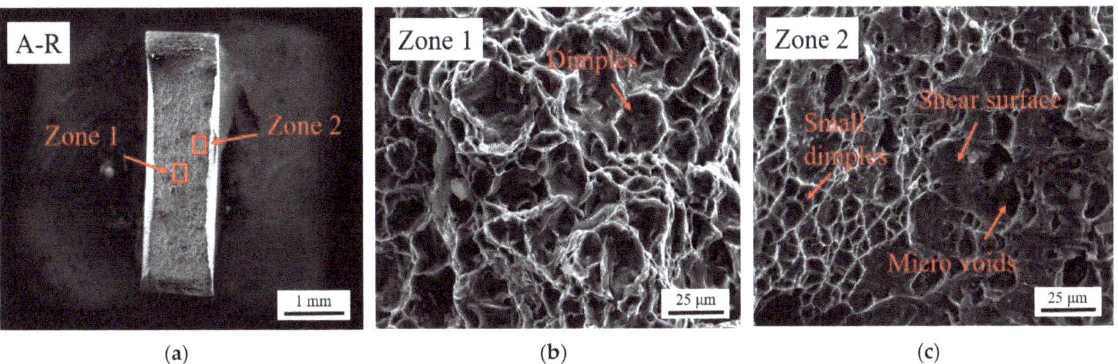

Figure 17. Fractography analysis of A-R material sample: (**a**) Selected areas showing type of fracture at middle (Zone 1) and edge of fracture face (Zone 2); (**b**) High magnification of Zone 1; (**c**) High magnification of Zone 2.

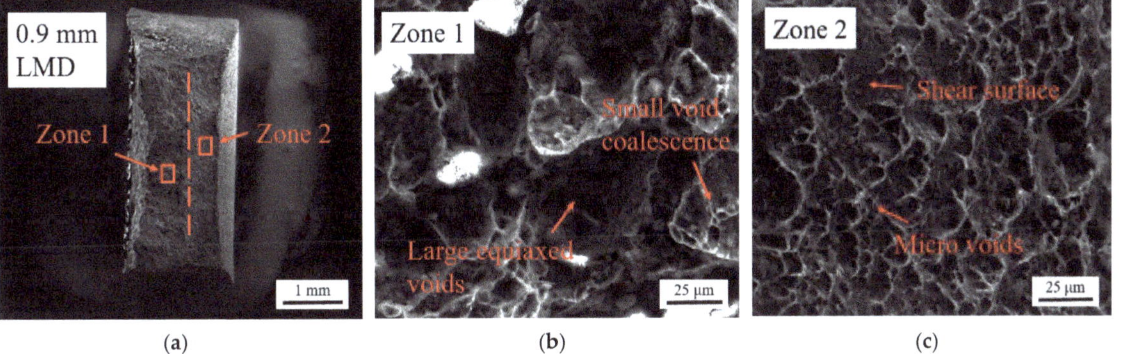

Figure 18. Fractography analysis of 0.9 mm LMD sample: (**a**) Differentiation between LMD added thickness (Zone 1) and non-transformed substrate (Zone 2); (**b**) High magnification of Zone 1; (**c**) High magnification of Zone 2.

Vickers hardness testing of the A-R material and LMD part was performed. For the in-plane measurements a depth of 280 μm was removed by grinding to measure the microhardness of the added LMD layer, Figure 19a. The hardness results for the A-R material are highly repeatable with a mean of 153 ± 3 HV. The hardness results for the tailored materials exhibit large scattering which is likely due to the inhomogeneous microstructure caused by the grain growth [18]. The mean hardness values for the tailored materials are in the range of 159–166 HV with significant overlap of the standard deviation error bars showing the difference is not statistically significant. This indicates the hardness is consistent across the tailored material despite the difference in LMD build thickness. Microhardness measurements were performed along the gauge length of the fractured tensile samples, from the top grip region towards the fracture site, Figure 19b. A rise in microhardness across the gauge of approximately 173–182 HV was observed across all material samples.

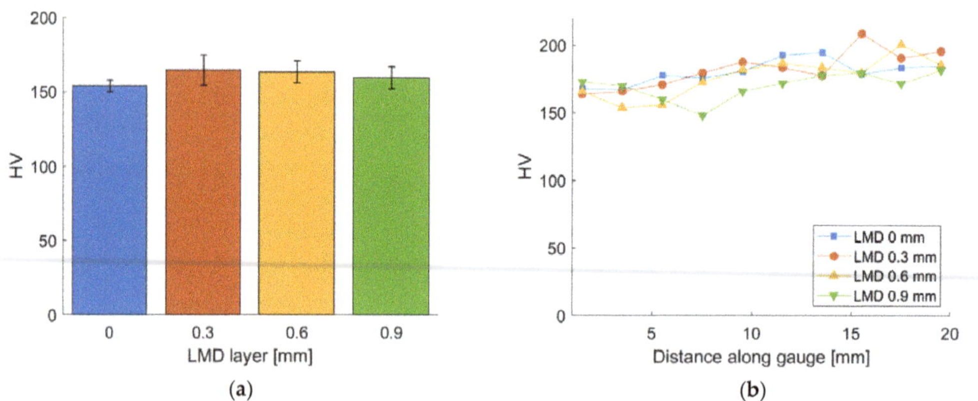

Figure 19. Vickers microhardness results: (**a**) In-plane microhardness of added LMD layer; (**b**) Microhardness along gauge length of fractured tensile specimens.

3.5. SPIF of LMD Tailored Preform

SPIF was performed on the LMD + SPIF preform until fracture. A three-dimensional representation was generated for the LMD + SPIF preform, Figure 20a, and the LMD + SPIF part, Figure 20b. The colour map shows thickness variation.

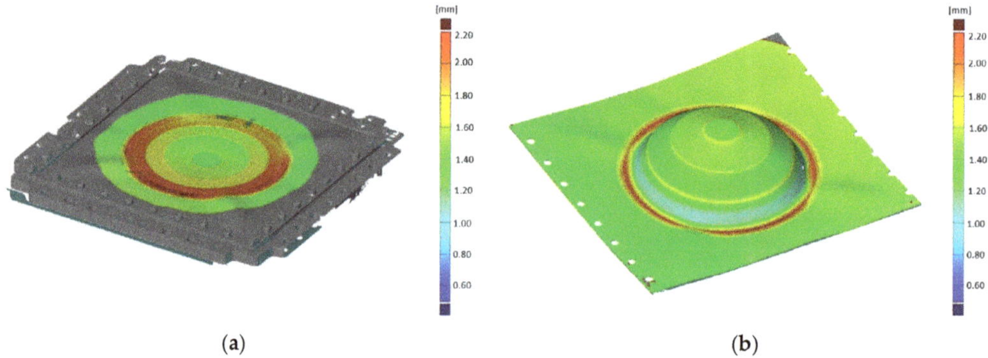

Figure 20. Three-dimensional scans of the LMD tailored SPIF preform and part: (**a**) Thickness distribution across LMD tailored SPIF preform; (**b**) Thickness distribution across LMD tailored SPIF part.

The thickness distribution was calculated across the SPIF part and LMD + SPIF parts, Figure 21a. The rise in thickness at the edge of the unformed section for the LMD + SPIF part is the unformed LMD thickened material. The results indicate the percentage (%) rate of thinning is equal for both formed sheets when measured directly from the as-received starting thickness, Figure 21b. However, when measured from the starting thickness of the reinforced wall sections, the % rate of thinning is lower for the LMD + SPIF part compared to the SPIF part, Figure 20c, suggesting greater thickness homogeneity in SPIF parts is achievable by using tailored preforms with additional thickness in regions of expected thinning.

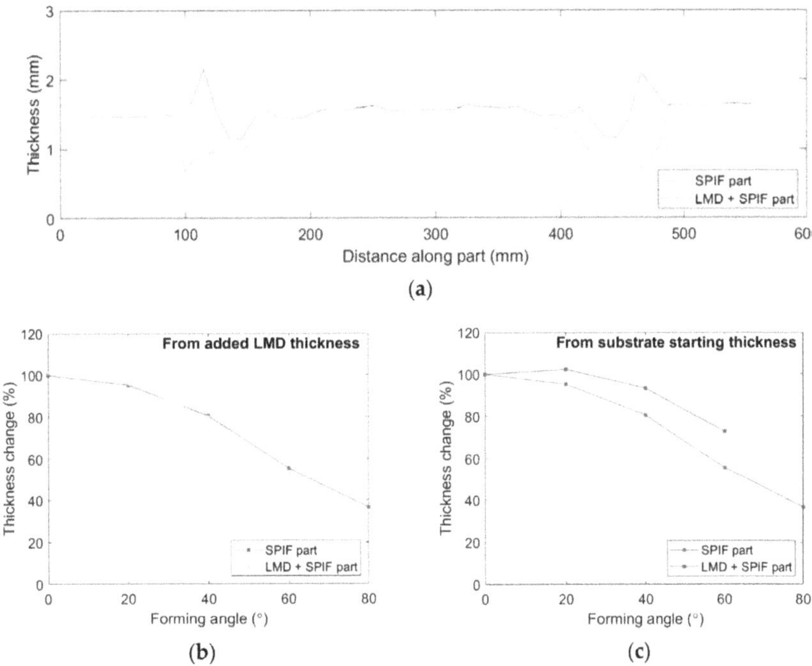

Figure 21. Thickness measurements of SPIF sheets with and without LMD added layer: (**a**) Thickness distribution across SPIF parts; (**b**) % change of each angled section from LMD added thickness; (**c**) % change of each angled section from substrate starting thickness.

Despite the results indicating improved thickness homogeneity, the LMD + SPIF preform exhibited cracks in the LMD surface and eventual fracture in the 60° wall angle section. In comparison, the SPIF part without LMD tailoring failed in the 80° wall angle section. Further analysis of the LMD + SPIF part is planned to better understand the cause of this failure with the goal of optimising the hybrid SPIF + LMD process.

4. Summary and Conclusions

The aim of this work was to generate a LMD tailored preform with variable thickness to mitigate thinning, a common defect in room temperature SPIF of titanium parts with high angled walls. An initial material study of a LMD tailored CP-Ti50A sheet with localised thickening was performed. Following this, SPIF was performed on a LMD tailored CP-Ti50A preform sheet. To facilitate the hybrid LMD + SPIF process a modular fixture was designed to constrain the tailored titanium sheet during LMD, post-processing, and SPIF.

The main findings of this work are summarised below:

- Microstructural analysis of the LMD tailored material showed distinct regions across the thickness of the material sample. This consisted of the LMD deposited layer, a HAZ interface zone and the non-transformed substrate material.
- EBSD analysis of the LMD tailored material showed grain growth in response to LMD. Fragmentation of large grains was observed with intragranular LAB misorientations and fully enclosed HAB grains.
- Isotropic mechanical properties were observed in the LMD tailored material in contrast to the anisotropic properties of the A-R CP-Ti50A sheet. The LMD build height appeared to have little influence on strength. As such, an SPIF preform can be designed without significant consideration of directionality, or concerns about the thickness of LMD material. The results indicate LMD reinforcement reduces material ductility and lowers formability of CP titanium.

- In-situ DIC analysis during tensile testing showed the LMD build geometry to alter effective strain distribution across the sample surface, with surface strain propagating parallel to the LMD beads. This differs from the homogeneous spread of surface strain across the A-R samples.
- Fractography analysis of the LMD thickened material fracture face showed a population of dimples and fine microscopic cracks, accompanied by large equiaxed voids with internal coalesced microscopic voids. No delamination of the deposit layer was evident suggesting good cohesion between the LMD layer and substrate.
- Microhardness values measured across the tailored material were statistically equal despite the difference in LMD build thickness.
- LMD tailoring was seen to achieve greater homogeneity in the SPIF part, however the part failed during forming the 60° wall angled section.

The result of this study indicates tailoring preforms with variable thickness offers a potential route to optimise room temperature SPIF of titanium parts. However, issues regarding this hybrid process have been discovered which must be addressed. One such issue is the limited formability of the LMD tailored material which limits the achievable wall angle for the SPIF part. A subsequent quality assessment and analysis of the fracture modes of the final sheet part is planned.

Author Contributions: Investigation, M.M.; formal analysis, M.M.; writing—original draft preparation, M.M.; writing—review and editing, E.Y. and P.B.; supervision, E.Y. and P.B.; funding acquisition, E.Y. All authors have read and agreed to the published version of the manuscript.

Funding: This research was funded by the Advanced Forming Research Centre, University of Strathclyde, and Faculty of Engineering under a Research Excellence Award.

Acknowledgments: The authors acknowledge the support of the Advanced Forming Research Centre (AFRC) and the Department of Design, Manufacturing & Engineering Management of the University of Strathclyde.

Conflicts of Interest: The authors declare no conflict of interest.

References

1. Allwood, J.M.; Shouler, D.R.; Tekkaya, A.E. The Increased Forming Limits of Incremental Sheet Forming Processes. *Key Eng. Mater.* **2007**, *344*, 621–628. [CrossRef]
2. More, K.R.; Sisodia, V.; Kumar, S. A Brief Review on Formability, Wall Thickness Distribution and Surface Roughness of Formed Part in Incremental Sheet Forming. In *Advances in Manufacturing Processes*; Lecture Notes in Mechanical Engineering; Springer: Singapore, 2021; pp. 135–149. [CrossRef]
3. Jackson, K.; Allwood, J. The mechanics of incremental sheet forming. *J. Mater. Process. Technol.* **2008**, *209*, 1158–1174. [CrossRef]
4. Oleksik, V.; Trzepieciński, T.; Szpunar, M.; Chodoła, Ł.; Ficek, D.; Szczęsny, I. Single-point incremental forming of titanium and titanium alloy sheets. *Materials* **2021**, *14*, 6372. [CrossRef] [PubMed]
5. Trzepiecinski, T.; Oleksik, V.; Pepelnjak, T.; Najm, S.; Paniti, I.; Maji, K. Emerging Trends in Single Point Incremental Sheet Forming of Lightweight Metals. *Metals* **2021**, *11*, 1188. [CrossRef]
6. Chen, F.K.; Chiu, K.H. Stamping formability of pure titanium sheets. *J. Mater. Process. Technol.* **2005**, *170*, 181–186. [CrossRef]
7. Jagtap, R.; Kumar, S. An experimental investigation on thinning and formability in hybrid incremental sheet forming process. *Procedia Manuf.* **2019**, *30*, 71–76. [CrossRef]
8. Behera, A.K.; de Sousa, R.A.; Ingarao, G.; Oleksik, V. Single point incremental forming: An assessment of the progress and technology trends from 2005 to 2015. *J. Manuf. Process.* **2017**, *27*, 37–62. [CrossRef]
9. Jeswiet, J.; Micari, F.; Hirt, G.; Bramley, A.; Duflou, J.; Allwood, J. Asymmetric single point incremental forming of sheet metal. *CIRP Ann.* **2005**, *54*, 88–114. [CrossRef]
10. Lu, H.; Liu, H.; Wang, C. Review on strategies for geometric accuracy improvement in incremental sheet forming. *Int. J. Adv. Manuf. Technol.* **2019**, *102*, 3381–3417. [CrossRef]
11. Shamsari, M.; Mirnia, M.J. Formability improvement in single point incremental forming of truncated cone using a two-stage hybrid deformation strategy. *Int. J. Adv. Manuf. Technol.* **2017**, *94*, 2357–2368. [CrossRef]
12. Ambrogio, G.; Gagliardi, F.; Serratore, G.; Ramundo, E.; Filice, L. SPIF of tailored sheets to optimize thickness distribution along the shaped wall. *Procedia Manuf.* **2019**, *29*, 80–87. [CrossRef]
13. Ambrogio, G.; Gagliardi, F.; Muzzupappa, M.; Filice, L. Additive-incremental forming hybrid manufacturing technique to improve customised part performance. *J. Manuf. Process.* **2018**, *37*, 386–391. [CrossRef]

14. Kumar, N.; Belokar, R.M.; Agrawal, A. Multi-objective optimization of quality characteristics in single point incremental forming process by response surface methodology. *IJEMS* **2019**, *26*, 254–264.
15. Raut, J.; Marathe, S.; Raval, H. Numerical Investigation on Single Point Incremental Forming (SPIF) of Tailor Welded Blanks (TWBs). In *Advances in Simulation, Product Design and Development*; Springer: Singapore, 2020; pp. 43–54.
16. Bambach, M.; Sviridov, A.; Weisheit, A.; Schleifenbaum, J.H. Case Studies on Local Reinforcement of Sheet Metal Components by Laser Additive Manufacturing. *Metals* **2017**, *7*, 113. [CrossRef]
17. Hama-Saleh, R.; Weisheit, A.; Schleifenbaum, J.H.; Ünsal, I.; Sviridov, A.; Bambach, M. Formability Analysis of Micro-Alloyed Sheet Metals Reinforced by Additive Manufacturing. *Procedia Manuf.* **2020**, *47*, 1023–1028. [CrossRef]
18. Yu, J.; Rombouts, M.; Maes, G.; Motmans, F. Material Properties of Ti6Al4V Parts Produced by Laser Metal Deposition. *Phys. Procedia* **2012**, *39*, 416–424. [CrossRef]
19. Barro, Ó.; Arias-González, F.; Lusquiños, F.; Comesaña, R.; Del Val, J.; Riveiro, A.; Badaoui, A.; Gómez-Baño, F.; Pou, J. Improved commercially pure titanium obtained by laser directed energy deposition for dental prosthetic applications. *Metals* **2020**, *11*, 70. [CrossRef]
20. Gu, D.D.; Meiners, W.; Wissenbach, K.; Poprawe, R. Wissenbach, Laser additive manufacturing of metallic components: Materials, processes, and mechanisms. *Int. Mater. Rev.* **2012**, *57*, 133–164. [CrossRef]
21. MatWeb, TIMET TIMETAL®50A CP Titanium (ASTM Grade 2). Available online: http://www.matweb.com/search/datasheet.aspx?matguid=247af0c81c8a4a9da817bafa791d20bf&ckck=1 (accessed on 8 February 2022).
22. AP&C, Advanced Powders. Available online: https://www.advancedpowders.com/powders/titanium/cp-ti2 (accessed on 31 March 2022).
23. Kusuma, C.; Ahmed, S.H.; Mian, A.; Srinivasan, R. Effect of Laser Power and Scan Speed on Melt Pool Characteristics of Commercially Pure Titanium (CP-Ti). *J. Mater. Eng. Perform.* **2017**, *26*, 3560–3568. [CrossRef]
24. Lutjering, G.; Williams, J.C. *Titanium*, 2nd ed.; Springer: Hamburg, Germany, 2007.
25. Qarni, M.J.; Sivaswamy, G.; Rosochowski, A.; Boczkal, S. On the evolution of microstructure and texture in commercial purity titanium during multiple passes of incremental equal channel angular pressing (I-ECAP). *Mater. Sci. Eng. A* **2017**, *699*, 31–47. [CrossRef]
26. Yamanaka, K.; Saito, W.; Mori, M.; Matsumoto, H.; Sato, S.; Chiba, A. Abnormal grain growth in commercially pure titanium during additive manufacturing with electron beam melting. *Materialia* **2019**, *6*, 100281. [CrossRef]

Article

Integrated Numerical Simulations and Experimental Measurements for the Sintering Process of Injection-Molded Ti-6Al-4V Alloy

Shaohua Su [1,2], Zijian Hong [1,3,*], Yuhui Huang [1,*], Peng Wang [2], Xiaobao Li [2], Junwen Wu [2] and Yongjun Wu [1,*]

1 School of Materials Science & Engineering, Zhejiang University, Hangzhou 310027, China
2 Jiangsu Gian Technology Co., Ltd., Changzhou 213016, China
3 Cyrus Tang Center for Sensor Materials and Applications, State Key Laboratory of Silicon Materials, Zhejiang University, Hangzhou 310027, China
* Correspondence: hongzijian100@zju.edu.cn (Z.H.); huangyuhui@zju.edu.cn (Y.H.); yongjunwu@zju.edu.cn (Y.W.)

Abstract: Metal injection molding (MIM) is an advanced manufacturing technology that enables the mass production of high-performance and complex materials, such as the Ti-6Al-4V alloy. The determination of the size change and deformation of the Ti-6Al-4V alloy after the sintering process is challenging and critical for quality control. The numerical simulation could be a fast and cost-effective way to predict size change and deformation, given the large degrees of freedom for the sintering process. Herein, a finite element method based on the thermal-elastic-viscoplastic macroscopic model is developed to predict the shrinkage, deformation, relative density, and crack of injection-molded Ti-6Al-4V after sintering, using the Simufact software. Excellent agreements between experimental measurements and numerical simulations of the size and deformation are demonstrated (within a 3% error), confirming the accuracy of the numerical model. This approach can serve as a guideline for the mold design and sintering optimization of the MIM process.

Keywords: Ti-6Al-4V alloy; metal injection molding; numerical simulation; sintering process

Citation: Su, S.; Hong, Z.; Huang, Y.; Wang, P.; Li, X.; Wu, J.; Wu, Y. Integrated Numerical Simulations and Experimental Measurements for the Sintering Process of Injection-Molded Ti-6Al-4V Alloy. *Materials* **2022**, *15*, 8109. https://doi.org/10.3390/ma15228109

Academic Editors: Jai-Sung Lee, Mateusz Kopec and Denis Politis

Received: 29 September 2022
Accepted: 12 November 2022
Published: 16 November 2022

1. Introduction

The Ti-6Al-4V alloy has advantages, including its low density, high strength, good biocompatibility, and excellent heat and corrosion resistance, as well as being non-magnetic and luxurious [1]. Such superior features promote its applications in various fields, such as aerospace, transportation, chemical, medical, and wearable industries [2–4]. Due to the relatively low thermal conductivity (about 1/5 of iron) and Young's modulus (about half of iron) [5–7], the Ti-6Al-4V alloy is difficult to be processed by traditional methods, such as machining or forging. As one of the state-of-the-art advanced manufacturing technologies, metal injection molding (MIM) has become the major manufacturing technique for complex Ti-6Al-4V materials, with a relatively low-cost and near-net-shape process [8–14]. MIM combines the shape-making complexity of plastic injection molding with the material flexibility of powder metallurgy, which is composed of four sequential steps: mixing, injection molding, debinding, and sintering [15,16].

Sintering is one of the key processes for MIM that determines the accuracy and performance of the materials. It is a thermal process below the melting point of the main constituent, which could facilitate particle bonding through solid-state diffusion that increases structural strength and density and reduces system energy [17–20]. The sintering phenomenon involves particle fusion, volume reduction, porosity reduction, and grain growth. This process is largely influenced by multiple factors, including particle size, particle shape, composition, initial density, heating rate, sintering peak temperature, sintering holding time, sintering pressure, sintering atmosphere, etc. [21–23]. For instance, the brown part before sintering is a porous packing of loose powder held together by weak surface bonds with initial porosity of ~35% to 50%. During sintering, the individual particles fuse to create a dense, strong monolithic part, which undergoes a large shrinkage during the sintering process, generally 10% to 20% [24].

Because of the large shrinkage in the sintering process, determination of the size change and deformation is necessary yet challenging for industrial mass production to produce the materials in near-net-shape and high quality. Given the complexity of the parameter space for the sintering process, as compared to the conventional trial and error method, numerical simulation of the sintering stage could be a fast and cost-effective alternative to predict the size change and deformation [25]. This could enable the accurate design of the furnace and sintering parameters (sintering temperature, heating rate, etc.) towards customized products with controlled quality while largely minimizing the experimental time and cost. Currently, there are different methodologies to model sintering processes, including continuum, micromechanical, multiparticle, and molecular dynamics approaches, which cover multiple length scales. Among these methodologies, continuum models, such as finite element analysis, could predict relevant attributes, such as shrinkage, deformation, and density profile, at a relatively large length scale [26–30]. Nosewicz et al. [31] presented a viscoelastic model of powder sintering developed within the discrete element framework This model has been applied to simulate the free and pressure sintering process of Ni-Al Mohsin et al. [18] used a finite element (FE) method based on a thermo-kinetic model to describe the densification process of MIM copper during sintering. The research focused on measuring thermos physical properties and numerical simulation associated with the sintering process. Furthermore, some enhancements were suggested in the temperature field calculation of the FE model to mimic a real furnace condition. Jeong et al. [32] developed a unified model for describing compaction and sintering based on plasticity theory. A method for predicting the final dimensions of a powder material was proposed. The proposed model simulated the powder process continuously and simultaneously and was more effective than previous models that treat compaction and sintering separately. Kwon et al. [16] studied the simulation of the sintering densification and shrinkage behavior of powder-injection-molded 17-4 PH stainless steel. The predictive capability of the model was verified by comparing the theoretical calculations with the experimentally observed variation in sintering shrinkage of the samples determined by dilatometry. Sahli et al. [21] studied the sintering of micro-parts using the powder injection molding process, with numerical simulations and experimental analysis. They concluded that the FE simulation results had better agreement with the experiments at high temperatures. Although the finite-element-based continuum simulation method has been widely employed in the modeling of the sintering process, there are barely studies on the numerical simulation and experimental analysis of the Ti-6Al-4V sintering process by MIM. As compared to other metal materials, Ti-6Al-4V by MIM has lower maturity, higher experimental cost, and greater difficulty in densification and dimensional control; the integrated theoretical and experimental understandings for the sintering process of injection-molded Ti-6Al-4V is, thus, greatly needed.

Herein, we aim to simulate the shrinkage, deformation, relative density, and crack of the injection-molded Ti-6Al-4V after sintering, using the finite element method. Thermo-elasto-viscoplastic constitutive law based on continuum mechanics was employed to describe the sintering process. Numerical simulations of the Ti-6Al-4V round and elongated specimens were carried out with the Simufact software and compared with experimental results. It is shown that our finite-element-based sintering model could accurately predict the densification behaviors of the specimens (within a 3% error, as compared to the experimental data). This study can serve as a guideline and example for the mold design and sintering optimization of the MIM process towards industrial applications.

2. Methods

The working flow of this research is shown in Figure 1. Numerical simulation and experiment of Ti-6Al-4V sintering by MIM were carried out in parallel to do a point-by-point comparison. For numerical simulation, based on the continuum mechanics theory, a macroscopic sintering model based on the viscoplastic constitutive law (details given in Section 3) was established first, and then the parameters for Ti-6Al-4V were determined. Finally, the numerical simulations for the Ti-6Al-4V round specimen and elongated specimen were carried out. The FE method is a numerical computational method for solving a system of differential equations through approximation functions applied to each element, called domain-wise approximation. This method is very powerful for the typical complex geometries encountered in powder metallurgy. The factors affecting the sintering shrinkage and deformation of MIM parts, such as gravity, friction, and temperature, were analyzed [26]. Simufact software with a sintering module was used for the FE simulation to avoid unnecessary cost and time expenditures and improve materials quality [33].

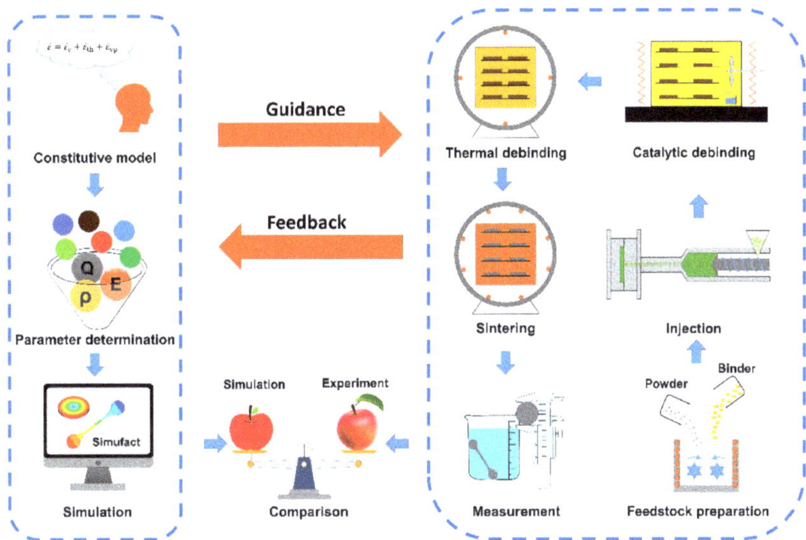

Figure 1. Schematic diagram of the workflow in this study.

On the other hand, Ti-6Al-4V was sintered experimentally using MIM. High pure atomized Ti-6Al-4V powder (Avimetal PM, Beijing, China) with a spherical shape was used as the raw material in this study. The feedstock was prepared with a polyformaldehyde (POM) -based binder system. The green parts were injected into an injection molding machine (NEX80IIIT, NISSEI, Nagano-ken, Japan) at a maximum injection pressure of 130 MPa. The injection temperatures from barrel to nozzle were 50 °C, 170 °C, 180 °C, 185 °C, 190 °C, and 195 °C. Catalytic debinding was performed in a catalytic debinding furnace (STZ-400L-OA, Sinterzone Company, Shenzhen, China) at 130 °C for 7 h with nitric acid as the catalytic medium. This process was mainly used to remove POM. Thermal debinding was performed in a graphite furnace (VM40/40/150, Hiper, Ningbo, China) at 700 °C for 2.5 h to remove the backbone binder. The sintering process was performed at 1100 °C for 6 h in a vacuum sintering furnace (pressure $< 10^{-3}$ Pa, VM42/45/125, Hiper, Ningbo, China). The sintering curve of Ti-6Al-4V is shown in Figure 2. The density of product specimens was measured by the Archimedes method with deionized water. The relative density was calculated as the ratio of the measured density to the theoretical density. Dimension was measured by vernier caliper.

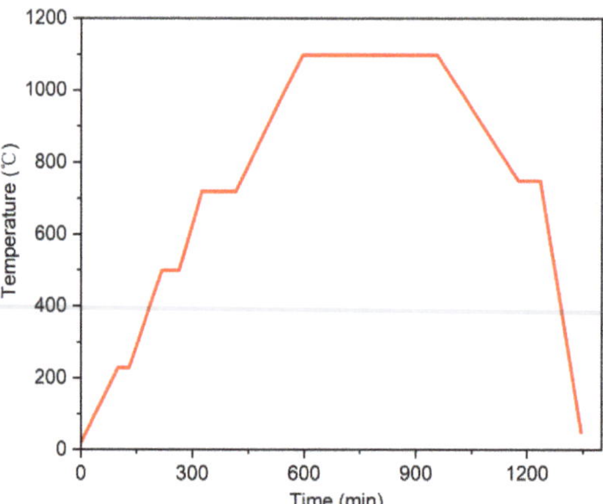

Figure 2. A typical sintering curve for Ti-6Al-4V.

Eventually, the numerical simulation and experiment results were compared apples to apples to verify the correctness of the sintering model.

3. Constitutive Model of Sintering Process

3.1. Principle

The sintering process provides energy to bond the individual powder particles together to eliminate the pores between particles. The sintering process can be divided into three stages: (1) aggregation of the particles and disappearance of the borders begin to produce a neck in the points of contact between particles. Grain boundaries are formed between two adjacent particles in the contact plane. The centers of particles are only slightly closer with very low shrinkage. (2) Pores are reduced due to the growth of the neck. These pores are reorganized and interconnected like a cylindrical channel to maximize the contact between the particles. The grain growth takes place later in this stage. (3) Pores are closed, isolated, and located principally at the boundaries or within the grains. This step is much slower than the first two stages. Additionally, the densification rate becomes slow and the grain growth is more evident [22].

The brown part composed of powders and pores after debinding is considered as a compressible continuum media in the macroscopic model. This porous particle has high surface energy, and, as the temperature increases, the movement of atoms intensifies, which decreases the surface energy. A viscoplastic constitutive law is adopted to describe the densification behavior of powders. In addition, the effects of elasticity and thermal expansion are taken into account. The densification process of sintered materials is affected by the evolution of the temperature field [34]. During the sintering process, elastic strain, thermal strain, and elastic–plastic strain are affected by the temperature change. Diffusion process, grain growth, creep, surface tension effect, gravity, friction, and thermal dependence are considered in the sintering process, with the coupling of several thermo-mechanical phenomena.

3.2. Thermo-Elasto-Viscoplastic Constitutive Law

Thermo-elasto-viscoplastic constitutive law is employed to describe the sintering process. The total strain rate $\dot{\varepsilon}$ consists of three parts: elastic strain rate $\dot{\varepsilon}_e$, thermal strain rate $\dot{\varepsilon}_{th}$, and viscoplastic strain rate $\dot{\varepsilon}_{vp}$, i.e., [35]

$$\dot{\varepsilon} = \dot{\varepsilon}_e + \dot{\varepsilon}_{th} + \dot{\varepsilon}_{vp} \quad (1)$$

The elastic strain rate and thermal strain rate are due to the change in furnace temperature [36]. The elastic strain rate is assumed to be linear and isotropic and can be expressed with Hooke's law [22]:

$$\dot{\sigma} = k_e \dot{\varepsilon}_e \tag{2}$$

where $\dot{\sigma}$ is the stress rate and k_e is the elastic stiffness matrix while the thermal strain rate caused by thermal expansion can be expressed as [21]:

$$\dot{\varepsilon}_{th} = \alpha \Delta \dot{T} I \tag{3}$$

where α is the material thermal expansion coefficient, $\Delta \dot{T}$ is the temperature change rate, and I is the second order identity tensor.

When the temperature exceeds a certain transition temperature, viscoplastic strain instead of elastic strain dominates the sintering process, and the specimen begins to densify. Viscoplastic strain refers to the creep of materials at high temperature due to grain boundary diffusion and lattice diffusion [17]. According to the linear viscoplastic theory, the viscoplastic strain rate can be expressed as [34]:

$$\dot{\varepsilon}_{vp} = \frac{\sigma'}{2G_p} + \frac{tr(\sigma) - 3\sigma_s}{9K_p} I \tag{4}$$

where σ is the stress tensor in sintered materials, $tr(\sigma)$ is the trace of the stress tensor, σ' is the deviatoric stress tensor, σ_s is the sintering stress, and G_p and K_p are the shear viscosity modulus and bulk viscosity modulus of the material, respectively. The first term on the right side of the equation determines deformation during sintering, and the second term that is originated from the hydrostatic stress determines the volume shrinkage during sintering [26].

Sintering stress, which changes with the evolution of microstructure during different sintering stages, is the driving force for densification due to interfacial energy reduction [16,24]. Initially, the sintering stress is related to the particle size and sintering neck size in the early stage. Then, the pores in the sintered part are interconnected and cylindrical, so sintering stress is related to the curvature of the pores. Finally, the grains grow up rapidly, and the pores are located at the intersection of grain boundaries. Sintering stress depends on the size of grains and pores. Since shrinkage mainly occurs in the middle stage of the sintering process where the grain growth is not obvious, the following equation can be used to approximate the sintering stress [35]:

$$\sigma_s = \frac{C\rho^2}{r_p} \tag{5}$$

where r_p is the radius of powder, ρ is the density of the sintered material, and C is a material parameter, which is a function of surface free energy.

Both shear viscosity modulus and bulk viscosity modulus depend on relative density, temperature, and microstructural factors, such as grain and pore size [16,22]. The shear and bulk viscosity modulus for the sintered materials are defined as the following [21,24]:

$$G_p = \frac{\eta_p}{2(1 + v_p)} \tag{6}$$

$$K_p = \frac{\eta_p}{3(1 - 2v_p)} \tag{7}$$

where η_p and v_p are the uniaxial viscosity and viscous Poisson's ratio of the material, respectively [17].

3.3. Steady Creep Law

Creep is considered the major source for the viscoplastic behavior [37]. Steady-state creep rate, which is affected by material properties and environment, is the most important index to evaluate the creep resistance of materials at high temperatures. It is generally expressed in terms of an Arrhenius-like relation [38]:

$$\dot{\varepsilon}_s = A\sigma^n exp\left(-\frac{Q}{RT}\right) \tag{8}$$

where $\dot{\varepsilon}_s$ is the steady-state strain rate (s^{-1}), A is the pre-exponential constant, Q is the creep activation energy, R is the gas constant, σ is the applied stress (in Pa), n is the stress exponent, and T is the temperature.

3.4. Relative Density

In a sintering process, particles connect and fuse together while the pores among the particles reduce gradually. Macroscopically, the sintered parts shrink with the volume reduction and increase in the density. The change of relative density during sintering conforms to the mass conservation law, where the mass of the powder conserves with the gradual disappearance of the pores in the brown parts [34]:

$$\dot{\rho} = -\rho tr(\dot{\varepsilon}) \tag{9}$$

where $tr(\dot{\varepsilon})$ is the trace of total strain rate.

3.5. Friction

Friction acts as an obstacle to shrinkage, and the greater the friction, the less sufficient the shrinkage. The friction (f) can be expressed as:

$$f = \mu N \tag{10}$$

where μ is the sliding friction coefficient, which is related to the material and contact surface roughness; N is the positive pressure. If N is in the vertical direction, it equals the gravity.

4. Simulation

Commercially available software Simufact (https://www.simufact.com/, accessed on 20 September 2022) was used for numerical simulations in this study, based on the above models.

4.1. Parameter

To simulate the process, it is necessary to specify the basic properties of the material, which are temperature-dependent [22,33]. Most of the material parameters for Ti-6Al-4V are already predefined by the software vendor. The properties for Ti-6Al-4V used in the simulation are tabulated in Table 1. Powder distribution is considered homogeneous.

Figure 3 shows the additional physical properties of Ti-6Al-4V used for the simulations. As shown in Figure 3a, as the temperature increases from room temperature, the thermal conductivity increases almost monotonically from 5 W/(m·K) to 35 W/(m·K), which is consistent with the previous report [39], while due to the thermal expansion, the density of Ti-6Al-4V decreases slightly with increasing temperature (Figure 3b). Notably, the density plateau around 985 °C can be attributed to the α + β ↔ β phase transformation. Whereas the Young's modulus gradually decreases with the increase in temperature, owing to the softening of the material at elevated temperatures, as shown in Figure 3c. The mechanical response of Ti-6Al-4V is expressed through flow stress curves at different temperatures (from 20 °C to 1600 °C), which is presented in Figure 3d. The material is easier to deform with the decrease in flow stress at higher temperatures.

Table 1. Materials parameters for Ti-6Al-4V used in the simulation.

Menu	Parameter	Value	Unit
General properties	Gravitational acceleration	9.8	m/s^2
	Initial relative density	60.1	%
	Friction coefficient	0.3	-
	Pre-exponential constant	2.36×10^{-36}	1/(Pa·s)
	Creep activation energy	277	kJ/mol
	Gas constant	8.31	J/(mol·K)
Thermal properties	Specific heat capacity	540	J/(kg·K)
	Melting temperature	1640	°C
	Latent heat for melting	419	J/g
	Thermal expansion coefficient	8.84×10^{-6}	1/K
Mechanical properties	Ultimate strain	12	%
	Stress exponent	5.64	-
	Poisson's ratio	0.34	-
	Yield strength	941	MPa
	Tensile strength	1000	MPa

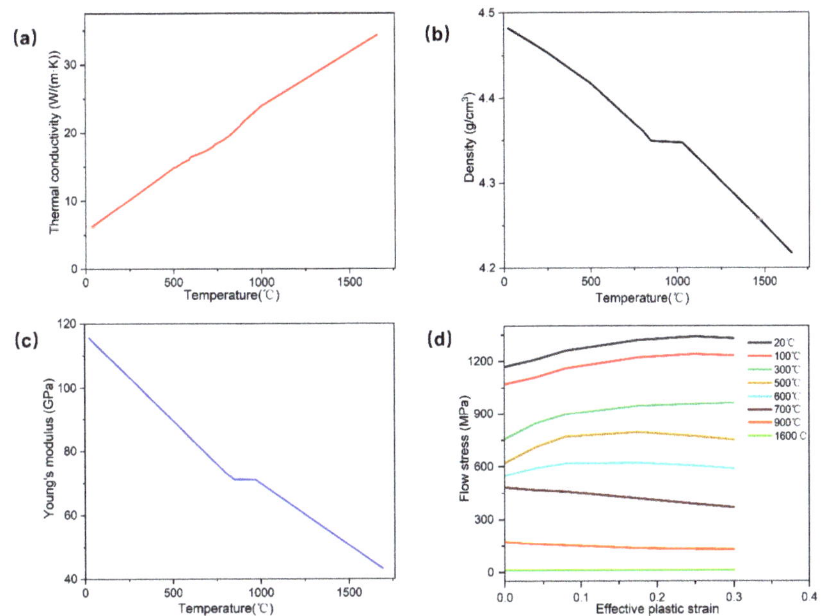

Figure 3. Physical properties of Ti-6Al-4V used in the simulations. (**a**) Thermal conductivity, (**b**) density, (**c**) Young's modulus, and (**d**) flow curves.

4.2. Round Specimen Simulation

As shown in Figure 4, a round specimen is selected for the first simulation. The numerical simulation aims to predict the density variations and shrinkages of the 3D specimens in sintering. The dimension of the ceramic plate is set as 100 mm × 100 mm × 2 mm (Figure 4a), and the round specimen is placed on top of a ceramic plate. It can be observed that after sintering, the round specimen shows a negative/positive displacement on the center/edge (Figure 4b), indicating a central symmetrical shrinking of the sample from the center of the sample. This can be understood given that the center of the round specimen is the gate for injection. As shown in Figure 4c, the relative density of the specimen is about 95.26%, and the sintering process makes the final density of sintered specimen almost uniform. The uniformity of sintered density is also related to factors, such as green density distribution, specimen shape, and sintering temperature field. A slightly lower density can be seen in the center of the round specimen that could be due to the pull effect from surrounding positions during sintering.

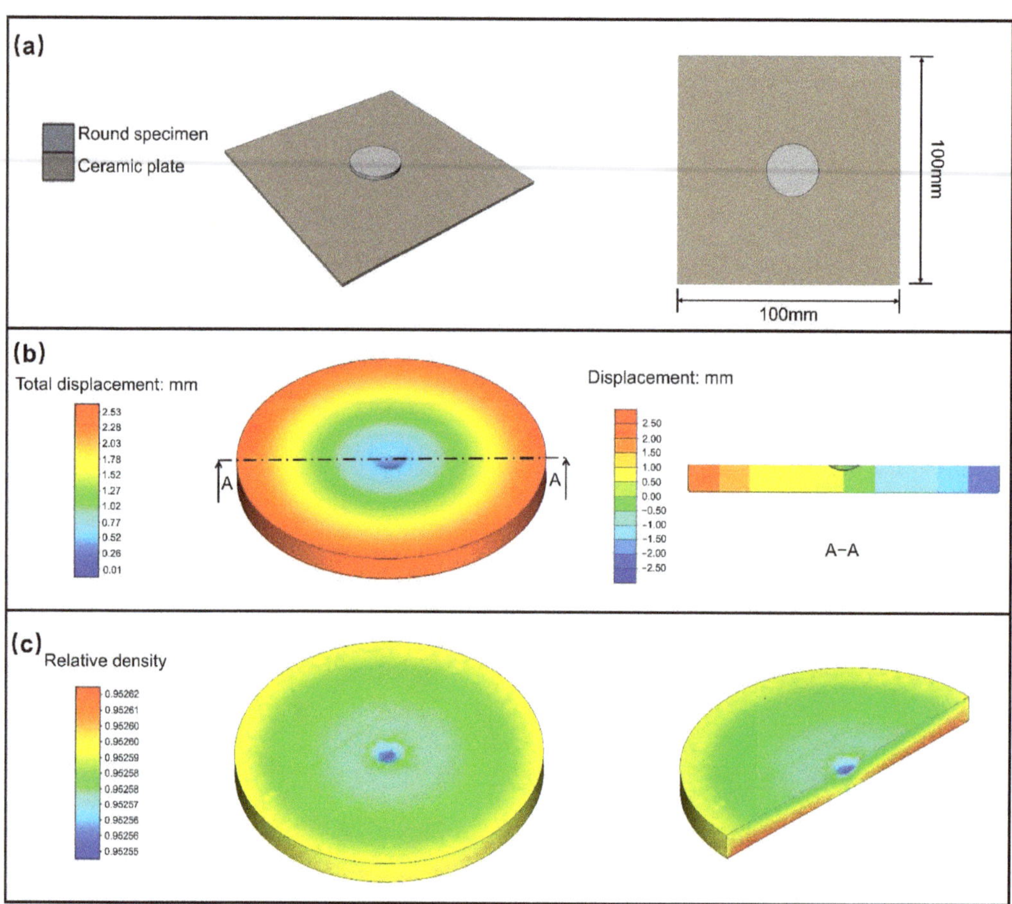

Figure 4. Simulation of round specimen for Ti-6Al-4V. (**a**) Layout, (**b**) total displacement, and (**c**) relative density.

The parameter comparison of the round specimen for Ti-6Al-4V is presented in Table 2. The errors are very small for dimension and density, which means the simulation values are in good agreement with the experimental measurements. Since the density simulation comprehensively considers the shrinkage in different directions, the density error is relatively larger than that of the dimension. Due to the regular shape of the round specimen, it exhibits linear shrinkage during sintering.

Table 2. Parameters comparison for round specimen of Ti-6Al-4V.

Parameter	Before Sintering	Experiment after Sintering	Simulation after Sintering	Error
Diameter (mm)	34.80	29.78	29.90	0.4%
Thickness (mm)	2.91	2.48	2.49	0.4%
Relative density (%)	60.10	97.80	95.26	−2.6%

4.3. Elongated Specimen Simulation

As shown in Figure 5, elongated specimens were carried out for the second simulation. There are two sintering modes, one is laid flat on the ceramic plate (Figure 5a) and the other is supported by ceramic blocks on both ends (Figure 5b). The overall length, width, and thickness of the green specimen are 114 mm, 23 mm, and 3.8 mm, respectively. For the second sintering mode, the span between ceramic blocks is 80.7 mm.

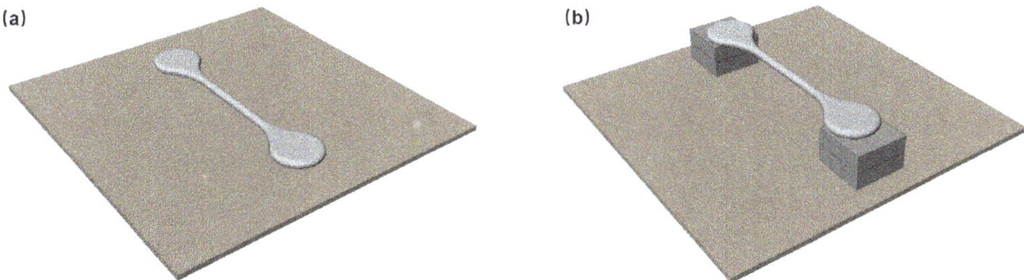

Figure 5. Elongated specimen of Ti-6Al-4V. (**a**) Laid and (**b**) supported.

Properties of the elongated specimen for the laid Ti-6Al-4V are shown in Figure 6. The gray and colored contours in Figure 6a represent the specimen before and after sintering, respectively. During thermal debinding and sintering, binder elimination and subsequent particles bonding take place, resulting in the dimensional change of the MIM parts. Compared with the green part, the dimensional change after debinding was not noticeable whilst the dimensional change after sintering was clearly evident [36,40]. The shape of the elongated specimen is more complex than that of the round specimen, which has multiple shrinkage centers during sintering, resulting in a higher nonlinearity of shrinkage. The variations in particle size, composition, mold wear, furnace temperature distribution, and other factors, such as reactions between particles during sintering, all affect the dimensional control of the MIM parts [26]. It can be seen that the simulated and experimental values have good consistency in length, width, and thickness direction (Figure 6b), which shows the accuracy of the numerical simulation. The final density is depicted in Figure 6c, which shows an almost homogenous distribution with a value of ~95.4%. In some cases, cracks can be formed in the sintered specimen due to the high thermal or stress gradients. In this study, no crack was found in the simulations (Figure 6d) or experiments, which shows that the rationality of the sintering curve had well-designed heating and cooling rates.

The deformation of the elongated specimen for the supported Ti-6Al-4V from the simulation and experiment is also shown in Figure 7. The gray and colored contours in Figure 7a represent the specimen after thermal debinding and sintering from simulation, respectively. On the other hand, Figure 7b,c show the experimental results of the specimen after thermal debinding and sintering, respectively. It is discovered that the specimen has almost no deformation and shrinkage after thermal debinding. For specimens after sintering, the farther away from the support block, the greater the deformation, which is due to the effect of the gravity. A maximum deformation of 9.26 mm can be observed after sintering, consistent with the experimentally measured value of 9.01 mm, giving rise to a relative error of only ~2.8%, showing the accuracy of the numerical model. In the sintering process, the specimen bended due to its own gravity. According to Equation (4), shrinkage in the sintering process depends on the ratio of sintering stress and bulk viscosity modulus, while the sintering deformation depends on the shear viscosity modulus. Shrinkage can be simulated with greater accuracy than that of deformation.

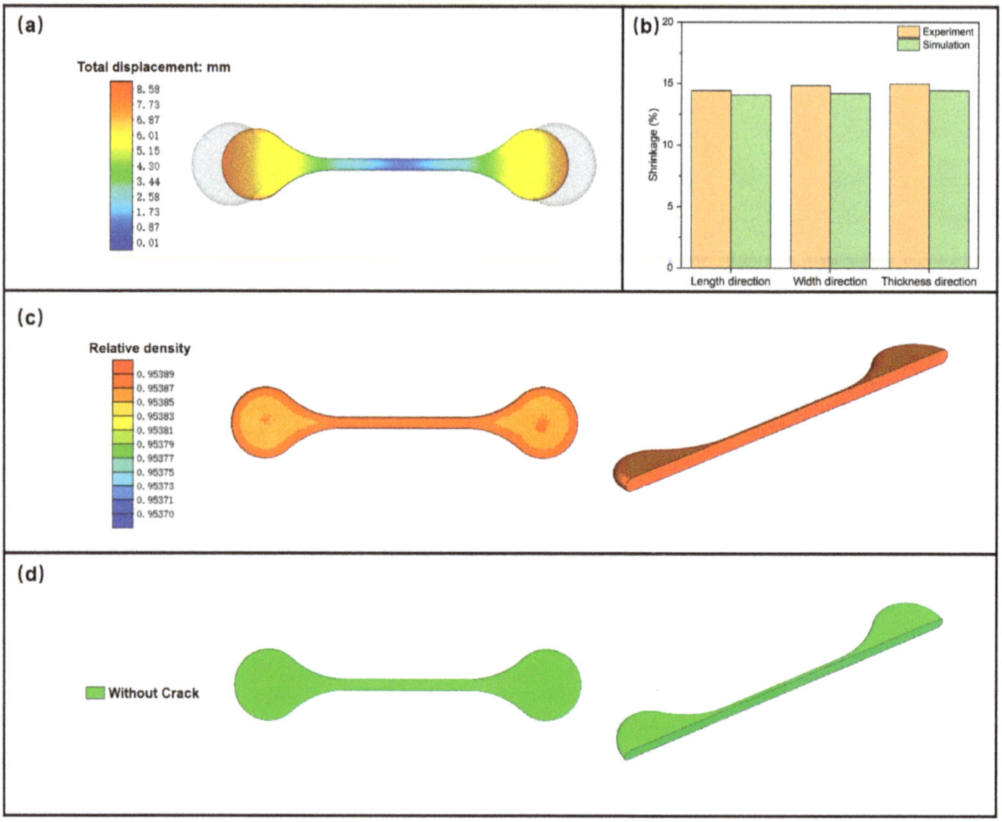

Figure 6. Properties of the elongated specimen for laid Ti-6Al-4V. (**a**) Total shrinkage, (**b**) shrinkage along different directions, (**c**) relative density, and (**d**) crack.

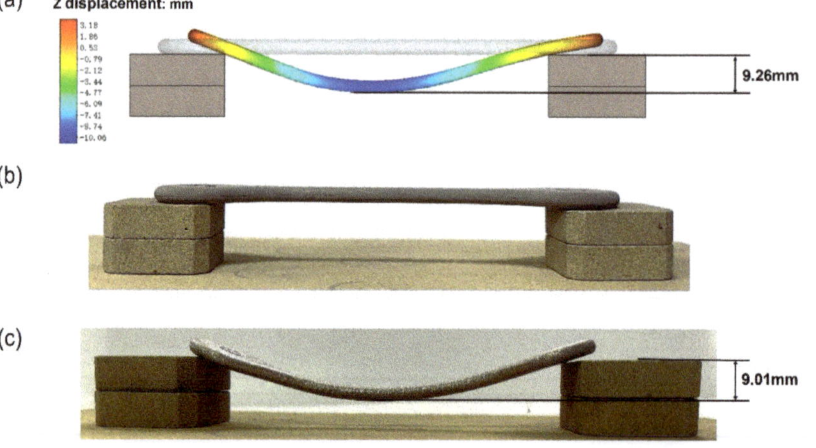

Figure 7. Deformation of elongated specimen for supported Ti-6Al-4V. (**a**) Numerical simulation, (**b**) specimen after thermal debinding, and (**c**) specimen after sintering.

MIM specimens reach very low strength levels during sintering. Accordingly, weak forces, such as gravity, friction, and non-uniform heating, induce deformation [26]. Since gravity only acts on the vertical direction, the specimen shrank unevenly and deformed in the sintering process. The effect of gravity on the shrinkage and deformation of sintered specimens is related to the size of the specimen and the type of ceramic block used in sintering. At the same time, the friction between porous specimens and ceramic blocks also led to the sintering deformation. The deformation of the specimen results in the change of stress distribution in which the compressive stress is beneficial to the densification of sintering, while the tensile stress is contrary.

The experimental relative density is very close to the simulated value, whether laid or supported. Interestingly, the experimental values of sintered density are slightly larger than the simulated ones, which is understandable given that the simulation is carried out under ideal vacuum conditions, where only the role of thermal radiation is considered. Although there is still appropriate heat conduction and convection, in addition to thermal radiation experimentally, the specimen is subjected to more heat action with an additional heat source.

5. Conclusions

In conclusion, we have developed an FE method based on the thermal-elastic-viscoplastic macroscopic model to predict the shrinkage, deformation, relative density, and crack of injection-molded Ti-6Al-4V after sintering, using the commercially available Simufact software. Experiments were simultaneously carried out to justify the accuracy of the sintering model and simulation method. The results showed a good agreement (within a 3% error) between the experimental measurements and numerical simulations. The slightly larger sintered density and shrinkage in the experiment than those in the simulation are due to additional thermal convection and conduction during the sintering experiments. The deformation was affected by gravity, friction, specimen shape, and support mode. This approach can serve as a guideline for mold design and sintering optimization and could be extended to other MIM materials while a follow-up study on the influence of sintering temperatures and material's compositions is needed to further extend this method.

Author Contributions: Conceptualization, S.S., Z.H. and Y.W.; methodology, S.S., Z.H. and Y.H.; software, S.S.; validation, S.S.; formal analysis, S.S., Y.W. and P.W.; investigation, S.S. and X.L.; data curation, S.S.; writing—original draft preparation, S.S. and Z.H.; writing—review and editing, S.S. and Z.H.; supervision, J.W.; funding acquisition, J.W. All authors have read and agreed to the published version of the manuscript.

Funding: This research was funded by National Key Research and Development Program of China, grant number 2021YFB3701900.

Institutional Review Board Statement: Not applicable.

Informed Consent Statement: Not applicable.

Data Availability Statement: The data are available from the corresponding author (hongzijian100@zju.edu.cn) upon reasonable request.

Acknowledgments: Z.H. would like to acknowledge a start-up grant from Zhejiang University.

Conflicts of Interest: The authors declare no conflict of interest.

References

1. Tong, J.; Bowen, C.; Persson, J.; Plummer, A. Mechanical properties of titanium-based Ti-6Al-4V alloys manufactured by powder bed additive manufacture. *Mater. Sci. Technol.* **2016**, *33*, 138–148. [CrossRef]
2. Moghadam, M.S.; Fayyaz, A.; Ardestani, M. Fabrication of titanium components by low-pressure powder injection moulding using hydride-dehydride titanium powder. *Powder Technol.* **2021**, *377*, 70–79. [CrossRef]
3. Ghanmi, O.; Demers, V. Molding properties of titanium-based feedstock used in low-pressure powder injection molding. *Powder Technol.* **2021**, *379*, 515–525. [CrossRef]

4. Suwanpreecha, C.; Alabort, E.; Tang, Y.B.; Panwisawas, C.; Reed, R.C.; Manonukul, A. A novel low-modulus titanium alloy for biomedical applications: A comparison between selective laser melting and metal injection moulding. *Mater. Sci. Eng. A* **2021**, *812*, 141081. [CrossRef]
5. Sales, W.F.; Schoop, J.; Jawahir, I.S. Tribological behavior of PCD tools during superfinishing turning of the Ti6Al4V alloy using cryogenic, hybrid and flood as lubri-coolant environments. *Tribol. Int.* **2017**, *114*, 109–120. [CrossRef]
6. Liang, X.L.; Liu, Z.Q. Tool wear behaviors and corresponding machined surface topography during high-speed machining of Ti-6Al-4V with fine grain tools. *Tribol. Int.* **2018**, *121*, 321–332. [CrossRef]
7. Liang, X.L.; Liu, Z.Q.; Liu, W.T.; Wang, B.; Yao, G.H. Surface integrity analysis for high-pressure jet assisted machined Ti-6Al-4V considering cooling pressures and injection positions. *J. Manuf. Process.* **2019**, *40*, 149–159. [CrossRef]
8. Dehghan-Manshadi, A.; Bermingham, M.J.; Dargusch, M.S.; StJohn, D.H.; Qian, M. Metal injection moulding of titanium and titanium alloys: Challenges and recent development. *Powder Technol.* **2017**, *319*, 289–301. [CrossRef]
9. Xu, P.; Pyczak, F.; Limberg, W.; Willumeit-Römer, R.; Ebel, T. Superior fatigue endurance exempt from high processing cleanliness of metal-injection-molded β Ti-Nb-Zr for bio-tolerant applications. *Mater. Des.* **2021**, *211*, 110141. [CrossRef]
10. Mahmud, N.N.; Azam, F.'A.A.; Ramli, M.I.; Foudzi, F.M.; Ameyama, K.; Sulong, A.B. Rheological properties of irregular-shaped titanium-hydroxyapatite bimodal powder composite moulded by powder injection moulding. *J. Mater. Res. Technol.* **2021**, *11*, 2255–2264. [CrossRef]
11. Hamidi, M.F.F.A.; Harun, W.S.W.; Samykano, M.; Ghani, S.A.C.; Ghazalli, Z.; Ahmad, F.; Sulong, A.B. A review of biocompatible metal injection moulding process parameters for biomedical applications. *Mater. Sci. Eng. C* **2017**, *78*, 1263–1276. [CrossRef] [PubMed]
12. Bootchai, S.; Taweejun, N.; Manonukul, A.; Kanchanomai, C. Metal injection molded titanium: Mechanical properties of debinded powder and sintered metal. *J. Mater. Eng. Perform.* **2020**, *29*, 4559–4568. [CrossRef]
13. Subaşıa, M.; Safarianb, A.; Karataş, Ç. An investigation on characteristics and rheological behaviour of titanium injection moulding feedstocks with thermoplastic-based binders. *Powder Metall.* **2019**, *62*, 229–239. [CrossRef]
14. Subaşi, M.; Safarian, A.; Karataş, Ç. The investigation of production parameters of Ti-6Al-4V component by powder injection molding. *Int. J. Adv. Manuf. Technol.* **2019**, *105*, 4747–4760. [CrossRef]
15. Meng, J.H.; Loh, N.H.; Fu, G.; Tay, B.Y.; Tor, S.B. Micro powder injection moulding of alumina micro-channel part. *J. Eur. Ceram. Soc.* **2011**, *31*, 1049–1056. [CrossRef]
16. Kwon, Y.S.; Wu, Y.X.; Suri, P.; German, R.M. Simulation of the sintering densification and shrinkage behavior of powder-injection-molded 17-4 PH stainless steel. *Metall. Mater. Trans. A* **2004**, *35*, 257–263. [CrossRef]
17. Kong, X.; Quinard, C.; Barriere, T.; Gelin, J.C. Micro powder injection moulding of 316L stainless steel feedstock and numerical simulation of the sintering stage. In Proceedings of the 10th International Conference on Numerical Methods in Industrial Forming Processes, Pohang, Republic of Korea, 13–17 June 2010.
18. Mohsin, I.U.; Lager, D.; Hohenauer, W.; Gierl, C.; Danninger, H. Finite element sintering analysis of metal injection molded copper brown body using thermo-physical data and kinetics. *Comput. Mater. Sci.* **2012**, *53*, 6–11. [CrossRef]
19. German, R.M. Thermodynamics of sintering. In *Sintering of Advanced Materials*, 1st ed.; Fang, Z.Z., Ed.; Woodhead Publishing: Sawston, UK, 2010; Volume 1, pp. 3–32.
20. Choi, J.-P.; Lee, G.-Y.; Song, J.-I.; Lee, W.-S.; Lee, J.-S. Sintering behavior of 316L stainless steel micro-nanopowder compact fabricated by powder injection molding. *Powder Technol.* **2015**, *279*, 196–202. [CrossRef]
21. Sahli, M.; Djoudi, H.; Gelin, J.C.; Barriere, T.; Assoul, M. Numerical simulation and experimental analysis of the sintered micro-parts using the powder injection molding process. *Microsyst. Technol.* **2018**, *24*, 1495–1508. [CrossRef]
22. Mamen, B. Experimental Investigation and Numerical Simulation of Thermal Debinding and Sintering Processes in Powder Injection Moulding. Ph.D. Thesis, University of Franche-Comté, Besançon, France, 2013.
23. Węglewski, W.; Basista, M.; Chmielewski, M.; Pietrzak, K. Modelling of thermally induced damage in the processing of Cr-Al_2O_3 composites. *Compos. Part B-Eng.* **2012**, *43*, 255–264. [CrossRef]
24. Braginsky, M.; Tikare, V.; Olevsky, E. Numerical simulation of solid state sintering. *Int. J. Solids Struct.* **2005**, *42*, 621–636. [CrossRef]
25. Oyama, K. Diplas, S.; M'hamdi, M.; Gunnæs, A.E.; Azar, A.S. Heat source management in wire-arc additive manufacturing process for Al-Mg and Al-Si alloys. *Addit. Manuf.* **2019**, *26*, 180–192.
26. Kang, T.G.; Ahn, S.; Chung, S.H.; Chung, S.T.; Kwon, Y.S.; Park, S.J.; German, R.M. Modeling and simulation of metal injection molding (MIM). In *Handbook of Metal Injection Molding*, 2nd ed.; Heaney, D.F., Ed.; Woodhead Publishing: Sawston, UK, 2019; Volume 2, pp. 219–252.
27. Martin, S.; Guessasma, M.; Léchelle, J.; Fortin, J.; Saleh, K.; Adenot, F. Simulation of sintering using a Non Smooth Discrete Element Method. Application to the study of rearrangement. *Comput. Mater. Sci.* **2014**, *84*, 31–39. [CrossRef]
28. Song, J.; Barriere, T.; Liu, B.; Gelin, J.C. Numerical simulation of sintering process in ceramic powder injection moulded components. In Proceedings of the 9th International Conference on Numerical Methods in Industrial Forming Processes, Porto, Portugal, 17–21 June 2007.
29. Behrensa, B.-A.; Chugreeva, A.; Chugreev, A. FE-simulation of hot forging with an integrated heat treatment with the objective of residual stress prediction. In Proceedings of the 21st International ESAFORM Conference on Material Forming, Palermo, Italy, 23–25 April 2018.

30. Bauer, A.; Manurung, Y.H.P.; Sprungk, J.; Graf, M.; Awiszus, B.; Prajadhiana, K. Investigation on forming–welding process chain for DC04 tube manufacturing using experiment and FEM simulation. *Int. J. Adv. Manuf. Technol.* **2019**, *102*, 2399–2408. [CrossRef]
31. Nosewicz, S.; Rojek, J.; Pietrzak, K.; Chmielewski, M. Viscoelastic discrete element model of powder sintering. *Powder Technol.* **2013**, *246*, 157–168. [CrossRef]
32. Jeong, M.-S.; Yoo, J.-H.; Rhim, S.-H.; Lee, S.-K.; Oh, S.-I. A unified model for compaction and sintering behavior of powder processing. *Finite Elem. Anal. Des.* **2012**, *53*, 56–62. [CrossRef]
33. Pagáč, M.; Hajnyš, J.; Halama, R.; Aldabash, T.; Měsíček, J.; Jancar, L.; Jansa, J. Prediction of model distortion by FEM in 3D printing via the selective laser melting of stainless steel AISI 316L. *Appl. Sci.* **2021**, *11*, 1656. [CrossRef]
34. Shi, J.; Cheng, Z.; Barriere, T.; Liu, B.; Gelin, J.C. Multiphysic coupling and full cycle simulation of microwave sintering applied to a ceramic compact obtained by ceramic injection moulding. *Powder Metall.* **2017**, *60*, 404–414. [CrossRef]
35. Song, J.; Gelin, J.C.; Barriere, T.; Liu, B. Experiments and numerical modelling of solid state sintering for 316L stainless steel components. *J. Mater. Process. Technol.* **2006**, *177*, 352–355. [CrossRef]
36. Sahli, M.; Lebied, A.; Gelin, J.C.; Barriere, T.; Necib, B. Numerical simulation and experimental analysis of solid-state sintering response of 316 L stainless steel micro-parts manufactured by metal injection molding. *Int. J. Adv. Manuf. Technol.* **2015**, *79*, 2079–2092. [CrossRef]
37. Aryanpour, G.; Mashl, S.; Warke, V. Elastoplastic–viscoplastic modelling of metal powder compaction: Application to hot isostatic pressing. *Powder Metall.* **2013**, *56*, 14–23. [CrossRef]
38. Gu, Y.; Zeng, F.H.; Qi, Y.L.; Xia, C.Q.; Xiong, X. Tensile creep behavior of heat-treated TC11 titanium alloy at 450–550 °C. *Mater. Sci. Eng. A* **2013**, *575*, 74–85. [CrossRef]
39. Villa, M. Metallurgical and Mechanical Modelling of Ti-6Al-4V for Welding Applications. Ph.D. Thesis, University of Birmingham, Birmingham, UK, 2016.
40. Sahli, M.; Gelin, J.C.; Barriere, T. Simulation and modelling of sintering process for 316L stainless steel metal injection molding parts. *Adv. Mater. Process. Technol.* **2015**, *1*, 577–585. [CrossRef]

Article

Orthotropic Behavior of Twin-Roll-Cast and Hot-Rolled Magnesium ZAX210 Sheet

Madlen Ullmann *, Christoph Kaden, Kristina Kittner and Ulrich Prahl

Institute of Metal Forming, Technische Universität Bergakademie Freiberg, Bernhard-von-Cotta Straße 4, 09599 Freiberg, Germany
* Correspondence: madlen.ullmann@imf.tu-freiberg.de

Abstract: Magnesium sheet metal alloys offer a deformation asymmetry, which is strongly related to grain size and texture. In order to predict deformation behavior as well as to provide methods to eliminate anisotropy and yield asymmetry, a lot of effort is invested in studying the tension–compression asymmetry of magnesium alloys. However, only a few studies deal with the characterization of the yield asymmetry of magnesium wrought alloys, especially Ca-containing alloys, related to temperature and strain. In this study, the orthotropic behavior of a twin-roll-cast, homogenized, rolled and finish-annealed Mg-2Zn-1Al-0.3Ca (ZAX210) magnesium alloy was investigated by tensile testing at room temperature, 150 °C and 250 °C. The *r*-values were determined and the Hill'48 yield criterion was used for the constitutive formulation of the plastic yielding and deformation. The yield loci calculated using Mises and Hill'48 as well as the determined *r*-values reveal an almost isotropic behavior of the ZAX210 alloy. The r-value increases with increasing logarithmic strain. At 0.16 logarithmic strain the *r*-values at room temperature vary between 1 (0°) and 1.5 (45° and 90°). At higher temperatures (250 °C), *r*-values close to 1 at all tested directions are attained. The enhanced yield asymmetry can be attributed to the weaker basal texture that arises during hot rolling and final annealing of the twin-roll-cast ZAX210 magnesium alloy. In comparison to AZ31, the ZAX210 alloy shows a yield behavior close to transversal isotropy. Finally, responsible mechanisms for this behavior are discussed.

Keywords: orthotropic; anisotropy; yield loci; Hill; Lankford; localization; AZ31; magnesium; sheet

Citation: Ullmann, M.; Kaden, C.; Kittner, K.; Prahl, U. Orthotropic Behavior of Twin-Roll-Cast and Hot-Rolled Magnesium ZAX210 Sheet. *Materials* **2022**, *15*, 6437. https://doi.org/10.3390/ma15186437

Academic Editor: Mateusz Kopec

Received: 16 August 2022
Accepted: 13 September 2022
Published: 16 September 2022

1. Introduction

The use of magnesium alloys has a positive impact on energy efficiency in the automotive and aerospace industries. In order to promote the industrial use of magnesium materials produced by forming technology, one focus of research in recent years has been on the development of magnesium alloys with improved formability. Sheet metal forming processes with multiaxial loading are used for manufacturing large-area products in mass production with tailored mechanical and technological property profiles. In order to describe the hardening behavior, for example as a basis for forming simulation, it is necessary to establish a suitable flow criterion. The basis for this is provided by investigations of magnesium sheets under biaxial loading to support the selection of the anisotropic yield criterion parameters [1,2].

The anisotropic nature of the hexagonal closed-packed structure of conventional magnesium alloys leads to the formation of strong basal textures during deformation or thermomechanical processing. Related studies have shown that these textures are mainly responsible for the formation of yield asymmetry, since during the deformation parallel to the c-axis is the formation of tensile twinning of the dominant deformation mechanism [3–6]. Davis et al. (2019) [3] showed the most effective strategies to reduce yield asymmetry and anisotropy by texture weakening and solid solution strengthening in Mg-Y alloys. Comparable effects were demonstrated by Kamrani and Fleck (2014) [5], showing that a pronounced weakening of the texture in ZEK100 alloyed by Ca goes along with a reduced yield asymmetry. Dobron et al. (2018) [7] describe that the formation of extension twins during extrusion along the extrusion direction leads to a significant reduction in the compressive yield strength, while the tensile yield strength is not influenced. The authors reveal that this behavior is related to twinning activity in the ZX10 magnesium alloy. The combination of pre-compression and subsequent heat treatment additionally causes the formation of a yield plateau. According to [7], this is due to the propagation of twins beyond grain boundaries, which is associated with the stress relaxation mechanism after strengthening of the material by the presence of a relatively high dislocation density and fine Mg_2Ca precipitates. A comparable behavior was presented by [8] by investigating yielding by twinning during compression of an extruded AZ31 magnesium alloy.

Calcium as an alloying element is known to weaken basal texture in magnesium alloys [5,9,10], which may result in a reduced yield asymmetry. Nakata et al. (2018) [11] presented a Mg-1.1Al-0.24Ca-1.0Mn magnesium alloy with improved compressive yield stress and a small yield asymmetry after extrusion and aging treatment. The increase in the strength was attributed to the grain size refinement and the uniform distribution of the Guinier Preston zones by aging. The addition of Mn supports the grain refinement by suppressing significant grain growth because of nano-scaled Mn-containing precipitates. Investigations of Song et al. (2020) [12] revealed that torsion deformation and aging treatment of an extruded AZ91 rod can enhance the yield strength effectively, while the yield asymmetry is eliminated. In summary, texture weakening [6,13,14], solid solution strengthening, grain size reduction [15] and precipitation strengthening [16,17] are effective strategies to develop magnesium alloys with reduced anisotropy and yield asymmetry. However, few studies address the yield asymmetry of Ca-containing magnesium alloy sheets; in particular, properties are often presented only at room temperature, not at elevated temperatures with coincident strain dependence.

The Mg-2Zn-1Al-0.3Ca (ZAX210) magnesium alloy was designed driven by the effort to develop an alloy with higher formability without using expensive Rare Earth elements. The hot rolling process leads to the formation of a fine-grained microstructure, weaker basal texture and a lower anisotropy of the mechanical properties [18]. As mentioned above, studies about the behavior of yield anisotropy are currently not available. As a result of the increasing importance of the alloy for industrial applications, the orientation dependent flow behavior of the ZAX210 magnesium alloy was investigated in order to provide basic yield model data for the numerical simulation.

2. Materials and Methods

2.1. Materials

Twin-roll-cast, rolled and finish-annealed magnesium sheets of Mg-2Zn-1Al-0.3Ca (ZAX210) alloy with a thickness of 1.5 mm were used for these investigations. The chemical composition is shown in the table below (Table 1).

Table 1. Chemical composition of the ZAX210 sheets (wt.%) measured by optical emission spectrometry [19].

Zn	Al	Ca	Mn	Cu	Fe	Ni	Others	Mg
2.290	0.920	<0.250	0.040	0.001	0.005	0.001	<0.045	Bal.

The sheets were produced as described in [19]. Microstructural and texture characterization were performed using ZEISS GeminiSEM 450 device at the Institute of Metal Forming, Freiberg, Germany. The accelerating voltage for electron backscatter diffraction (EBSD) analysis was between 15 kV and 20 kV. A step size of 0.5 µm was selected. The AZtec software was used to analyze the recorded data.

2.2. Experimental Setup

The tensile tests to determine the LANKFORD parameters (r-values) were carried out on the biaxial testing device BTA-840 (TA Instruments former BÄHR Thermoanalyse GmbH, Hüllhorst, Germany) at the Institute of Metal Forming, TU Bergakademie Freiberg (Figure 1). The system is equipped with a total of four hydraulic cylinders, which are aligned at right angles in a cross shape on one plane and can be controlled individually. For the uniaxial tensile test only two cylinders are required. Induction coils are installed above and below the center of the specimen to heat the specimens. This allows a very precise specification of temperature–time curves.

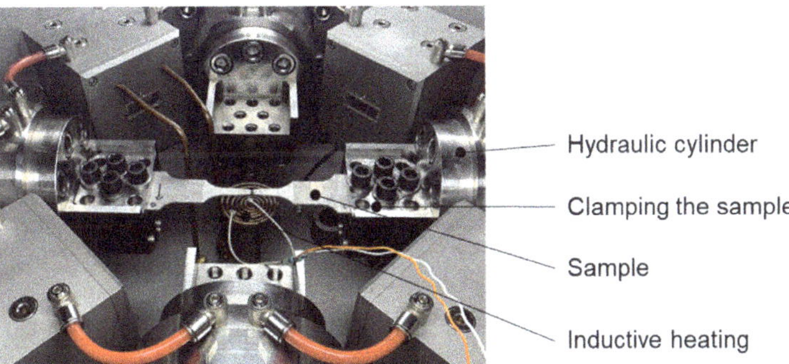

Figure 1. Biaxial testing device BTA-840 (BÄHR Thermoanalyse GmbH) at the Institute of Metal Forming, TU Bergakademie Freiberg.

To determine the r-values, tensile specimens were produced in the rolling direction ($r_{0°}$), transverse to the rolling direction ($r_{90°}$) and at an angle of 45° to the rolling direction ($r_{45°}$) according to the specifications of DIN EN ISO 6892 (Figure 2). The tests were carried out at room temperature, 150 °C and 250 °C, the feed rate of the hydraulic cylinders was 0.35 mm/s, and the heating rate was 2 K/s. For a better statistical validation, three samples for each temperature and direction were tested. For the subsequent evaluation, cameras were installed above the specimens to film the tensile test. A grid was applied to the surface of the specimen to improve the detection of length and width strain.

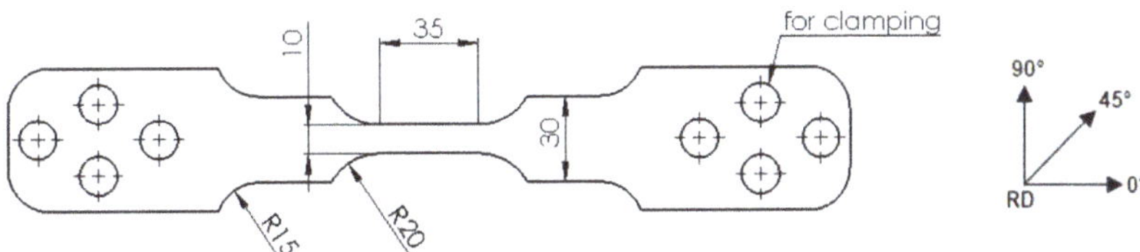

Figure 2. Sample dimensions and sampling directions for the uniaxial tensile tests (dimensions in mm).

2.3. Graphical Evaluation

Digital image correlation software, ZEISS GOM Correlate Pro, was used to determine the deformation and calculate the r-values. To use the software, the filmed videos of the individual tensile tests were broken down into individual frames and were uploaded into the program. Next, virtual extensometers were placed on the specimen within the measurement range (Figure 3). A total of three marks were used for length and five for width. The program then measured the deformation of the individual images, the data were uploaded into a calculation program and the r-values were calculated and output as a function of the local logarithmic. Strain was calculated automatically.

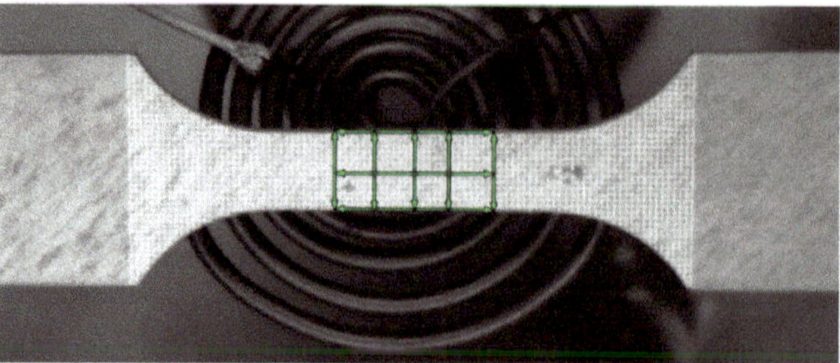

Figure 3. Virtual extensometer over measuring range (20 × 10 mm), GOM Correlate.

2.4. Hill'48 Yield Criterion

Hill [20] proposed a constitutive formulation for the plastic yielding and deformation of anisotropic metals. The quadratic yield criterion is given by Equation (1):

$$2f(\sigma) = F\left(\sigma_{yy} - \sigma_{zz}\right)^2 + G(\sigma_{zz} - \sigma_{xx})^2 + H\left(\sigma_{xx} - \sigma_{yy}\right)^2 + 2\left(L\tau_{yz}^2 + M\tau_{zx}^2 + N\tau_{xy}^2\right) = 1 \quad (1)$$

where f is the yield function; σ_{xx}, σ_{yy} and σ_{zz} are the stresses in the rolling, transverse and thickness directions; τ_{xy}, τ_{yz} and τ_{zx}, are the shear stresses in the xy, yz and zx planes, respectively. The Hill'48 material parameters F, G, H, L, M and N are the constants that define the anisotropy of the material. When applying this yield criterion to sheet metal, it can be simplified under the assumption of plane stress condition ($\sigma_{xx} = \tau_{yz} = \tau_{zx} = 0$):

$$2f(\sigma) = (G + H)\sigma_{xx}^2 + (F + H)\sigma_{yy}^2 - 2H\sigma_{xx}\sigma_{yy} + 2N\tau_{xy}^2. \quad (2)$$

In this study, the identification of the anisotropy parameters for the Hill'48 model was carried out by applying the classical approach, which uses the r-values from three uniaxial tension tests (0°, 45° and 90° to the rolling direction) at different temperatures:

$$F = \frac{r_0}{r_{90}(1 + r_0)} \quad (3)$$

$$G = \frac{1}{2}\left(1 + \frac{1 - r_0}{1 + r_0}\right) \quad (4)$$

$$H = \frac{1}{2}\left(1 + \frac{r_0 - 1}{1 + r_0}\right) \quad (5)$$

$$L = M = \frac{3}{2} \quad (6)$$

$$N = \frac{3}{2}\left(\frac{(1 + 2r_{45})(r_0 + r_{90})}{3r_{90}(1 + r_0)}\right) \quad (7)$$

In the present work, the tension–compression asymmetry method is not used because:

(1) When the sheet metal is deep drawn, the biaxial compressive stress in the sheet metal plane will not cause the so-called "thickening" of the sheet metal. If there is surface compressive stress, only wrinkles are formed, which macroscopically combine to form a larger sheet cross-section due to the high pressure in the press.

(2) In the special case of the inner bending radius, the small appearance of compressive stress can be neglected in the modeling.

(3) The manufacturing tolerance of a cylindrical compression specimen in the plane of the plate (direction of compression equal to direction of sheet thickness) is too large compared to the plate with a thickness of 1.5 mm, and the calculation of the flow curve will be accompanied by a large error.

3. Results and Discussion

3.1. Microstructure and Texture of the Initial State

Twin-roll-cast, rolled and finish-annealed magnesium sheets of Mg-2Zn-1Al-0.3Ca (ZAX210) alloy with a thickness of 1.5 mm were used for the investigations in this study. Figure 4 presents the EBSD map and the texture of the ZAX210 sheets. After final annealing, the sheets offer a microstructure with recrystallized equiaxed grains exhibiting an average grain size of 8.3 µm (±4.0 µm). The hot rolling and final annealing lead to the formation of a basal texture. The texture reveals for Ca-containing magnesium alloys a characteristic basal pole split in the rolling direction (RD) as well as a broadening of the texture components in the RD and transverse direction (TD), the latter being less pronounced. In general, the texture components have a low intensity. The c-axis of the hexagonal cells is tilted by 35° to 45° from the normal direction (ND) to RD. The prismatic pole intensity is aligned in the transverse direction without showing a six-fold symmetry of the maximum pole intensities across ND. This can be attributed to the pronounced basal pole broadening in RD.

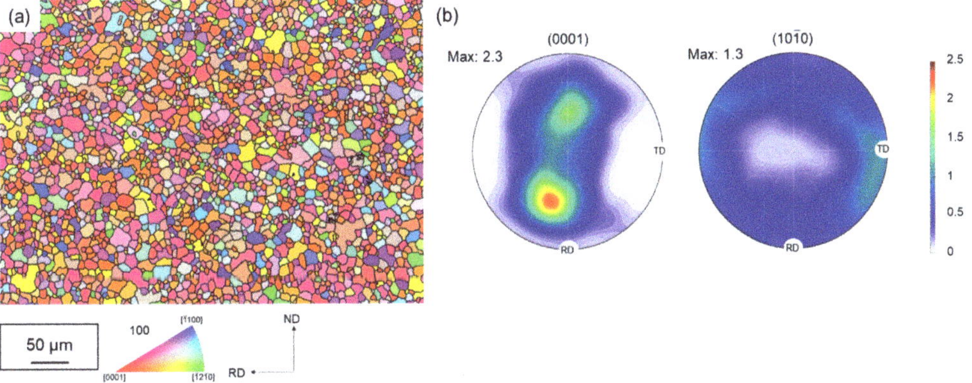

Figure 4. Microstructure and texture of the ZAX210 sheets: (**a**) EBSD-Map and (**b**) pole figures.

3.2. Variation in r-Values with Logarithmic Strain and Temperature and Hill'48 Coefficients

For the calculation of the *r*-values, the logarithmic strain up to the necking was used, whereby this range becomes smaller with increasing temperature. The *r*-values as a function of the local logarithmic strain are shown in Figure 4. The *r*-values are small (0.5 to 0.7) during the early stage of deformation and increase with local logarithmic strain. At all tested temperatures, there is a tendency that $r_{90^\circ} > r_{45^\circ} > r_{0^\circ}$. However, the difference between r_{90° and r_{45° is very small (maximum deviation is 10%). At room temperature, the *r*-value at a logarithmic strain of 0.16 is 1 (0°) and 1.5 (45°, 90°), respectively. With increasing the temperature to 150 °C the *r*-value also rises to 1.2 (0°) and 1.5 to 1.65 (45°, 90°). The *r*-values at 250 °C are generally lower and reach values close to 1. The obtained *r*-values are a direct result of the crystallographic texture present in the ZAX210 alloy, combined with the relative resolved shear strengths of the slip and twinning systems.

The plane anisotropy Δr is between −0.2 and 0 at all temperatures. These values indicate a slight anisotropic behavior in the sheet plane and lead to earing formation at an angle of 45° to the rolling direction. The maximum values for the mean vertical anisotropy r_m are 1.4 for RT, 1.5 for 150 °C and 1.1 for 250 °C. This gradual increase in *r*-value from the RD (rolling direction, 0°) to the TD (transverse direction, 90° to RD) is consistent with results for the AZ31 alloy from the literature, e.g., [21]. Yi et al. (2010) [22] explain the strength dependence of the *r*-value in AZ31 with the tilting of the basal planes. The more they are tilted in the direction of deformation, the more likely the activation of <a>-dislocations of the basal system can occur and strain in the c-axis direction can be realized, since the *r*-value is inversely proportional to the strain in the thickness direction. The variation in the *r*-values (0°, 45° and 90°) of the ZAX210 sheets can also be explained by their texture (see texture of ZAX210 sheets in [18] and Figure 4). Loading in the TD is advantageous for basal <a>-slip, since the tilted basal planes and the tilt angle is higher in the TD than in the RD. Therefore, the yield strength of TD is lower than that of RD. Compared with the *r*-value published for AZ31, the *r*-value close to 1 in alloy ZAX210 indicates that the cross-sectional contraction occurs isotropically, i.e., the strains in the width and thickness directions are similar. The achievement of *r*-values close to 1 is due to the activation of non-basal slip systems. Already, Duygulu et al. (2003) [23] showed in their simulations that only the activation of additional non-basal slip systems, especially the <c + a> pyramidal planes, leads to an increasing uniformity of the deformation, so that the *r*-value drops to the value close to isotropy (r~1).

With help of the *r*-values, the coefficients for the Hill'48 criteria could be determined. The graphs of the Hill'48 coefficients as a function of logarithmic strain are comparable for all temperatures and are shown for room temperature, 150 °C and 250 °C for the ZAX210 alloy in Figure 5. The specific values of the Hill'48 coefficients at room temperature are shown in Table 2.

Table 2. Anisotropic parameters in the Hill'48 yield function for rolled ZAX210 alloys at room temperature.

Log. Strain	F	G	H	L	M	N
0.015	0.5079	0.6511	0.3489	1.5	1.5	1.2977
0.05	0.4043	0.5123	0.877	1.5	1.5	1.4878
0.1	0.3499	0.4724	0.5276	1.5	1.5	1.5791
0.18	0.3222	0.4872	0.5128	1.5	1.5	1.6454

3.3. *Yield Loci Calculated Using Mises and Hill'48*

Figure 6 shows the calculated yield loci using von Mises and Hill'48 yield functions depending on logarithmic strain and temperature for the ZAX210 sheets. As the temperature increases, the *r*-values approach 1, resulting in isotropic behavior and thus Hill'48 and Mises converge. Likewise, the influence of logarithmic strain onto the *r*-value decreases with increasing temperature and almost disappears at 250 °C. Compared to AZ31 [24], ZAX210 shows a significantly more isotropic behavior. In the case of ZAX210, an approximately planar isotropic material behavior at a low temperature of 150 °C can be assumed even for moderately higher strains compared to AZ31.

Ultimately, it can be stated that the in-plane material flow behavior can be identified as orthotropic with decreasing anisotropy when forming temperature is rising. In comparison to AZ31, the ZAX210 alloy shows a yield behavior close to transversal isotropy, in particular for higher temperatures, when additional shear planes might be assumed to be activated.

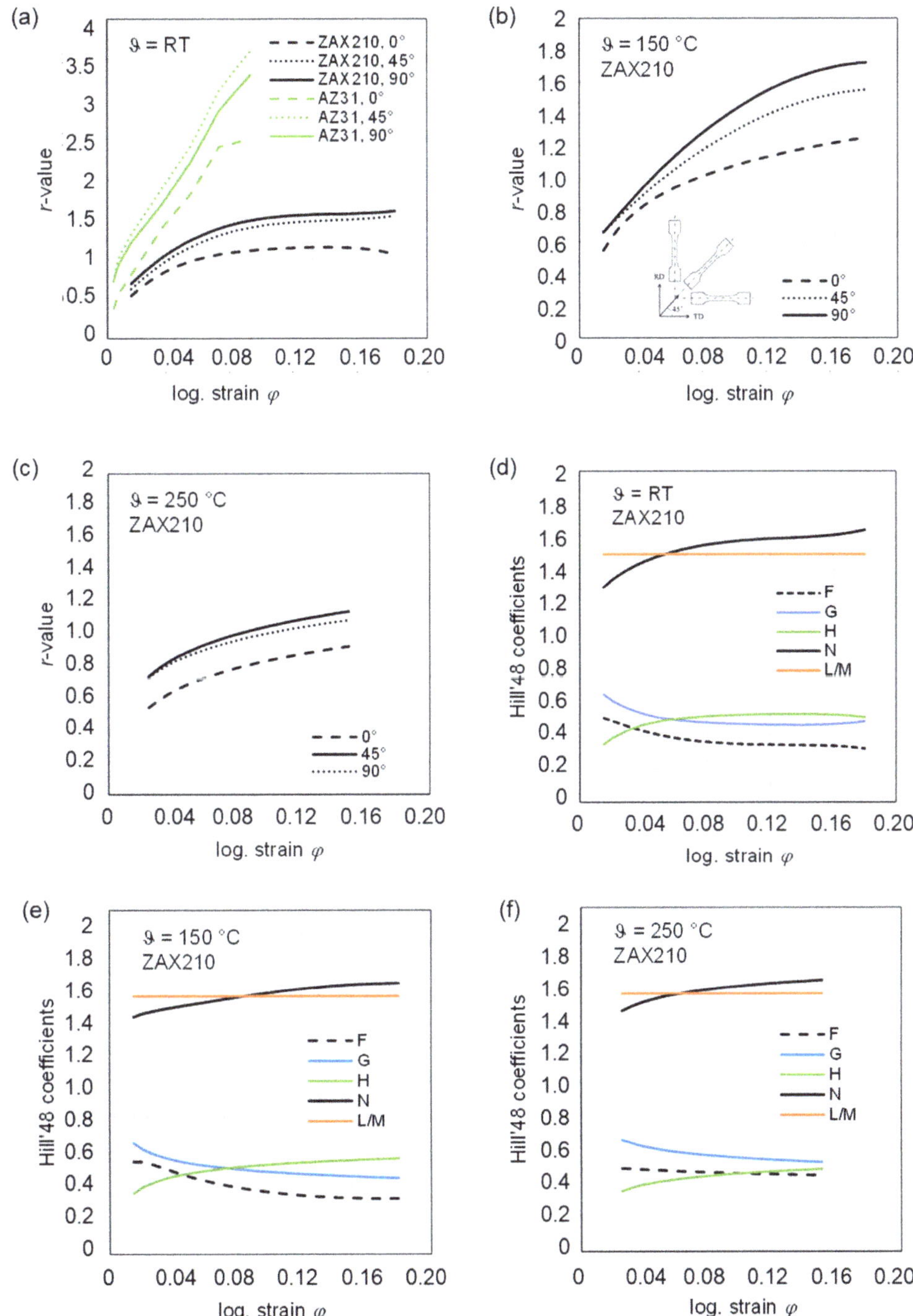

Figure 5. Variation in r-values with log. strain used to determine the material parameters for the anisotropic yield functions and Hill'48 coefficients: (**a**) ZAX210 in comparison with AZ31 (data from [24]) at RT, (**b**) ZAX210 at 150 °C, (**c**) ZAX210 at 250 °C, (**d**) Hill'48 coefficients for ZAX210 at RT, (**e**) Hill'48 coefficients for ZAX210 at 150 °C and (**f**) Hill'48 coefficients for ZAX210 at 250 °C.

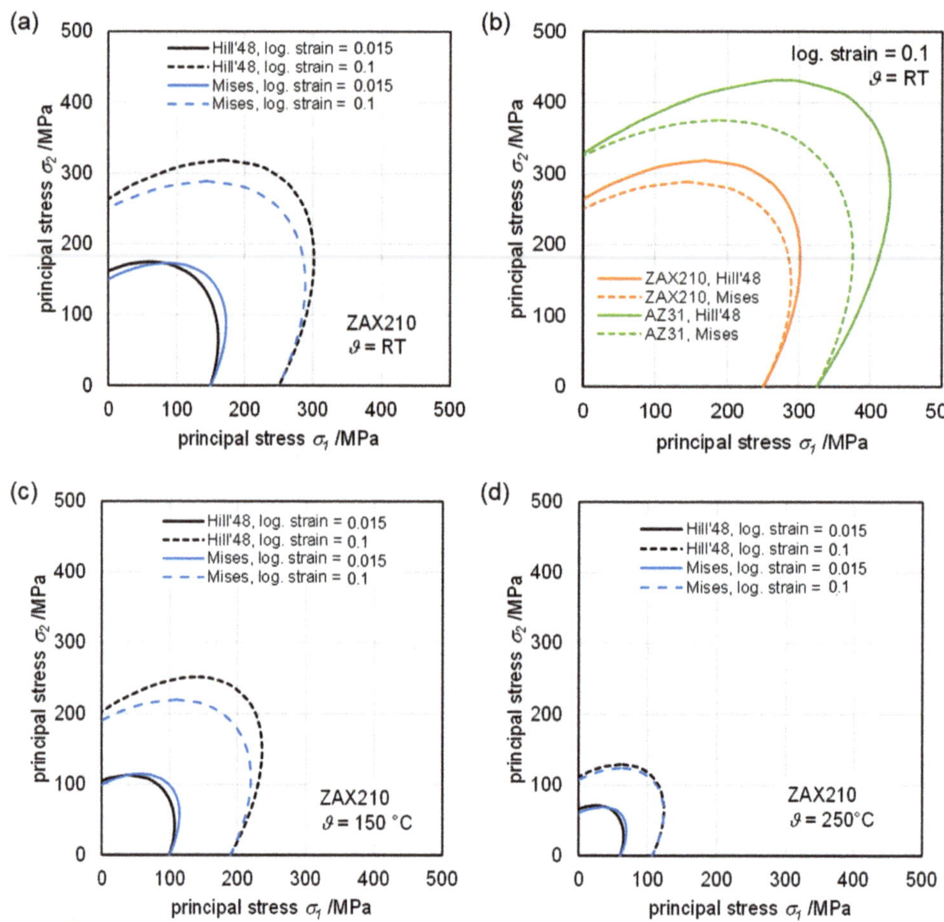

Figure 6. Yield loci of ZAX210 calculated with Hill'48 considering the r-values at different logarithmic strains, partly in comparison with Mises: (**a**) at room temperature, (**b**) at room temperature in comparison with AZ31, (**c**) at 150 °C, (**d**) at 250 °C.

4. Conclusions

In this study the orientation dependent flow behavior of the ZAX210 magnesium alloy was investigated in order to provide basic yield model data for the numerical simulation. The following conclusions were obtained.

(1) The r-values are small (0.5 to 0.7) during the early stage of deformation and increase with local logarithmic strain. At all tested temperatures, there is a tendency that $r_{90^\circ} > r_{45^\circ} > r_{0^\circ}$.
(2) The obtained r-values are a direct result of the crystallographic texture present in the ZAX210 alloy, combined with the relative resolved shear strengths of the slip and twinning systems.
(3) The plane anisotropy Δr is between 0 and–0.2 at all tested temperatures. These values indicate a slight anisotropic behavior in the sheet plane.
(4) The calculated yield loci using von Mises and Hill'48 yield functions depending on logarithmic strain and temperature for the ZAX210 sheets reveal an isotropic behavior, where Hill'48 and Mises converge and an approximately planar isotropic material

behavior at low temperature 150 °C can be assumed even for moderately higher strains compared to AZ31.

(5) The in-plane material flow behavior can be identified as orthotropic with decreasing anisotropy at elevated temperatures.

Author Contributions: Conceptualization, K.K. and M.U.; methodology, C.K. and M.U.; formal analysis, M.U.; investigation, C.K.; resources, U.P.; data curation, M.U.; writing—original draft preparation, C.K. and M.U.; writing—review and editing, K.K. and U.P.; visualization, M.U.; supervision, U.P.; project administration, K.K.; funding acquisition, U.P and M.U. All authors have read and agreed to the published version of the manuscript.

Funding: This research was funded within the project "Digitale Technologien für hybride Leichtbaustraukturen (dahlia)" by the European Union (European Regional Development Fund) and by the Free State of Saxony. Grant Number: 100383375.

Institutional Review Board Statement: Not applicable.

Informed Consent Statement: Not applicable.

Data Availability Statement: Not applicable.

Conflicts of Interest: The authors declare no conflict of interest.

References

1. Hama, T.; Takuda, H. Crystal plasticity finite-element simulation of work-hardening behavior in a magnesium alloy sheet under biaxial tension. *Comput. Mater. Sci.* **2012**, *51*, 156–164. [CrossRef]
2. Shi, B.; Yang, C.; Peng, Y.; Zhang, F.; Pan, F. Anisotropy of wrought magnesium alloys: A focused overview. *J. Magnes. Alloy.* **2022**, *10*, 1476–1510. [CrossRef]
3. Davis, A.E.; Robson, J.D.; Turski, M. Reducing yield asymmetry and anisotropy in wrought magnesium alloys—A comparative study. *Mater. Sci. Eng. A* **2019**, *744*, 525–537. [CrossRef]
4. Drozdenko, D.; Bohlen, J.; Yi, S.B.; Minárik, P.; Chmelík, F.; Dobroň, P. Investigating a twinning–detwinning process in wrought Mg alloys by the acoustic emission technique. *Acta Mater.* **2016**, *110*, 103–113. [CrossRef]
5. Kamrani, S.; Fleck, C. Effects of calcium and rare-earth elements on the microstructure and tension–compression yield asymmetry of ZEK100 alloy. *Mater. Sci. Eng. A* **2014**, *618*, 238–243. [CrossRef]
6. Wang, J.; Li, X.; Jin, P.; Li, S.; Ma, G.; Zhao, L. Reducing the tension-compression yield asymmetry in an extruded ZK60 alloy by ultrafine grains. *Mater. Res. Express* **2018**, *5*, 116518. [CrossRef]
7. Dobroň, P.; Drozdenko, D.; Olejňák, J.; Hegedüs, M.; Horváth, K.; Veselý, J.; Bohlen, J.; Letzig, D. Compressive yield stress improvement using thermomechanical treatment of extruded Mg-Zn-Ca alloy. *Mater. Sci. Eng. A* **2018**, *730*, 401–409. [CrossRef]
8. Barnett, M.R.; Nave, M.D.; Ghaderi, A. Yield point elongation due to twinning in a magnesium alloy. *Acta Mater.* **2012**, *60*, 1433–1443. [CrossRef]
9. Langelier, B.; Nasiri, A.M.; Lee, S.Y.; Gharghouri, M.A.; Esmaeili, S. Improving microstructure and ductility in the Mg–Zn alloy system by combinational Ce–Ca microalloying. *Mater. Sci. Eng. A* **2015**, *620*, 76–84. [CrossRef]
10. Stanford, N. The effect of calcium on the texture, microstructure and mechanical properties of extruded Mg–Mn–Ca alloys. *Mater. Sci. Eng. A* **2010**, *528*, 314–322. [CrossRef]
11. Nakata, T.; Xu, C.; Ajima, R.; Matsumoto, Y.; Shimizu, K.; Sasaki, T.T.; Hono, K.; Kamado, S. Improving mechanical properties and yield asymmetry in high-speed extrudable Mg-1.1Al-0.24Ca (wt%) alloy by high Mn addition. *Mater. Sci. Eng. A* **2018**, *712*, 12–19. [CrossRef]
12. Song, B.; Wang, C.; Guo, N.; Pan, H.; Xin, R. Improving Tensile and Compressive Properties of an Extruded AZ91 Rod by the Combined Use of Torsion Deformation and Aging Treatment. *Materials* **2017**, *10*, 280. [CrossRef] [PubMed]
13. Agnew, S.R.; Brown, D.W.; Tome, C.N. Validating a polycrystal model for the elastoplastic response of magnesium alloy AZ31 using in situ neutron diffraction. *Acta Mater.* **2006**, *54*, 4841–4852. [CrossRef]
14. Homma, T.; Mendis, C.L.; Hono, K.; Kamado, S. Effect of Zr addition on the mechanical properties of as-extruded Mg–Zn–Ca–Zr alloys. *Mater. Sci. Eng. A* **2010**, *527*, 2356–2362. [CrossRef]

15. Robson, J.D.; Twier, A.M.; Lorimer, G.W.; Rogers, P. Effect of extrusion conditions on microstructure, texture, and yield asymmetry in Mg–6Y–7Gd–0.5wt%Zr alloy. *Mater. Sci. Eng. A* **2011**, *528*, 7247–7256. [CrossRef]
16. Robson, J.D.; Stanford, N.; Barnett, M.R. Effect of precipitate shape on slip and twinning in magnesium alloys. *Acta Mater.* **2011**, *59*, 1945–1956. [CrossRef]
17. Sasaki, T.T.; Elsayed, F.R.; Nakata, T.; Ohkubo, T.; Kamado, S.; Hono, K. Strong and ductile heat-treatable Mg–Sn–Zn–Al wrought alloys. *Acta Mater.* **2015**, *99*, 176–186. [CrossRef]
18. Ullmann, M.; Kittner, K.; Henseler, T.; Stöcker, A.; Prahl, U.; Kawalla, R. Development of new alloy systems and innovative processing technologies for the production of magnesium flat products with excellent property profile. *Procedia Manuf.* **2019**, *27*, 203–208. [CrossRef]
19. Kittner, K.; Ullmann, M.; Henseler, T.; Kawalla, R.; Prahl, U. Microstructure and Hot Deformation Behavior of Twin Roll Cast Mg-2Zn-1Al-0.3Ca Alloy. *Materials* **2019**, *12*, 1020. [CrossRef]
20. Hill, R. A theory of the yielding and plastic flow of anisotropic metals. *Proc. R. Soc. Lond. A* **1948**, *193*, 281–297. [CrossRef]
21. Agnew, S.R. Plastic anisotropy of magnesium alloy AZ31B sheet. In *Essential Readings in Magnesium Technology*; Springer: Cham, Switzerland, 2002; pp. 169–174.
22. Yi, S.; Bohlen, J.; Heinemann, F.; Letzig, D. Mechanical anisotropy and deep drawing behaviour of AZ31 and ZE10 magnesium alloy sheets. *Acta Mater.* **2010**, *58*, 592–605. [CrossRef]
23. Duygulu, Ö.; Agnew, S.R. The effect of temperature and strain rate on the tensile properties of textured magnesium alloy AZ31B sheet. *Magnes. Technol.* **2003**, *237*, 242.
24. Andar, M.O.; Kuwabara, T.; Steglich, D. Material modeling of AZ31 Mg sheet considering variation of r-values and asymmetry of the yield locus. *Mater. Sci. Eng. A* **2012**, *549*, 82–92. [CrossRef]

Article

Unusual Spreading of Strain Neutral Layer in AZ31 Magnesium Alloy Sheet during Bending

Chao He [1], Lintao Liu [1], Shengwen Bai [1,2,*], Bin Jiang [1,2,*], Hang Teng [3], Guangsheng Huang [1], Dingfei Zhang [1] and Fusheng Pan [1]

1 National Engineering Research Center for Magnesium Alloys, College of Materials Science and Engineering, Chongqing University, Chongqing 400044, China
2 State Key Laboratory of Mechanical Transmissions, Chongqing University, Chongqing 400044, China
3 Nanjing Yunhai Special Metal Co., Ltd., Nanjing 211212, China
* Correspondence: baisw@cqu.edu.cn (S.B.); jiangbinrong@cqu.edu.cn (B.J.)

Abstract: In this work, we reported an unusual phenomenon of strain neutral layer (SNL) spreading in an as-rolled AZ31B magnesium alloy sheet during V-bending. The SNL on the middle symmetrical surface perpendicular to the transverse direction (TD) of the sheet extended to the compression region and was accompanied by a mound-like feature. However, the SNL on the side surface perpendicular to the TD was distributed with a parallel band feature. The underlying mechanism was revealed by the finite element (FE) analysis. The results indicate that the three-dimensional compressive stresses in the compression region of the bending samples were responsible for the above phenomenon. Moreover, the area of the SNL in the middle position gradually decreased as the bending test progressed. The findings in this study provide some new insights into the bending deformation behavior of magnesium alloy.

Keywords: magnesium; V-bending; strain neutral layer; microstructure

Citation: He, C.; Liu, L.; Bai, S.; Jiang, B.; Teng, H.; Huang, G.; Zhang, D.; Pan, F. Unusual Spreading of Strain Neutral Layer in AZ31 Magnesium Alloy Sheet during Bending. *Materials* **2022**, *15*, 8880. https://doi.org/10.3390/ma15248880

Academic Editors: Mateusz Kopec and Denis Politis

Received: 27 October 2022
Accepted: 8 December 2022
Published: 12 December 2022

1. Introduction

As a potential lightweight alternative that substitutes for aluminum and high-strength steel alloys, magnesium (Mg) alloys have been greatly developed in recent decades [1–4]. However, their inherent hexagonal close-packed (HCP) structure would restrict the number of activated slip systems at room temperature (RT) [5,6]. In addition, as one of the most important products of wrought Mg alloys, the sheets prepared by extrusion and rolling always exhibit a strong basal texture [7–9]. These factors result in the low formability of Mg alloy sheets at RT, which greatly limits the wide industrial application of Mg alloys [10–12]. Therefore, how to improve the formability of magnesium alloys has become a meaningful and urgent topic. During sheet forming, such as deep drawing and hemming, the material at the corner of the part undergoes bending deformation [13,14]. Therefore, bendability is an essential indicator that represents the formability of Mg alloy sheets. The evaluation and improvement of the bendability of Mg alloy sheets has attained adequate attention [15–21].

During the bending process, there exists a stress gradient through-thickness direction of the bending samples. The tensile stress is distributed in the outer region, while compressive stress is distributed in the inner region of bending samples [17,19,22]. Additionally, the bending samples can be divided into three parts along the thickness direction according to the different strain states, namely compressive strain region (CSR), strain neutral layer (SNL) and tensile strain region (TSR) [17,23,24]. A lot of efforts have been made to investigate the evolution of SNL during the bending of magnesium alloy. For example, Li et al. [16] demonstrated that the SNL tended to first shift toward the outer tensile region, and then shift toward the inner compressive region during the three-point bending of Mg alloy sheets, which is mainly attributed to the asymmetric response of the Mg alloy in tension and compression. Wang et al. [25] investigated the evolution of springback and neutral for AZ31 magnesium alloy by V-bending tests. The results indicate that the SNL shifted to the tension zone of the sample. The offset of SNL decreased with increasing the bending temperature because of the weakening of asymmetry between the tensile and compressive regions. Huang et al. [26] also found that the SNL of the AZ31 magnesium alloy sheet shifted toward the tension region during the V-bending at the temperature of 150 °C. Bai et al. [27] reported that the greater the shift of the neutral layer during the V-bending of the AZ31/Mg-Gd laminated composite sheet, the higher the strain on the outer layer, which makes the sample more prone to fracture. Most of the current studies focus on the shifting of SNL along the thickness direction during the bending. However, the importance of the morphology of SNL is ignored.

The present study observed an unusual phenomenon about the SNL spreading toward the CSR during the V-bending test of AZ31B Mg alloy. Optical microscopy (OM) and finite element (FE) analysis were utilized to investigate the morphology and stress state of the bending samples. It was found that the three-dimensional compress stresses were responsible for the SNL expansion. The underlying mechanisms are discussed based on the FE simulation results.

2. Experiments and Simulations

The material used in this study was a hot-rolled AZ31B sheet with a thickness of 3 mm. Tables 1 and 2 present the chemical compositions and tensile properties along the rolling direction (RD) of the sheet, respectively. Bending samples were machined into strips with the dimension of 50 mm × 12 mm × 3 mm from the as-rolled sheets along the RD. The V-bending test was operated on a universal testing machine (Instron 5985, Boston, MA, USA) at a speed of 1.5 mm/min. The distance between the two supports of the bending tooling was 32 mm. The die angle was 60°, and the radius of the punch was 2 mm. The texture of the as-rolled sheet was examined by the Electron Back-Scattered Diffraction (EBSD, NordlysMax2, Oxford, UK) technique. The microstructures of bending samples were characterized by the optical microscope (OM, ZEISS Axiovert 40 MAT, Oberkochen, Germany). The normal direction (ND)–rolling direction (RD) surfaces of the samples for EBSD and OM measurement were mechanically ground and polished. Thereafter, the OM sample was etched with a solution composed of 1.5 g picric acid, 5 mL acetic acid and 25 mL ethyl alcohol. The EBSD sample was electrochemically polished with the electrolyte AC2 at a voltage of 20 V and a temperature of −25 °C for 90 s.

Table 1. Chemical compositions of the AZ31B Mg alloy.

Elements	Al	Zn	Mn	Mg
Content (wt.%)	2.87	0.87	0.21	Bal.

Table 2. Tensile properties along the RD of the AZ31B Mg alloy sheet.

Yield Strength (MPa)	Ultimate Tensile Strength (MPa)	Elongation (%)
153	303	19.7

FE simulations were performed by the commercial software Deform 3D to analyze the distributions of stress and strain during the V-bending (Figure 1a). In the FE model, the bending sample was regarded as a rigid–plastic body, while both the punch and die were set to be rigid bodies. The bending sample was meshed into tetrahedral elements, and the number of elements was 50,000. The Coulomb friction model was employed at the interface between the sample and tooling, and the friction coefficient was set to 0.3 [28]. The stress–strain data obtained through the tensile test along the RD were input into the model for calculation (Figure 1b).

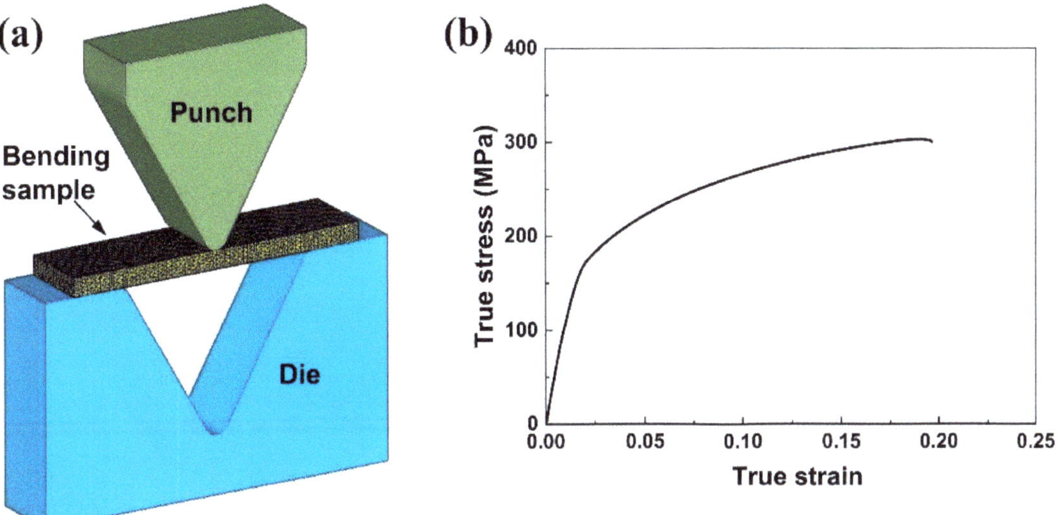

Figure 1. FE model of the V-bending (**a**) and stress–strain data used in the FE simulation (**b**).

3. Results and Discussion

3.1. Bending Behavior of the As-Rolled AZ31 Sheet

The initial microstructures of the as-rolled AZ31 sheet are shown in Figure 2a. The completed recrystallized grains are observed, and the average grain size is ~12 μm. In the (0002) pole figure, the *c*-axis of numerous grains are parallel to the ND of the sheet, and the maximum pole intensity is about 12.1 M.U.D (Figure 2b). The typical basal texture is usually formed during the rolling of the Mg alloy [29,30]. Tam et al. [31], for example, also found a strong basal texture with the *c*-axis of most grains parallel to the ND in the rolled AZ31 sheet. The bending load–stroke curve and the corresponding image of the bent sample are shown in Figure 2c,d, respectively. The bending load increases sharply in the initial stage, corresponding to the elastic deformation of the Mg alloy. With further increase in the stroke, the load rises slowly and reaches the peak value due to the work hardening during the plastic deformation. Thereafter, the load is gradually reduced until the test is stopped when the fracture occurs. The maximum stroke before cracking arrived at 3.0 ± 0.2 mm, and the final bending angle is $136 \pm 3°$. Furthermore, the cracks are located at the middle position of the outer tensile region, as shown in Figure 2d.

3.2. Microstructure Evolution during Bending

Figure 3 shows the microstructures when the bending test is interrupted at the punch stroke of 1.5 mm. The OM images of the side surface (position I in Figure 3i) of the sample are presented in Figure 3a–d. Figure 3e–h reveals the OM images on the middle symmetrical surface (position II in Figure 3i) perpendicular to the transverse direction (TD). In the TSR, a large number of deformation bands (marked with red arrows) is observed (Figure 3e). Twinning morphologies (marked with green arrows) appear frequently in the CSR (Figure 3b). Furthermore, the SNL without deformation bands or twinning morphologies is located between the TSR and CSR (see Figure 3c). For the as-rolled Mg alloys with strong basal texture, it is widely accepted that the slip dominates the tensile strain, and the {10−12} extension twinning dominates the compressive strain during the bending process. The reason for this phenomenon is that tensile stress in the TSR is perpendicular to the c-axes, which is favorable for the activation of basal and prismatic <a> slips. The {10−12} extension twinning is more likely to be activated under the compression stress state in the CSR [19]. Bai et al. [27] found that extension twinning was activated in the

CRS during the bending of the AZ31/Mg–Gd-laminated composite sheet. In the research of Tang et al. [18], a lot of twinning bands were observed in the CSR during the four-point bending of the extruded AZ31 plate, while few twinning bands were formed in the TSR. The phenomena observed in the current study are consistent with previous studies.

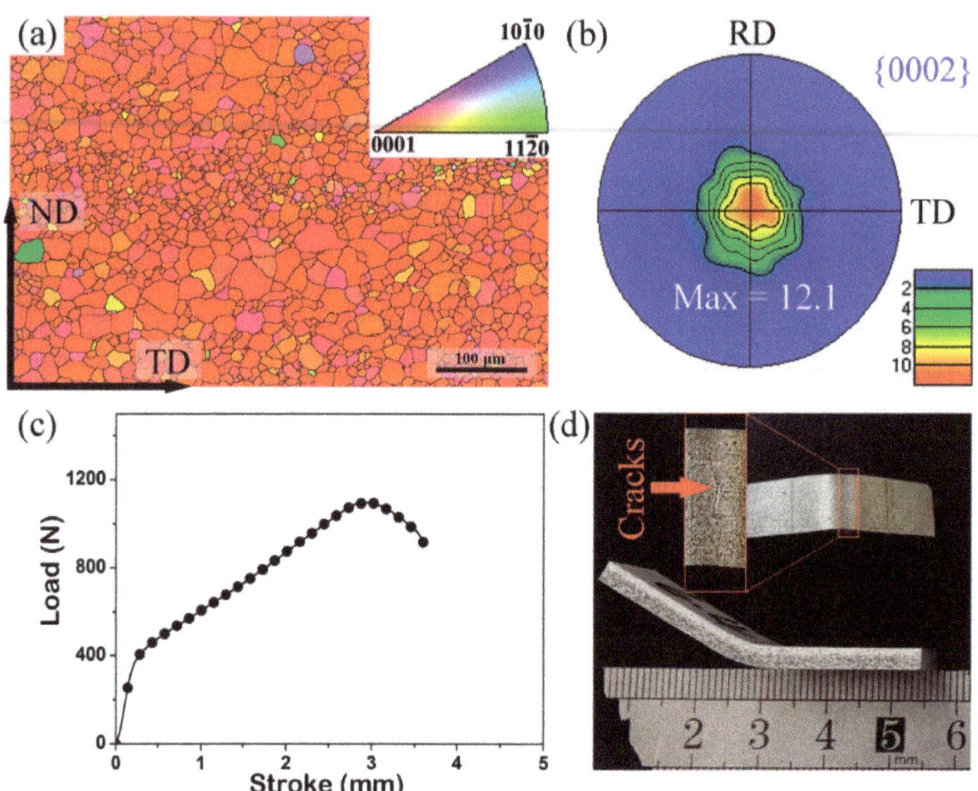

Figure 2. Results of EBSD measurement and bending experiment: (**a**) IPF coloring map, (**b**) (0002) pole figure of the as-rolled AZ31 alloys sheet, (**c**) bending load-stroke curve and (**d**) images of the bent sample.

Comparing the microstructures at position I and position II, the number of deform bands in the TSR of position II is significantly more than that of position I, which means that the tensile strain at position II is more severe than that at position I under the same punch stroke. The formation of deform bands during the plastic forming of the as-rolled Mg alloy sheets has been reported in many studies. The possible formation mechanisms of these deform bands have been proposed to be related to twinning-induced shear banding (TISB) [32,33] and weak texture-induced shear banding (WTISB) [34–36]. The boundaries of SNL are highlighted by the red dotted lines, as shown in Figure 3a,e. The SNL morphologies in these two positions are markedly different. In position I, the boundaries of SNL in Figure 3a are nearly parallel with each other, which is in the region between the CSR and the TSR. This phenomenon is consistent with the results reported by Lee et al. [13], Singh et al. [37] and Jin et al. [23]. For example, Lee et al. [13] observed the side surface of the bent sample using an optical microscope and also found that the boundaries of SNL are nearly parallel with each other. The SNL boundary extends to the compressed region in position II, as shown in Figure 3e. Compared with the position I, the SNL in position II is mound-like, which has never been reported before.

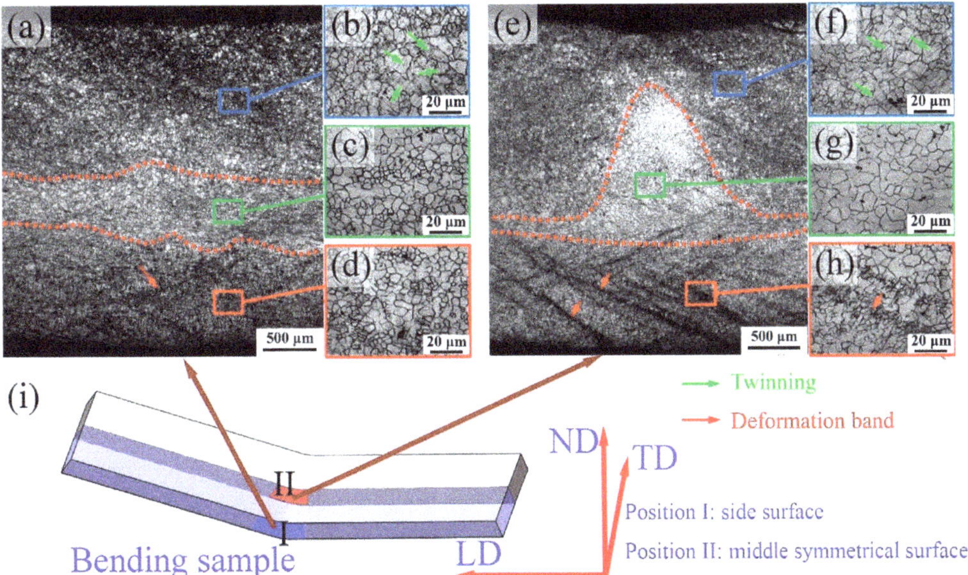

Figure 3. Optical micrographs (**a**–**d**) of the side surface (position I in Figure 3i) and (**e**–**h**) middle symmetrical surface (position II in Figure 3i) perpendicular to the transverse direction (TD) of the sample when the bending test interrupts at the punch stroke of 1.5 mm and (**i**) schematic diagram of the bending sample.

Figure 4 shows the microstructures when the bending test is interrupted at the punch stroke of 3.0 mm, in which the macrocracks are observed. More deform bands are formed in the TSR as the punch stroke increases from 1.5 mm to 3 mm, no matter whether in position I or position II. This is because the strain in the TSR increases as the bending proceeds. The macrocracks are formed and extended in position II. However, there is no sign of cracks in the side surface (position I) of the bending samples, which is consistent with the images of the bent sample shown in Figure 2d. According to Figures 3 and 4, the tensile strain in the TSR of position II is larger than that of position I during the whole bending test. For the distribution of SNL, a similar phenomenon that parallels the band-like SNL in position I and the mound-like SNL in position II is observed. However, the area of mound-like SNL in position II is decreased with the punch stroke increased from 1.5 mm to 3.0 mm, as shown in Figure 4b.

3.3. Simulated Results

To analyze the difference in the SNL between positions I and II, the stress component (σ_x) and strain component (ε_x) in the length direction of the bending sample are extracted from the FE simulated results, as shown in Figure 5. Four paths along the thickness of the bending AZ31B sample are defined to better understand the stress distribution during the bending test. Paths 1 and 3 are located at the middle position in positions I and II of the bending samples, respectively. Paths 2 and 4 are situated at the site away from the middle position, about 3 mm from paths 1 and 3, respectively. Figure 5a–c shows the distribution maps and the corresponding detailed curves of σ_x along the four paths. All the stress curves along the four paths show an 'S' shape. This type of stress distribution has been reported by Ren et al. [24]. The higher slope interval that is highlighted by the green shaded area corresponds to the SNL, and the low slope intervals represent the TSR and CSR, as shown in Figure 5c. It is revealed that the stress distribution curves of positions I and II are similar in the TSR. In the CSR, the stress values in position I (paths 1 and 2) are significantly lower than those in position II (paths 3 and 4). This means that the stress distribution along the transverse direction (TD) is not uniform during the bending process. The stress in the middle position is higher than that in the side position, which is consistent with the results observed in Figures 3 and 4. Figure 5d,e show the simulated strain when the punch stroke is 1.5 mm. It is revealed that the boundaries of SNL are nearly parallel with each other in the side position (position I) and arched upward in the middle position (position II), which is consistent with the results of OM observation (Figure 3). Figure 5f shows the strain distribution curves at the two positions along the four paths. Here, we define the elastic strain as the strain with the range of −0.002 ~ 0.002, such that the SNL can be marked by the green shaded area in Figure 5f. In CSR, the strain along path 3 is significantly lower than those along other paths. This means that the SNL extends to the CSR in position II. The simulation results (Figure 3) are highly consistent with the OM observed results.

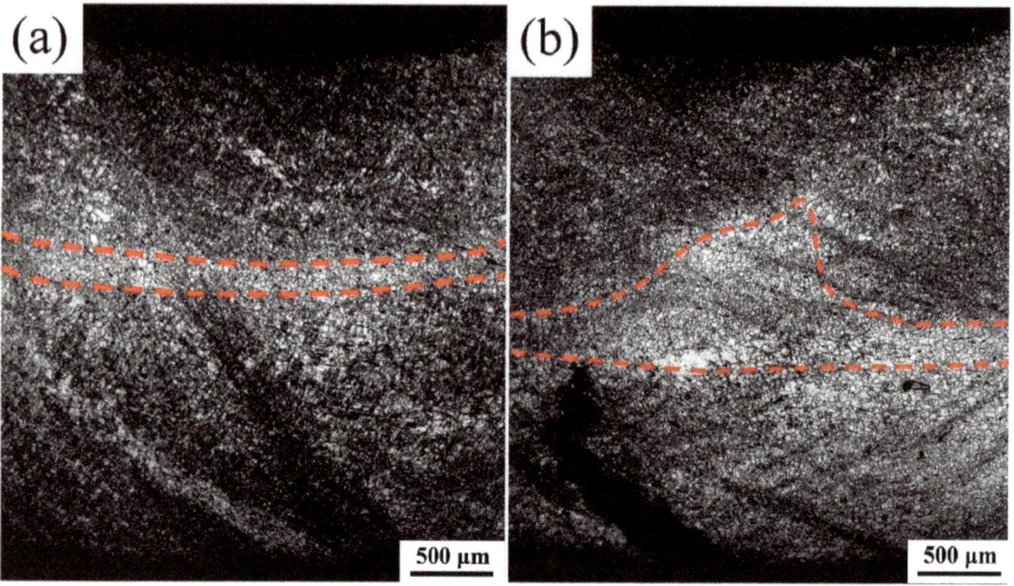

Figure 4. Optical micrographs (**a**) of the side surface (position I in Figure 3i) and (**b**) middle symmetrical surface (position II in Figure 3i) perpendicular to the transverse direction (TD) of the sample when the bending test interrupts at the punch stroke of 3.0 mm.

The analytical results indicate that the SNL spreading phenomenon of AZ31B magnesium alloy is closely related to the distribution of compression stress during bending. The material is in a relatively stable state under triaxial compression or tension stresses [13,37]. The same idea is used to understand the spreading phenomenon of the SNL in this study. Figure 6 presents the stress components in the length direction (LD), transverse direction (TD) and normal direction (ND) of the bending sample along paths 1 and 3, denoted as σ_x, σ_y and σ_z, respectively. Along path 1, σ_x is significantly greater than σ_y and σ_z. Consequently, the stress component in LD is responsible for the deformation behavior in the position I, and the boundaries of SNL are parallel with each other. For path 3, the difference among the stress components in LD, TD and ND is smaller than that along path 1. In the CSR of position I (corresponding to path 1), the material undergoes a stress state close to the uniaxial compression. The stress state in the CSR of position II is triaxial compression (Figure 6b). In position II, the highest compressive stress components in LD, TD and ND are 298, 199 and 175 MPa, respectively. In the grains under the triaxial compressive stress state, the strain coordination modes, such as basal <a> slip and {10−12} extension twinning, are hard to be activated. The area of the SNL extends to the CSR and forms a mound-like elastic strain region on the middle symmetrical surface perpendicular to TD.

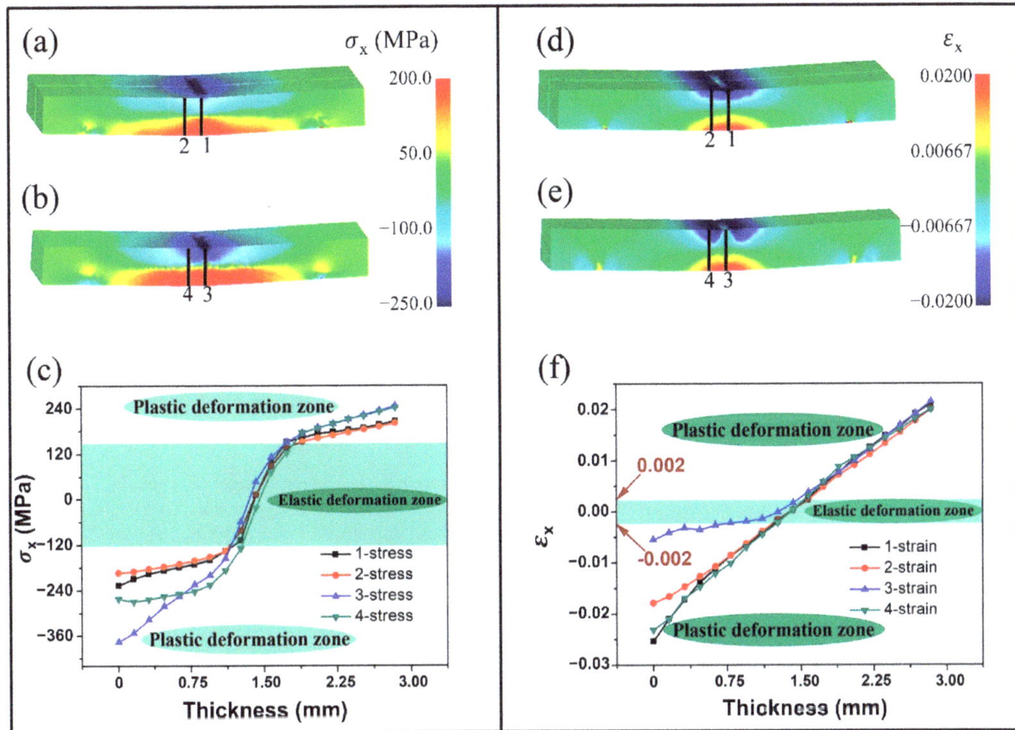

Figure 5. Simulated stress component σ_x (**a**,**b**) and strain component ε_x (**d**,**e**) in the side position (**a**,**d**) and middle position (**b**,**e**); distributions of stress (**c**) and strain (**f**) along the paths 1, 2, 3 and 4.

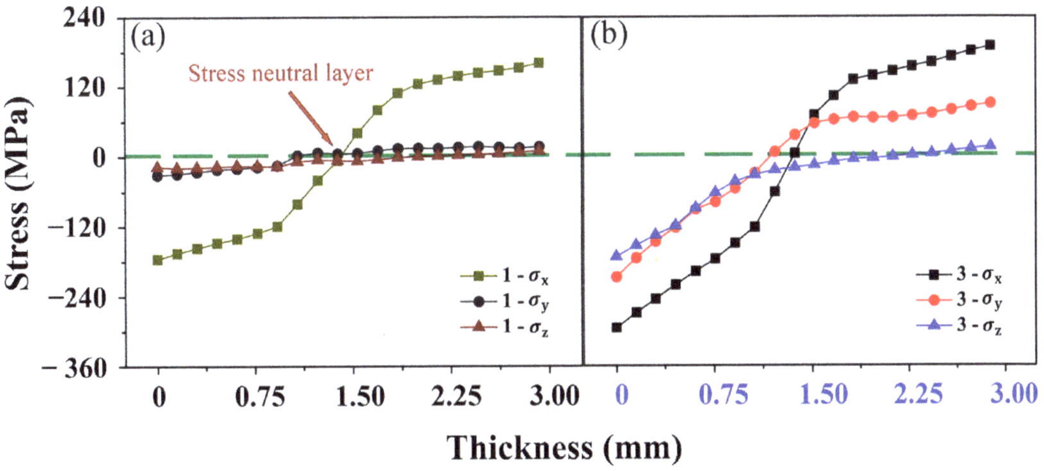

Figure 6. Distributions of σ_x in the LD, TD and ND along path 1 (**a**) and path 3 (**b**).

It is defined that the region with a stress value of 0 MPa at the junction of the tensile stress region and compressive stress region of the bending sample is the stress neutral layer. The elastic strain region is defined as the strain neutral layer. However, many reports only considered the stress component in the LD [37–41] and ignored the stress components in the TD and ND. The stress state in the bending sample is much more complex than that of uniaxial tension or compression. Only the three-dimensional stress state is considered, rather than the uniaxial stress separately; the phenomenon of strain SNL spreading can be explained legitimately. The results are important for better understanding the bending behavior of Mg alloy sheets.

4. Conclusions

In this work, an unusual phenomenon of SNL spreading is revealed during the V-bending test. The SNL on the middle symmetrical surface perpendicular to TD extends to the compression region with a mound-like boundary. The SNL in the side position is distributed with a parallel band feature. This difference in SNL distribution is mainly attributed to the difference in three-dimensional stress distributions between the side position and the middle position of the bending sample. The three-dimensional compressive stresses in the compressed region are responsible for the SNL spreading phenomenon. An improved FE model, in which the effects of anisotropy and tension–compression asymmetry of the material are taken into consideration, should be constructed in future studies to obtain more insights into the evolution of SNL during the bending of magnesium alloys.

Author Contributions: Funding acquisition, S.B. and B.J.; investigation, C.H., L.L. and H.T.; validation, G.H., D.Z and F.P.; writing—original draft, C.H.; writing—review and editing, S.B., L.L., B.J., H.T., G.H., D.Z. and F.P. All authors have read and agreed to the published version of the manuscript.

Funding: This research was funded by the National Natural Science Foundation of China (52101124, 51971044, U1910213, 52001037 and U2037601), Qinghai Scientific & Technological Program (2018-GX-A1), and Independent Research Project of State Key Laboratory of Mechanical Transmissions (SKLMT-ZZKT-2022M12).

Institutional Review Board Statement: Not applicable.

Informed Consent Statement: Not applicable.

Data Availability Statement: Not applicable.

Conflicts of Interest: The authors declare no conflict of interest.

References

1. Zeng, Z.; Nie, J.F.; Xu, S.W.; Davies, C.H.J.; Birbilis, N. Super-formable pure magnesium at room temperature. *Nat. Commun.* **2017**, *8*, 972. [CrossRef] [PubMed]
2. Fu, H.; Ge, B.; Xin, Y.; Wu, R.; Fernandez, C.; Huang, J.; Peng, Q.J.N.l. Achieving high strength and ductility in magnesium alloys via densely hierarchical double contraction nanotwins. *Nano Lett.* **2017**, *17*, 6117–6124. [CrossRef] [PubMed]
3. Hu, Y.J.; Wang, J.Y.; Xu, N.X.; Zhai, W.; Wei, B. Three orthogonally arranged ultrasounds modulate solidification microstructures and mechanical properties for AZ91 magnesium alloy. *Acta Mater.* **2022**, *241*, 118382. [CrossRef]
4. Hu, Y.J.; Zhou, Q.; Zhai, W.; Wang, J.Y.; Wei, B. Improved mechanical performances of dynamically solidified $Mg_{97.7}Y_{1.4}Al_{0.9}$ alloy by three dimensional ultrasounds. *Mater. Sci. Eng. A* **2022**, *860*, 144153. [CrossRef]
5. Bian, M.; Huang, X.; Chino, Y. Substantial improvement in cold formability of concentrated Mg–Al–Zn–Ca alloy sheets by high temperature final rolling. *Acta Mater.* **2021**, *220*, 117328. [CrossRef]
6. Nakata, T.; Xu, C.; Kaibe, K.; Yoshida, Y.; Yoshida, K.; Kamado, S. Improvement of strength and ductility synergy in a room-temperature stretch-formable Mg-Al-Mn alloy sheet by twin-roll casting and low-temperature annealing. *J. Magnes. Alloy.* **2022**, *10*, 1066–1074. [CrossRef]
7. Sabbaghian, M.; Fakhar, N.; Nagy, P.; Fekete, K.; Gubicza, J. Investigation of shear and tensile mechanical properties of ZK60 Mg alloy sheet processed by rolling and sheet extrusion. *Mater. Sci. Eng. A* **2021**, *828*, 142098. [CrossRef]
8. Wang, Y.; Li, F.; Bian, N.; Du, H.Q.; Da Huo, P. Mechanism of plasticity enhancement of AZ31B magnesium alloy sheet by accumulative alternating back extrusion. *J. Magnes. Alloy.* 2021; in press. [CrossRef]
9. Lee, G.M.; Lee, J.U.; Park, S.H. Effects of surface roughness on bending properties of rolled AZ31 alloy. *J. Magnes. Alloy.* 2022; in press. [CrossRef]
10. Javaid, A.; Czerwinski, F. Effect of hot rolling on microstructure and properties of the ZEK100 alloy. *J. Magnes. Alloy.* **2019**, *7*, 27–37. [CrossRef]
11. Lee, J.H.; Kwak, B.J.; Kong, T.; Park, S.H.; Lee, T. Improved tensile properties of AZ31 Mg alloy subjected to various caliber-rolling strains. *J. Magnes. Alloy.* **2019**, *7*, 381–387. [CrossRef]
12. Xu, J.; Guan, B.; Xin, Y.; Wei, X.; Huang, G.; Liu, C.; Liu, Q. A weak texture dependence of Hall–Petch relation in a rare-earth containing magnesium alloy. *J. Mater. Sci. Technol.* **2022**, *99*, 251–259. [CrossRef]

13. Lee, J.U.; Kim, S.-H.; Kim, Y.J.; Park, S.H. Improvement in bending formability of rolled magnesium alloy through precompression and subsequent annealing. *J. Alloy. Compd.* **2019**, *787*, 519–526. [CrossRef]
14. Lee, J.U.; Kim, S.-H.; Lee, D.H.; Kim, H.J.; Kim, Y.M.; Park, S.H. Variations in microstructure and bending formability of extruded Mg–Al–Zn–Ca–Y–MM alloy with precompression and subsequent annealing treatment conditions. *J. Magnes. Alloy.* **2022**, *10*, 2475–2490. [CrossRef]
15. Ning, F.; Zhou, X.; Le, Q.; Li, X.; Ma, L.; Jia, W. Investigation of microstructure and texture during continuous bending of rolled AZ31 sheet by experiment and FEM. *J. Mater. Res. Technol.* **2019**, *8*, 6232–6243. [CrossRef]
16. Li, F.-F.; Fang, G. Modeling of 3D plastic anisotropy and asymmetry of extruded magnesium alloy and its applications in three-point bending. *Int. J. Plast.* **2020**, *130*, 102704. [CrossRef]
17. Singh, J.; Kim, M.-S.; Choi, S.-H. The effect of initial texture on micromechanical deformation behaviors in Mg alloys under a mini-V-bending test. *Int. J. Plast.* **2019**, *117*, 33–57. [CrossRef]
18. Tang, D.; Zhou, K.; Tang, W.; Wu, P.; Wang, H. On the inhomogeneous deformation behavior of magnesium alloy beam subjected to bending. *Int. J. Plast.* **2022**, *150*, 103180. [CrossRef]
19. Wang, W.; Zhang, W.; Chen, W.; Cui, G.; Wang, E. Effect of initial texture on the bending behavior, microstructure and texture evolution of ZK60 magnesium alloy during the bending process. *J. Alloys Compd.* **2018**, *737*, 505–514. [CrossRef]
20. Zhang, J.; Liu, H.; Chen, X.; Zou, Q.; Huang, G.; Jiang, B.; Tang, A.; Pan, F. Deformation Characterization, Twinning Behavior and Mechanical Properties of Dissimilar Friction-Stir-Welded AM60/AZ31 Alloys Joint During the Three-Point Bending. *Acta Metall. Sin.* **2021**, *35*, 727–744. [CrossRef]
21. Bai, S.; Liu, L.; Li, K.; Jiang, B.; Huang, G.; Zhang, D.; Pan, F. Investigation into the microstructure, tensile properties and bendability of Mg–Al–Zn/Mg-xGd laminated composite sheets extruded by porthole die. *J. Mater. Res. Technol.* **2022**, *21*, 12–29. [CrossRef]
22. Lee, J.U.; Lee, G.M.; Park, S.H. Bending properties of extruded AZ91–0.9Ca–0.6Y alloy and their improvement through precompression and annealing. *J. Magnes. Alloy.* **2022**, *10*, 2238–2251. [CrossRef]
23. Jin, L.; Dong, J.; Sun, J.; Luo, A.A. In-situ investigation on the microstructure evolution and plasticity of two magnesium alloys during three-point bending. *Int. J. Plast.* **2015**, *72*, 218–232. [CrossRef]
24. Ren, W.; Li, J.; Xin, R. Texture dependent shifting behavior of neutral layer in bending of magnesium alloys. *Scr. Mater.* **2019**, *170*, 6–10. [CrossRef]
25. Wang, L.; Huang, G.; Zhang, H.; Wang, Y.; Yin, L. Evolution of springback and neutral layer of AZ31B magnesium alloy V-bending under warm forming conditions. *J. Mater. Process. Technol.* **2013**, *213*, 844–850. [CrossRef]
26. Huang, G.; Wang, L.; Zhang, H.; Wang, Y.; Shi, Z.; Pan, F. Evolution of neutral layer and microstructure of AZ31B magnesium alloy sheet during bending. *Mater. Lett.* **2013**, *98*, 47–50. [CrossRef]
27. Bai, S.; Wei, L.; He, C.; Liu, L.; Dong, Z.; Liu, W.; Jiang, B.; Huang, G.; Zhang, D.; Xu, J.; et al. Effects of layer thickness ratio on the bendability of Mg-Al-Zn/Mg-Gd laminated composite sheet. *J. Mater. Res. Technol.* **2022**, *21*, 1013–1028. [CrossRef]
28. Song, X.; Tian, Y.; Quan, J.; Cui, S.; Li, J.; Fang, Q. V-shaped bending of Ti-6Al-4V titanium alloy sheet with elliptical hole. *Mater. Res. Express* **2020**, *6*, 1265j2. [CrossRef]
29. Guo, F.; Zhang, D.; Fan, X.; Jiang, L.; Yu, D.; Pan, F. Deformation behavior of AZ31 Mg alloys sheet during large strain hot rolling process: A study on microstructure and texture evolutions of an intermediate-rolled sheet. *J. Alloys Compd.* **2016**, *663*, 140–147. [CrossRef]
30. Huang, W.; Huo, Q.; Fang, Z.; Xiao, Z.; Yin, Y.; Tan, Z.; Yang, X. Damage analysis of hot-rolled AZ31 Mg alloy sheet during uniaxial tensile testing under different loading directions. *Mater. Sci. Eng. A* **2018**, *710*, 289–299. [CrossRef]
31. Tam, K.J.; Vaughan, M.W.; Shen, L.; Knezevic, M.; Karaman, I.; Proust, G. Modelling the temperature and texture effects on the deformation mechanisms of magnesium alloy AZ31. *Int. J. Mech. Sci.* **2020**, *182*, 105727. [CrossRef]
32. Basu, I.; Al-Samman, T.; Gottstein, G. Shear band-related recrystallization and grain growth in two rolled magnesium-rare earth alloys. *Mater. Sci. Eng. A* **2013**, *579*, 50–56. [CrossRef]
33. Yan, H.; Xu, S.W.; Chen, R.S.; Kamado, S.; Honma, T.; Han, E.H. Twins, shear bands and recrystallization of a Mg–2.0%Zn–0.8%Gd alloy during rolling. *Scr. Mater.* **2011**, *64*, 141–144. [CrossRef]
34. Chun, Y.B.; Davies, C.H.J. Texture effects on development of shear bands in rolled AZ31 alloy. *Mater. Sci. Eng. A* **2012**, *556*, 253–259. [CrossRef]
35. Guan, D.; Rainforth, W.M.; Gao, J.; Ma, L.; Wynne, B. Individual effect of recrystallisation nucleation sites on texture weakening in a magnesium alloy: Part 2—Shear bands. *Acta Mater.* **2018**, *145*, 399–412. [CrossRef]
36. Üçel, İ.B.; Kapan, E.; Türkoğlu, O.; Aydıner, C.C. In situ investigation of strain heterogeneity and microstructural shear bands in rolled Magnesium AZ31. *Int. J. Plast.* **2019**, *118*, 233–251. [CrossRef]
37. Singh, J.; Kim, M.-S.; Choi, S.-H. The effect of strain heterogeneity on the deformation and failure behaviors of E-form Mg alloy sheets during a mini-V-bending test. *J. Alloys Compd.* **2017**, *708*, 694–705. [CrossRef]
38. Baird, J.C.; Li, B.; Yazdan Parast, S.; Horstemeyer, S.J.; Hector, L.G.; Wang, P.T.; Horstemeyer, M.F. Localized twin bands in sheet bending of a magnesium alloy. *Scr. Mater.* **2012**, *67*, 471–474. [CrossRef]
39. Habibnejad-korayem, M.; Jain, M.K.; Mishra, R.K. Large deformation of magnesium sheet at room temperature by preform annealing, part II: "Bending". *Mater. Sci. Eng. A* **2014**, *619*, 378–383. [CrossRef]

40. Aslam, I.; Li, B.; McClelland, Z.; Horstemeyer, S.J.; Ma, Q.; Wang, P.T.; Horstemeyer, M.F. Three-point bending behavior of a ZEK100 Mg alloy at room temperature. *Mater. Sci. Eng. A* **2014**, *590*, 168–173. [CrossRef]
41. Hu, K.; Le, Q.; Zhou, W.; Li, C.; Liao, Q.; Yu, F. Three-point bending behavior and microstructural evolution of ZM20E, ZM20EX and AZ31 tubes. *Mater. Charact.* **2020**, *159*, 110065. [CrossRef]

Article

Multistage Tool Path Optimisation of Single-Point Incremental Forming Process

Zhou Yan [1], Hany Hassanin [2,*], Mahmoud Ahmed El-Sayed [3], Hossam Mohamed Eldessouky [3], JRP Djuansjah [4], Naser A. Alsaleh [4], Khamis Essa [1] and Mahmoud Ahmadein [4,5]

1 School of Engineering, University of Birmingham, Birmingham B152TT, UK; Zhou.Yan@bham.ac.uk (Z.Y.); k.e.a.essa@bham.ac.uk (K.E.)
2 School of Engineering, Technology, and Design, Canterbury Christ Church University, Canterbury CT1 1QU, UK
3 Department of Industrial Engineering, Arab Academy for Science Technology and Maritime, Alexandria 21599, Egypt; m_elsayed@aast.edu (M.A.E.-S.); hossam.eldessouky@aast.edu (H.M.E.)
4 College of Engineering, Imam Mohammad Ibn Saud Islamic University, Riyadh 11564, Saudi Arabia; jrdjuansjah@imamu.edu.sa (J.R.P.D.); naalsaleh@imamu.edu.sa (N.A.A.); maahmadein@imamu.edu.sa (M.A.)
5 Department of Production Engineering and Mechanical Design, Tanta University, Tanta 31512, Egypt
* Correspondence: hany.hassanin@canterbury.ac.uk

Abstract: Single-point incremental forming (SPIF) is a flexible technology that can form a wide range of sheet metal products without the need for using punch and die sets. As a relatively cheap and die-less process, this technology is preferable for small and medium customised production. However, the SPIF technology has drawbacks, such as the geometrical inaccuracy and the thickness uniformity of the shaped part. This research aims to optimise the formed part geometric accuracy and reduce the processing time of a two-stage forming strategy of SPIF. Finite element analysis (FEA) was initially used and validated using experimental literature data. Furthermore, the design of experiments (DoE) statistical approach was used to optimise the proposed two-stage SPIF technique. The mass scaling technique was applied during the finite element analysis to minimise the computational time. The results showed that the step size during forming stage two significantly affected the geometrical accuracy of the part, whereas the forming depth during stage one was insignificant to the part quality. It was also revealed that the geometrical improvement had taken place along the base and the wall regions. However, the areas near the clamp system showed minor improvements. The optimised two-stage strategy successfully decreased both the geometrical inaccuracy and processing time. After optimisation, the average values of the geometrical deviation and forming time were reduced by 25% and 55.56%, respectively.

Keywords: SPIF; sheet metal; forming; incremental; FEA; tool path; optimisation

Citation: Yan, Z.; Hassanin, H.; El-Sayed, M.A.; Eldessouky, H.M.; Djuansjah, J.R.P.; A. Alsaleh, N.; Essa, K.; Ahmadein, M. Multistage Tool Path Optimisation of Single-Point Incremental Forming Process. *Materials* **2021**, *14*, 6794. https://doi.org/10.3390/ma14226794

Academic Editors: Mateusz Kopec, Denis Politis and Filippo Berto

Received: 28 August 2021
Accepted: 4 November 2021
Published: 11 November 2021

1. Introduction

Conventional sheet metal-forming (SMF) techniques such as drawing, stamping, rolling, and stretch forming are well-established mass production processes for a wide range of sheet metal products, especially in the automotive and aerospace industries. The main disadvantage of conventional SMF is the need to design and manufacture special dies with the required geometry of the product. The cost of the SMF dies is directly proportional to the complexity of the geometry [1]. Therefore, these techniques are not compatible with the customisation or personalisation of products. On the other hand, additive manufacturing (AM) techniques offer the possibility of customising products with complex geometries, allowing them to find applications in several sectors [2–6]. However, additive manufacturing for automotive and aerospace industries is incapable of manufacturing large and curved sheet metals, such as vehicle body panels. Other challenges restricting AM products in automotive and aerospace industries are the poor surface finish, limited part size and disparities in the production quality [7,8].

The demand for low/medium mass customisation is continuously growing as a promising approach for fabricating products with a high degree of customisation at a low production cost. However, the current capabilities of additive manufacturing and conventional sheet metal-forming are incapable of satisfying the needs for mass customisation [9]. The SPIF process can reduce production costs by eliminating the need for die manufacturing. It also offers a high degree of flexibility in customisation, especially for medium-sized production, resulting in die-less metal forming [10]. Single-point incremental forming is based on the conventional sheet metal spinning commonly used for the production of axisymmetric shape complex parts, without the need for using dies [11–13]. In SPIF, the metal sheet is plastically deformed into the desired shape through an incremental localised progressive manner by using a spherical head tool. The concept was first introduced by Leszak et al. [14], and the first trials were carried out in Matsubara labs, Japan. However, the process was advanced further and is currently used in several industries, such as aerospace [15], automotive [16] and marine [17]. The drawbacks of SPIF are around the fabricated parts' poor dimensional and geometrical precision due to the absence of dies in the process, making it challenging to ensure tight dimensional tolerance. Other geometrical precision issues typically arise from sheet thinning and elastic spring back [18].

Recently, significant advancements have been achieved in the equipment front, such as using individual or in sync robots to improve the flexibility and precision of the process. Several researchers have studied the SPIF experimentally, analytically and/or numerically. The selection of the process parameters was shown to significantly affect the thickness distribution of the metal sheet. Duflou and colleagues [19] had experimentally investigated the effect of different parameters such as tool diameter, sheet thickness, part wall angle and vertical step size on the forming force in the SPIF process. In their study, they had used a simple truncated cone part and contour forming strategy. Their results demonstrated a solid relationship between the induced forming forces and the selected process parameters. Bambach et al. and Dejardin et al. reported the significance of four SPIF process parameters which define the formability of the metal sheet. These were sheet thickness, tool speed, tool diameter and forming strategy [20,21]. Different forming strategies lead to different strain distribution and, hence, variation in the formed part thickness. The tool path is the route through which the tool travels in order to deform the metal sheet to the required shape. Therefore, both the formed component geometry and the process tolerance strongly depend on the tool path. Contour and spiral tool path strategies are two forming strategies that are typically applied in the SPIF process [22]. Arfa et al. [23] compared the two tool path strategies, and they stated several advantages associated with adopting the spiral tool path strategy. This includes uniform formed part thickness, homogeneous strain distribution and reduced defects on the surface of the formed part. Other researchers had also reported the remarkable influence of the forming strategy on the quality of the formed part. Therefore, optimising the SPIF tool path has become an interesting research area [18,20,22].

Generally, most of the research investigating the tool path optimisation of the SPIF process focused on improving the forming strategy. The optimisation of the SPIF tool path initially concentrated on minimising the spring-back phenomenon and cutting down the tool trajectory. Although the experimental approach usually offers better accuracy, it is also much more time- and money-consuming. Essa [18] studied the use of a backing plate, tool path modification and a kinematic supporting tool to improve geometrical accuracy. Azaouzi and Lebaal [22] examined the spiral tool path strategy by optimising the weighting factor and the tool vertical tour number with the objective of controlling the tool path. They used the statistical response surface technique to optimise the spiral tool path. Through this optimisation, they successfully reduced the tool path length while improving the uniformity of the part thickness. Reese and Ruszkiewicz [24] realised that the spring-back of an aluminium truncated cone could be reduced by increasing the incremental step size in the z-direction. Recently, Maaß et al. [25] studied the influence of the step-down size on the characteristics of the forming mechanisms via numerical simulation. Their results indicated that increasing the step-down size reduced the part thickness and the geometrical accuracy. However, it was also shown to increase the waviness of the formed part [22,25]. Due to the dynamic and incremental concept of SPIF, it was reported that the process simulation often consumes considerable CPU time. Maidagan et al. and [26] Meier et al. [27] introduced the double-side incremental forming (DSIF) to improve the geometrical accuracy of the formed sheet. In this technique, two forming tools simultaneously work on the two sides of the sheet metal. The DSIF was found effective at improving geometric accuracy [28].

Gonzalez et al. [29] investigated the multistage incremental sheet forming to improve the formability and accuracy of SPIF of conical geometry. The results showed that multistage forming enhanced the geometric accuracy in the unformed areas. However, the sheet thickness was deteriorated compared to those formed using single-stage forming. Suresh et al. [30] managed to implement multistage incremental sheet forming experimentally to produce geometries with steep walls that were difficult to achieve using SPIF. Recent research introduced multistage forming as a technique to improve geometrical accuracy, spring-back and thickness distribution [29,31]. However, literature on optimising this technique statistically is lacking. The research question of this paper is related to how effective is the use of the design of experiments to optimise a multistage single incremental forming and what are the effects and interactions of multistage forming parameters on the geometrical accuracy and forming time?

2. Methodology

Finite element analysis was initially used and validated using experimental literature data. Step-down tool size and the forming depth have been varied in order to provide some insight. Initially, an FEA model of SPIF was built using an explicit solver and validated with experimental data in the literature. The part shape was a truncated aluminium alloy cone with a 180 mm diameter, 40 mm depth and 50° wall angle. The experiments were modelled using an ABAQUS/Explicit solver. Finally, the simulation results, such as the average geometrical deviation and the forming time in various simulation scenarios, have been used to analyse variance and identify the optimum parameters utilising the design of experiments (DoE) method and analysis of variance. The results obtained using the optimised multistage tool path parameters were compared to a single-stage one introduced by Maaß et al. [25].

2.1. FEA Modelling of SPIF Process

Finite element modelling (FEM) has been a powerful process modelling tool allowing the study of the deformation mechanism of the SPIF process and by implication to avoid restrictions associated with facilities and location. The sheet metal part is typically modelled using shell or solid elements. Solid elements are preferred for meshing the thickness direction to improve the accuracy of the results. However, this increased the computational time [21]. Cocchetti et al. [32] recommended the use of shell elements to save computational time. Explicit and implicit FEM solver techniques were investigated in the literature, and it was found that the explicit solvers were time-efficient, whereas the implicit method was accurate. Researchers have explored the use of implicit and explicit solvers aiming to reduce computational time. A simultaneous solution of equations in each time increment was used to solve the problem in the implicit solver. In the explicit FEA technique, the solution of the preceding step was employed to solve the succeeding increment. As a result, the implicit technique was reported to be more precise than the explicit solver due to eliminating error accumulation [23]. Therefore, using the explicit method primarily to explore the optimal range of process parameters, followed by employing the implicit solvers, could permit the prediction of more accurate results in a reasonable time. A schematic diagram of SPIF is illustrated in Figure 1. The spherical tip tool makes a sequence of contours and forms the desired shape incrementally into the metal sheet.

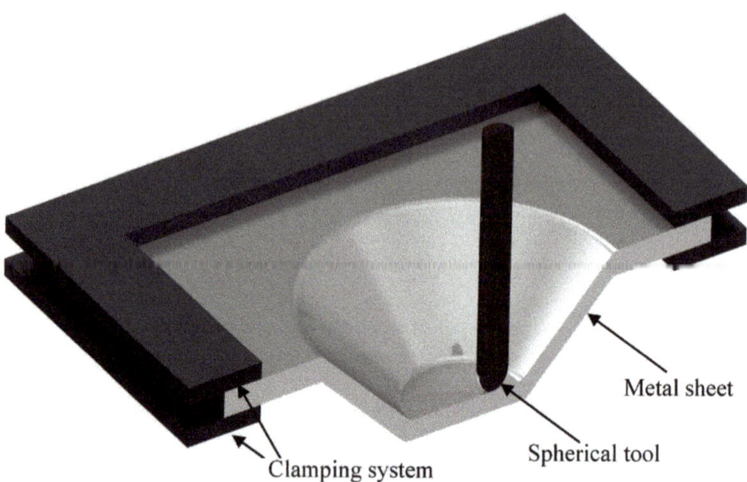

Figure 1. A schematic diagram of the SPIF system.

An asymmetric geometry represented the truncated cone, with a diameter of 180 mm, a depth of 40 mm and a wall angle of 50°, and it was constructed to simulate the SPIF in ABAQUS software. The sheet metal was defined as a (200 × 200 mm) blank sheet of 1.2 mm in thickness. The blank sheet was held between two blank holders and deformed using a spherical tool of 10 mm in diameter. Figure 2 shows the FEA model in which the metal sheet is considered a deformable body, whereas rigid bodies were assigned to the tool and the two clamping holders. The sheet material was aluminium 3003-O, with the following physical and mechanical properties: density ρ = 2700 kg/m^3, Poisson's ratio ν = 0.33, Young's modulus E = 70 GPa, ultimate strength σ_u = 95–135 MPa and yield stress σ_y = 35 MPa [19]. The flow stress equation of the material can be calculated using the Swift-type hardening law, as follows:

$$\overline{\sigma} = k(\varepsilon_0 + \overline{\varepsilon})^n \tag{1}$$

where k is the hardening coefficient, $\overline{\varepsilon}$ is the effective accumulated plastic strain and n is the strain hardening exponent [23]. For aluminium 3003-O: $k = 184$ MPa, ε_0 = 0.00196 and n = 0.224.

The velocity of the tool feed was maintained at 2000 mm/min to eliminate the effect of the tool speed on the process. It is typical to use a lubricant covering the metal sheet in the SPIF process, and this has been set as a friction coefficient of 0.09 between the forming tool and the metal sheet. In addition, the friction coefficient between the clamp, sheet and backing was set at 0.015. The tool path was initially defined by the movement in x, y and z directions. At each time increment, the coordinate of the tool reference point was initially generated using MATLAB and converted into displacement. The tool step-down was identified as 0.5 mm, and the explicit time integration was set as a nonlinear analysis in which the sheet metal was continuously loaded. In order to stabilise the time discretisation, the time increment was set as small as possible [32]. The maximum time increment was calculated by the smallest mesh size and the sound speed, as in Equation (2) [33]:

$$\Delta t \approx \frac{L_{min}}{S} \tag{2}$$

where Δt is maximum stable time increment, S is the sound speed and L_{min} is the smallest mesh size. For any isotropic shell element, S can be calculated using:

$$S = \sqrt{\frac{E}{(1-\nu^2)\rho}} \tag{3}$$

where ρ is the material density (ton/mm^3), ν is the Poisson's ratio and E is the Young's modulus (MPa). As shown from the above equation, as the density increases, the computational time decreases. However, further increases in mass scaling can result in inaccurate results. Therefore, the mass scaling factor was adapted according to the minimum mesh size in order to stabilise the time integration and save computational time. This resulted in adjusting the step time increment to around 10^{-5}, and the use of the shell element as the mass scaling was found suitable. The critical time step was controlled by the size of the in-plane elements rather than the thickness of the sheet metal [32]. After several simulation convergence trials, the simulation results were comparable to the results by Duflou et al. [19] and Arfa et al. when the mesh size of the part was 4 mm, the mesh size of the clamp and backing were 12 mm and the mass scaling was 4×10^4 [23]. Moreover, the kinetic energy was less than 5% of the model internal energy, indicating a negligible effect of the mass scaling on the results' accuracy. Therefore, a conservation analysis was carried out to reduce the computational time while not compromising the model accuracy. It was suggested that the suitable mesh size for the part would be 4 mm, with a mass scaling of 4×10^4 and 2500 elements.

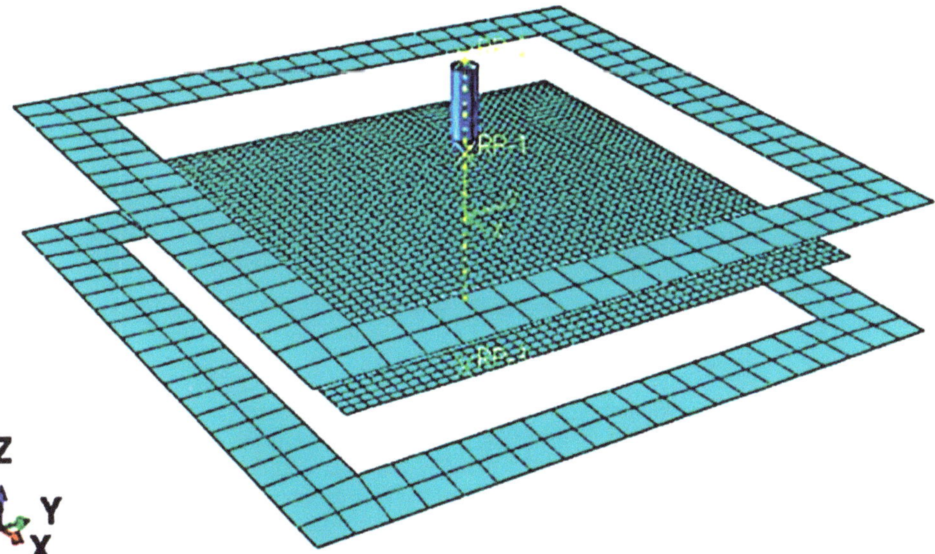

Figure 2. The simulation model of the SPIF process.

2.2. Model Validation

The model's reliability was assessed by comparing the current modelling results of a single-stage model with the experimental and simulation data in the literature [19].

The wall thickness of the part can be calculated using the sine principle, as shown in Equation (4) [23]:

$$T_{part} = T_0 \sin(90 - \alpha) \tag{4}$$

where T_{part} is the formed part thickness, T_0 is the thickness of the metal sheet and α is the cone angle.

The estimated thickness of the formed part was found to be ≈0.77 mm. The range of the formed part thickness, obtained from the present simulation and shown in Figure 3a, was from 0.6954 to 1.200 mm, which was comparable to that reported in the literature (0.7216–1.200 mm) [23].

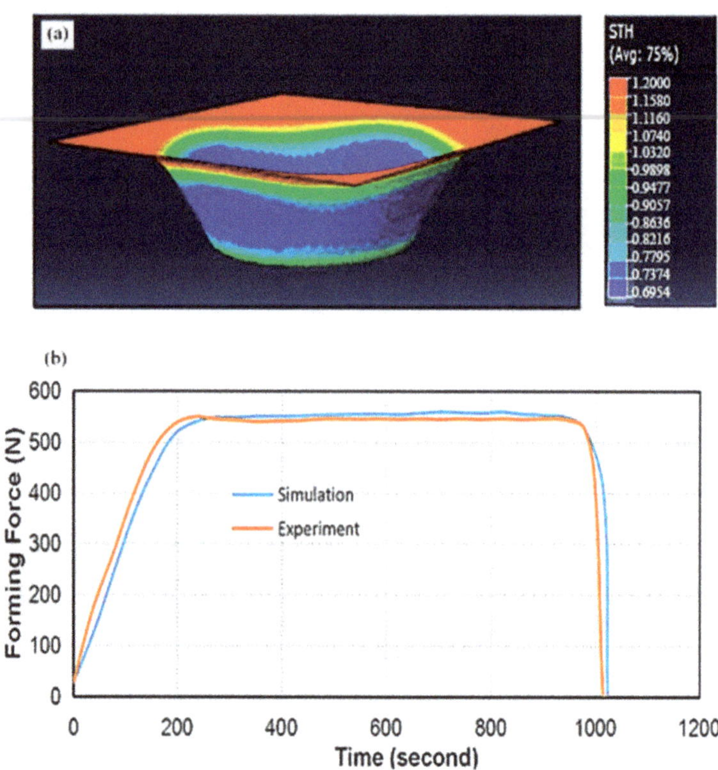

Figure 3. (**a**) The formed part thickness distribution, and (**b**) the forming force plot from the present simulation and the experimental results from the literature. The experimental graph was reused with permission from [19]. Elsevier.

Duflou et al. [19] used single-point incremental forming forces using a table-type dynamometer. To ensure an accurate comparison, the current study used the same setup parameters as those introduced by Duflou et al. [19], such as sheet material Al 3003-O, the thickness of 1.2 mm, a cone of 180 mm in diameter, 40 mm depth, 10 mm tool diameter and 50° wall angle. They optimised four different process parameters and their effect on the forming forces: the tool diameter, the sheet metal thickness, the vertical step size and the parts' wall angle. The approximate mean thickness of the formed part in the current study and in the literature [19], represented by the blue areas in Figure 3a, were 0.7374 and 0.7615 mm, respectively. Both values are considered accurate compared to the theoretical value (0.77 mm) with an error of 1.4% and 5.5%, respectively.

Figure 3b compares the forming force versus time in the current study with corresponding values found in the literature [19]. As shown, the two curves have a similar trend and coincide at the same maximum value. Therefore, the current simulation can be considered a valid model and can be reliably employed to investigate the multistage tool path optimisation.

3. Optimisation of the Multistage SPIF Process

3.1. Multistage Parameters

Two spiral tool paths were considered in this study (one for each forming strategy). The paths' parameters are the depths (h, H) and the maximum diameters (d, D). Lowercase and capital letters are associated with forming strategies 1 and 2 respectively, see Figure 4. The maximum depth and diameter during the second stage (D, H) must be the same as the part dimensions, which are 40 and 180 mm, respectively. In addition, the part should have the same wall angle as the desired geometry after the first stage to ensure high geometrical accuracy. Earlier studies had explored the effect of the step size in the range of 1.875 to 5.625 mm and found that a step size of 5 mm was optimal [22,25]. Generally, increasing the step-down size reduces the geometrical accuracy and the forming time. The step-down size of forming stages 1 and 2 (denoted as $\Delta z1$ and $\Delta z2$) have been considered in a range from 1 to 5 mm. Accordingly, the range of d and the value of h were calculated through Equations (5) and (6) and were found to be 35–39 and 178 mm, respectively.

$$H - h \leq \Delta z \tag{5}$$

$$d \leq D - \frac{2\Delta z}{\tan 50} \tag{6}$$

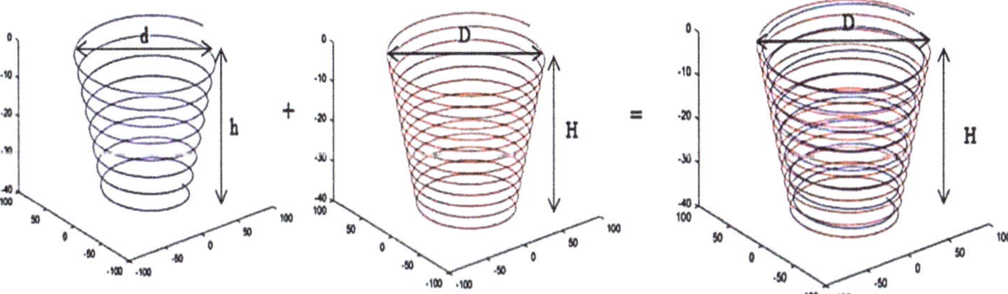

Figure 4. The two forming stages of the proposed SPIF process.

The spiral tool path parametric equation can be written as follow:

$$\begin{cases} X(\beta) = R(\beta)\cos\beta \\ Y(\beta) = R(\beta)\sin\beta \\ Z(\beta) = \frac{Z_{max}}{2\pi} \end{cases} \tag{7}$$

where $R(\beta)$ is the radius as a function of the spiral angle β, n is the vertical increment $= \frac{Z_{max}}{\Delta z}$, $0 \leqq \beta \leqq 2\pi n$ and Z_{max} is the maximum forming depth, which is calculated using Equation (8):

$$R(\beta) = r + \frac{Z(\beta)}{\tan(\theta)} \tag{8}$$

where r is the spherical tool radius and θ is the wall angle. Three parameters have been considered. Those are the vertical step-down sizes during the first and the second forming stages ($\Delta z1$ and $\Delta z2$, respectively), and the depth of the first forming stage (h), in which the ranges are $1 \leqq \Delta z1$ and $\Delta z2 \leqq 5$, and $35 \leqq h \leqq 39$.

Maaß et al. [25] investigated the relationship between the tool path parameters such as the vertical step size, Δz, and the part geometry of a similar truncated cone. The authors found that the optimum step size ratio for a single-stage incrementing forming Δz/tool radius is 0.25 for an accurate part geometry. Therefore, an initial value for Δz of 1.25 mm was employed in the current study, in comparison to the multistage optimised values.

3.2. The Response Surface Method (RSM)

The design of experiments is a statistical approach for designing and optimising processes with multiple input parameters, and it has been used extensively in sheet metal forming as well [34,35]. The response surface method was used to statistically investigate the effect of the process parameters using the two forming stages' spiral tool paths and to find the optimised parameters, as shown in Figure 4. The aim was to deform the sheet metal evenly and reduce spring-back with a minimum forming time. To achieve this aim, the Box–Behnken DoE was used to create an experimental plan with minimum trials. The Box–Behnken method is a statistical design used for the response surface approach to ensure that each parameter is placed at one of three equally spaced levels, such as −1, 0 and +1 [36].

A second-order regression equation can express the response surface function "Y". The order of the regression equation is typically kept as low as possible. Typically, the accuracy of second-order regression is the lowest accurate order, see Equation (9):

$$Y = b_0 + \sum b_i x_i + \sum b_{ii} x_i^2 + \sum b_{ij} x_i x_j \quad (9)$$

where the factors x_i are the process parameters; on the other hand, b_0, b_i, b_{ii} and b_{ij} are the regression equation coefficients determined using the least-square technique. Design-Expert V 7.0.0 (Stat-Ease Inc., Minneapolis, MN, USA) was applied to implement the DoE approach.

Three parameters were deemed appropriate in the study: vertical tool increment of the first forming stage, vertical tool increment of the second forming stage and the depth of the first forming stage. According to the Box–Behnken design, three levels of each parameter were considered, see Table 1. As shown, the 3 levels were 0 as the middle level, 1 as the high level and −1 as the lower level [32]. Furthermore, three centre points were considered (to permit the determination of the experimental error). This resulted in 15 parametric combinations, see Table 2. In this study, two responses: the average geometric deviation of the formed part and the forming time, were optimised, knowing that the geometric deviation of the part was calculated using the average distance between the formed and designed shapes [23,24].

Table 1. Factors and levels used in the DoE.

Factors	Process Parameters	−1	0	+1
Δz1 (mm)	Tool vertical increments of the first forming stage	1	3	5
Δz2 (mm)	Tool vertical increments of the second forming stage	1	3	5
h (mm)	Depth of the first forming stage	35	37	39

Table 2. Box–Behnken design parameters with calculated average deviation and forming time.

Run	Δz1 mm	Δz2 mm	*h* mm	Average Derivation mm	Forming Time (s)
1	5	3	39	1.6594	278
2	3	1	39	0.9792	658
3	3	5	35	2.077	258
4	3	3	37	1.5967	345
5	1	1	37	0.8773	1016
6	5	1	37	1.1333	633
7	3	3	37	1.5997	345
8	3	3	37	1.6056	345
9	5	5	37	2.2779	221

Table 2. *Cont.*

Run	Δz1 mm	Δz2 mm	*h* mm	Average Derivation mm	Forming Time (s)
10	5	3	35	1.8586	258
11	1	3	39	1.4476	689
12	1	5	37	1.5239	580
13	1	3	35	1.5967	645
14	3	1	35	1.3892	645
15	3	5	39	2.0536	261

4. Results

4.1. Analysis of Geometric Deviation

Three distinct geometric deviations were found at three regions, see Figure 5a. The deviation in region A was created by bending the metal sheet, while the ones found in region B were due to the spring-back effect. Finally, the pillow effect deviation at the base of the part (region C) was a result of the change in the transverse stress/strain [32,36]. The sheet metal parameters such as the internal bend angle (*A*), internal bend radius (*R*) and the sheet thickness (*T*) are shown in Figure 5b.

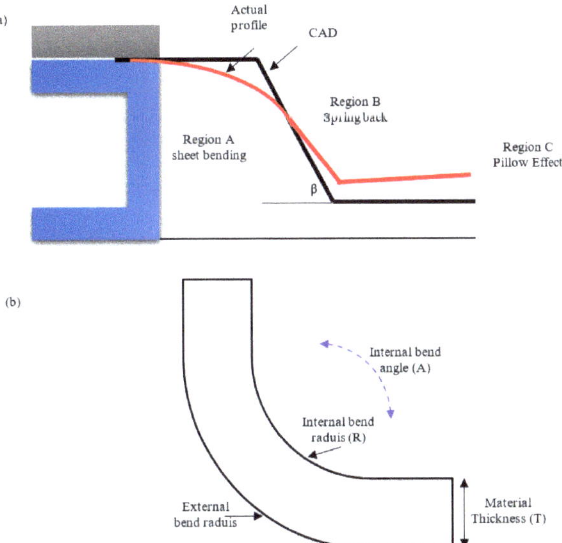

Figure 5. (**a**) A schematic diagram of the actual and ideal geometries. (**b**) Sheet metal parameters.

To assess the part quality in the fifteen simulation trials, the maximum error in each of the three regions was measured, see Figure 6. It was found that the maximum geometrical deviation in region A was about 4.70 mm for all simulation scenarios, which did not improve compared to the single stage (4.67 mm). There was a significant variation in the error values in region B, which ranged from 1.04 to 4.75 mm. The maximum geometrical deviation in region C varied from 0.7688 to 2.6546 mm. In conclusion, the second forming stage improved the geometric accuracy in regions B and C while not deteriorating in region A.

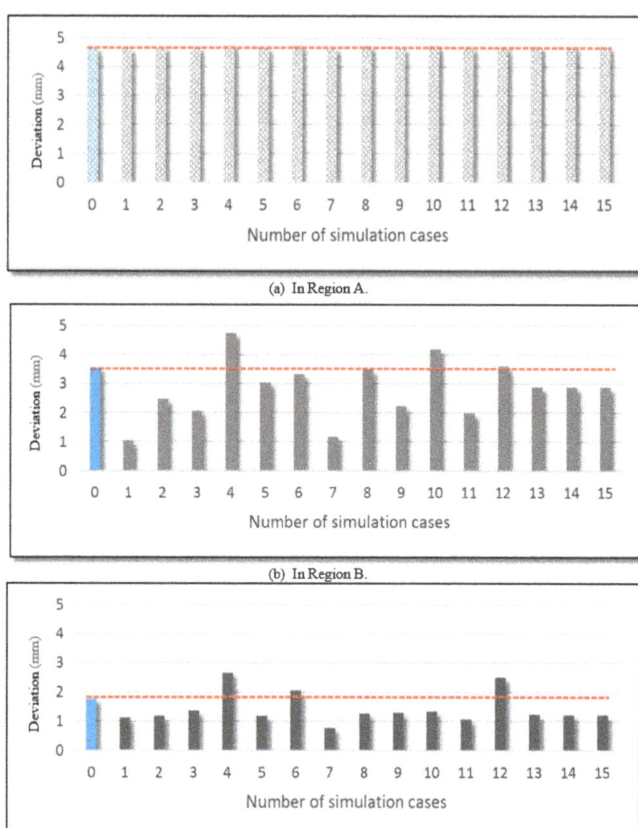

Figure 6. The maximum geometric deviation in region A, region B and region C.

4.2. Analysis of Variance (ANOVA)

Results of the average deviation and forming time are listed in Table 2. The least-square fitting *R*2 was employed to define the model fit [37]. According to the Box–Behnken design, the average geometrical deviation and forming time fit quadratic models with the least-square fitting *R*2 of 97% and 95%, respectively. This suggests that the models accurately describe the relationship between the input parameters and the outputs. The two models are functions of the vertical step-down sizes during the first forming stage (Δz1), the second forming stage (Δz2) and the depth during the first forming stage (*h*), see Equation (10).

$$\begin{aligned} Response = \; & b_0 + b_1(\Delta z1) + b_2(\Delta z2) + b_3(h) + b_4(\Delta z1\Delta z2) + b_5(\Delta z1h) \\ & + b_6(\Delta z2h) + b_7(\Delta z1)^2 + b_8(\Delta z2)^2 + b_9(h)^2 \end{aligned} \tag{10}$$

where b_0 is the average of the levels, and b_1, b_2 … , b_9 are the models' coefficients. Least-squares fitting, an approach for best curve fitting by minimising the total squares of the errors, was used to analyse the equation data shown in Table 3 and define the coefficients. The coefficients of the surface response model for the two outputs are listed in Table 3.

Table 3. Coefficients of the response surface models of the average deviation and forming time.

Coefficient	Average Deviation Model	Forming Time Model
$b0$	+40.94938	−4769.56250
$b1$	+0.21402	−254.87500
$b2$	−0.64276	−275.00000
$b3$	−2.06886	+330.50000
$b4$	+0.031125	+1.50000
$b5$	−0.00313125	−1.50000
$b6$	+0.024162	−0.62500
$b7$	−0.016468	+34.93750
$b8$	−0.020424	+31.93750
$b9$	+0.026445	−4.31250

The null hypothesis, which presumes that the process parameters have no effect, is rejected when the p-value is less than 0.05 (95% confidence level). As a result, parameters with p-values ≤ 0.05 are considered significant. The calculated p-values of all the parameters and interactions are listed in Table 4. The ANOVA results show that the most significant parameters are the vertical increments during stages one and two, the interaction between both increments and the depth of forming during stage one. Furthermore, the forming time was found to be significantly affected by the vertical increments of both forming stages.

Table 4. The p-values of the individual parameters and the interactions for the two outputs.

Model Parameter	p-Value	
	Average Deviation	Forming Time
$\Delta z1$	**0.0017**	**<0.0001**
$\Delta z2$	**<0.0001**	**<0.0001**
h	**0.0237**	0.1265
$\Delta z1\Delta z2$	**0.0340**	0.4722
$\Delta z1h$	0.7826	0.4722
$\Delta z2h$	0.0746	0.7592
$(\Delta z1)^2$	0.2012	<0.0001
$(\Delta z2)^2$	0.1276	<0.0001
h^2	0.0645	0.0846

Bold values indicate statistically significant process parameters (p-value < 0.05).

Figure 7a–c show the effect of the increment of the first forming stage (FS1), increment of the second forming stage (FS2) and the depth of the first forming stage on the average deviation using a quadratic model. It can be noted that the average deviation increased consistently with increasing any of the vertical increments of stages 1 and 2. In addition, the depth of the forming stage 1 was shown to have a slight adverse effect on the average deviation. Finally, the model shows that the interaction of the increments during the two forming stages was also significant. The geometric deviation significantly increases as the increment of any of the forming stages increases, Figure 7d. On the other hand, and as shown in Figure 8a,b, each of the increments of forming stages 1 and 2 were significant on the forming time, and the relationship between them follows a quadratic model. Both factors were shown to have virtually the same inverse effect on the forming time. At a depth of stage 1 of 37 mm, increasing the increment of forming stage 1 from 1 to 5 mm (at a constant value of the increment of forming stage 2 of 3 mm) resulted in a reduction of the forming time from 677 to 292 s, while the same increase of the increment of forming stage 2 (at a constant value of the increment of forming stage 1 of 3 mm) caused the forming time to decrease from 677 to 269 s.

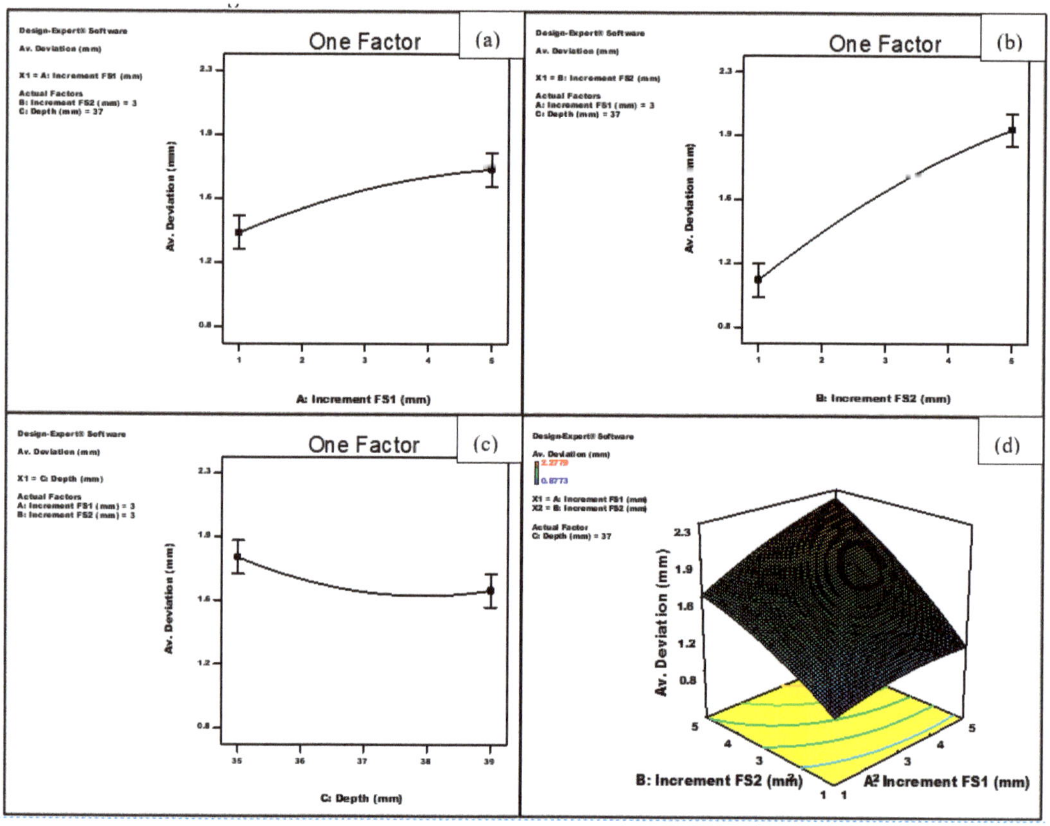

Figure 7. Effect of (**a**) increment of forming stage 1 (FS1), (**b**) increment of forming stage 2 (FS2), (**c**) depth of forming stage 1 and (**d**) the interaction between the increments of both forming stages on the average deviation.

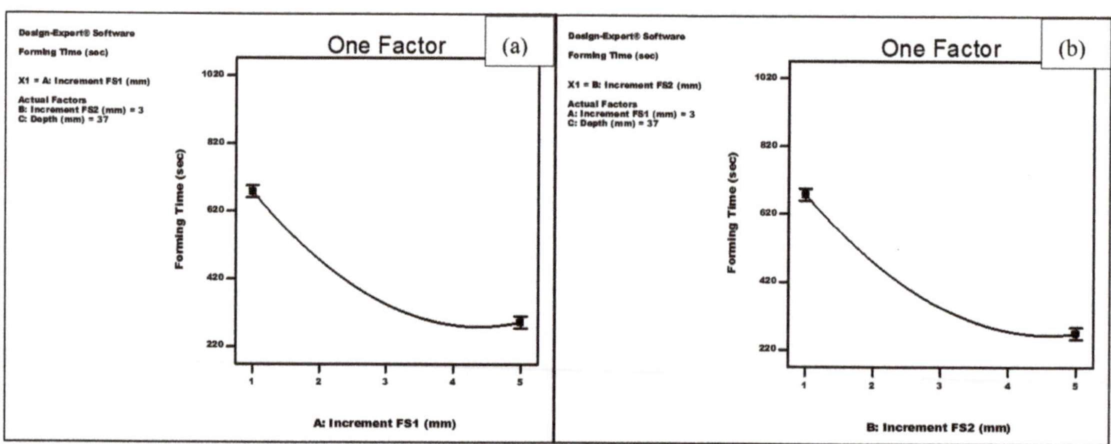

Figure 8. Effect of (**a**) increment of forming stage 1 (FS1) and (**b**) increment of forming stage 2 (FS2) on the forming time.

4.3. Optimisation of Process Parameters

The process parameters of the two-stage process were optimised. The objective function is to minimise both the geometrical deviation and the forming time. However, two different priorities were used in the objective function. The geometrical accuracy was defined with high priority, while the forming time was defined with low priority. Therefore, if both objectives have an inverse relationship, a trade-off will be made. In this case, the geometrical accuracy objective function will be met at the expense of the forming time to obtain a valid solution.

The experimental data were analysed, and the genetic algorithm (GA) was employed to calculate the process parameters [38]. The genetic algorithm is a search optimisation technique that is inspired by natural evolution. The geometrical deviation and forming time shown in Equation (10) and the corresponding constants listed in Table 3 were simultaneously solved, and the contour plot of the optimisation is shown in Figure 9. The model shows that the optimum values of the vertical increment of stage 1, vertical increment of stage 2 and the depth of forming stage 1 would be 4.5, 1.59 and 39 mm, respectively. At these values, the predicted average deviation and forming time were 1.21 mm and 473 s, respectively.

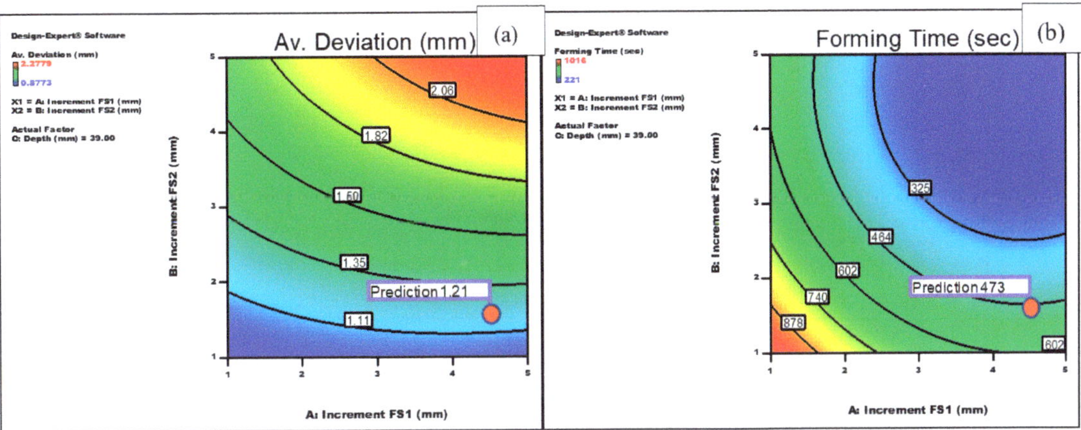

Figure 9. Predicted optimum process parameters for minimising the process responses (**a**) average deviation and (**b**) forming time.

4.4. Validation and Comparison

The aim of using the DoE optimisation was to minimise the geometrical deviation and forming time. The optimum process parameters were $\Delta z1 = 4.5$ mm, $\Delta z2 = 1.6$ mm and $h = 39$ mm. This set was employed in the FE model, and its effect on the part geometrical deviation and forming time is listed in Table 5. It was found that both the forming time and geometrical deviation were reduced by about 56% and 25%, respectively. Figure 10 shows the part geometry created by the optimised tool path parameters of a single-stage model created by Maaß et al. [25] and the current optimised process. The geometry of the formed part obtained using the current study's optimised parameters was found more accurate to the ideal profile than the one obtained using the single-stage process [25], especially the regions along the part wall and the base. The deviations in the geometry caused by the spring-back effect and the transverse strain/stress in regions B and C were slightly penalised. However, the change in the geometrical deviation in region A, caused by sheet bending, was found negligible. Figure 11 demonstrates a comparison of the part thickness distribution between the single-stage and double-stage optimised processes. The minimal part thickness of the optimised formed part was decreased by 1.6%, resulting in a more uniform thickness compared with the single-stage process.

Table 5. Comparison between the forming time, average deviation and minimum sheet thickness.

Parameters	Single Stage Using Δz = 1.25 mm, [25]	Optimised Solution	Reduction
Forming time (s)	1024	455	55.56%
Average deviation (mm)	1.5438	1.2351	25%
Minimum sheet thickness (mm)	0.6954	0.6770	1.6%

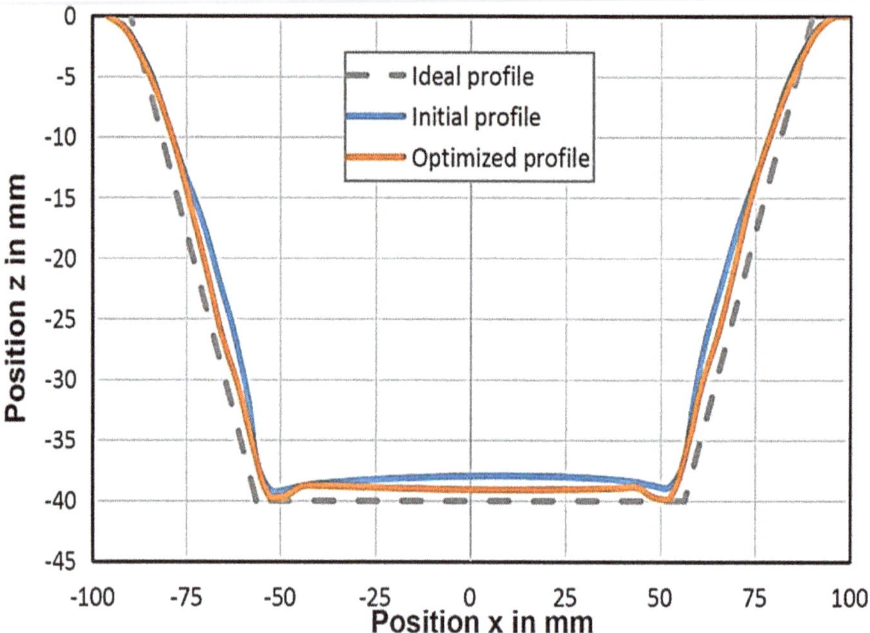

Figure 10. Section of the formed part produced using the single-stage process [25] and the optimised multistage process and ideal part.

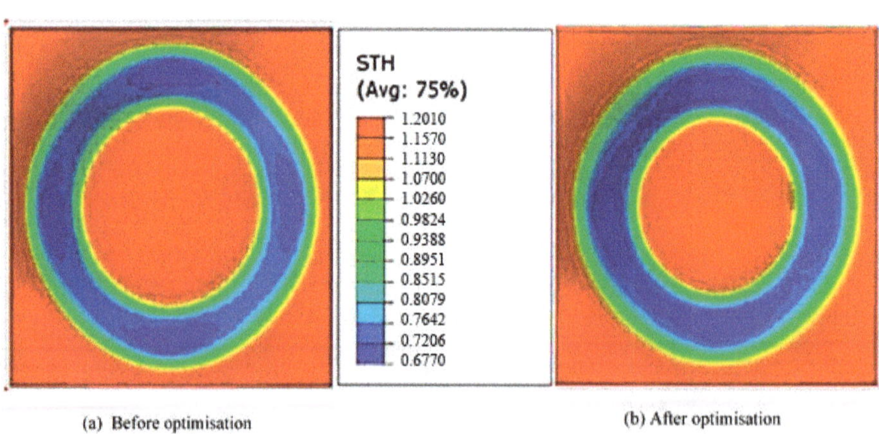

Figure 11. Thickness distribution of the (**a**) single-stage and (**b**) the optimised multistage SPIF process.

4.5. Forming Force

The forming force calculated using the optimised SPIF process parameters, shown in Figure 12, increases during the first stage then breaks at ~120 s as the tool trajectory moves between the two forming stages. The figure also shows that the forming force during the first forming stage of the part is more significant than that of the second stage, as the deformation primarily occurs during the first stage. In the second stage, the tool deforms the sheet metal slightly compared to the first stage. Hence, the small reaction force in the second forming stage works better in improving the part geometry and the wall thickness. This is because, after the first stage, there is still an undeformed depth, in which the forming force increases. As the forming depth in the first stage increases, the peak of the forming force in the second stage decreases. In addition, the forming force in the first stage rises rapidly and then experiences almost a flat plateau, as shown in Figure 12. On the other hand, the forming force in the second stage initially increases, stabilises for a short time and then increases rapidly at the end of the second stage. When the step size of the second stage and the undeformed depth are small, the strain and stress near the part base become relatively small by the end of the second forming stage, making the pillow effect phenomenon negligible.

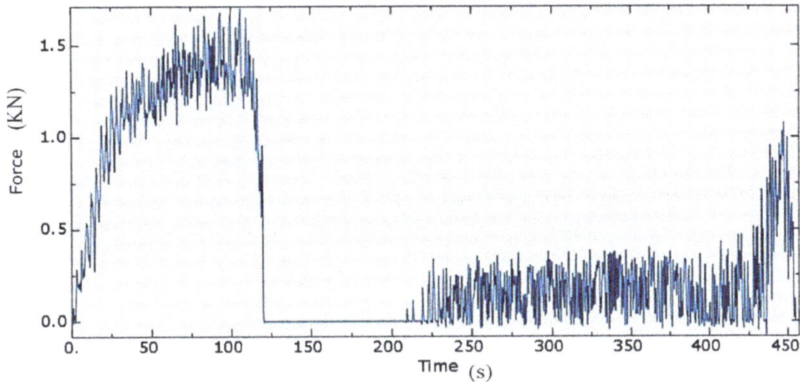

Figure 12. The forming force plot after optimisation.

5. Discussion

The contact between the tool and the metal sheet occurs within region B, where the metal sheet has been forced into the ideal shape by a tool progressing according to the designed path. With the tool movement, a sheet metal part is continuously free from constraints, and the residual stress makes it deform without a forming force. Hence, the actual part deflects further from the ideal profile, and the build-up of the local spring-back increases the geometrical deviation. Typically, the increase in the forming force increases the residual stress. For parts created in a single stage, local deformation becomes more significant as the residual stresses increase. However, after introducing the second forming stage, the developed local spring-back is small, which reduces the geometric deviation in region B [20]. The forming force in the second stage is smaller than in the first stage, though it is still sufficient to deform the metal sheet into the desired profile. As the forming force and the associated residual stress in the second stage become small, the deformation also becomes small, and the part thickness deforms more uniformly than the first stage.

Although the tool and sheet metal do not directly connect in region A, the bending occurs due to deformation in region B. The length of the neutral line in region A can be denoted by the bend allowance, as shown in Equation (11):

$$BA = A\left(\frac{\pi}{180}\right)(R + (KT)) \tag{11}$$

where R is the inside bend radius, A is the bend angle in degrees, T is the material thickness and K is the K-factor, which typically has a value between about 0.3 and 0.5. From Equation (11), it can be assumed that the bend allowance does not depend on the forming force. Therefore, the geometric deviation in region A is less dependent on the process parameters. On the other hand, the pillow effect that develops in region C develops from the bending of the metal sheet (due to in-plane stresses) [39]. This is because the material is largely deformed in the transverse direction within the tool vicinity. Consequently, it flows toward the metal sheet centre, which causes a significant pillow effect. The undeformed depth from the first stage can reduce the development of the pillow effect by restricting material plastic deformation near the base of the part.

Although the part dimensional accuracy is enhanced by using the proposed two-stage forming strategy, there is still a deviation between the formed part and the ideal shape, particularly near the area between the formed part and the clamping. As a result, the pillow effect becomes more evident as the size of the step-down increases. Further improvements in the design of the process can be considered in future work. For example, the tool path can be extended to the sheet centre point, reducing the pillow effect. In addition, the tool path start-points and endpoints of the two forming stages can be the same in order to reduce the spring-back effect due to the repeated forming. Furthermore, the distance between the forming and the clamped area can be reduced to minimise the sheet bending at the maximum diameter [40].

6. Conclusions

A two-stage forming strategy in SPIF was introduced and optimised to reduce the geometrical deviation and the processing time compared to those manufactured using a single forming tooling. A simulation model of the SPIF has been developed and solved using an explicit finite element analysis to study the optimal tool path for a truncated cone. The design of experiments using a response surface method was used to optimise the proposed two-stage forming strategy. The simulation results showed that the two-stage forming technique could significantly reduce both the geometrical deviation and the forming time. The step-down size in each forming stage was found to be the most significant parameter that affects the SPIF process' formability. Meanwhile, the step size of the second stage affected the part accuracy more than the step size of the first stage. The proposed and optimised two-stage forming strategy can reduce the geometric deviation caused by the spring-back and pillow effect while having an insignificant effect on those caused by sheet bending near the part and the clamp. The forming time and part geometric deviation were reduced by 56% and 25%, respectively. In addition, the part thickness distribution was found more uniform after optimisation, and the minimal thickness decreased by 1.6%.

Author Contributions: Conceptualization, H.H.; Formal analysis, Z.Y., H.M.E. and K.E.; Funding acquisition, M.A.; Investigation, H.H., M.A.E.-S. and N.A.A.; Methodology, Z.Y., H.H. and K.E.; Project administration, J.R.P.D.; Resources, N.A.A. and M.A.; Software, M.A.E.-S. and H.M.E.; Supervision, K.E.; Validation, Z.Y. and J.R.P.D.; Writing—review & editing, H.H., M.A.E.-S., H.M.E., J.R.P.D., N.A.A. and M.A. All authors have read and agreed to the published version of the manuscript.

Funding: This research was funded by the Deanship of Scientific Research at Imam Mohammad Ibn Saud Islamic University for funding this work through Research Group No. RG-21-12-02.

Institutional Review Board Statement: Not applicable.

Informed Consent Statement: Not applicable.

Data Availability Statement: The data underlying this article will be shared on reasonable request from the corresponding author.

Acknowledgments: The authors extend their appreciation to the Deanship of Scientific Research at Imam Mohammad Ibn Saud Islamic University for funding this work through Research Group No. RG-21-12-02.

Conflicts of Interest: The authors declare no conflict of interest.

References

1. Trzepieciński, T. Recent developments and trends in sheet metal forming. *Metals* **2020**, *10*, 779. [CrossRef]
2. Hassanin, H.; Alkendi, Y.; Elsayed, M.; Essa, K.; Zweiri, Y. Controlling the Properties of Additively Manufactured Cellular Structures Using Machine Learning Approaches. *Adv. Eng. Mater.* **2020**, *22*, 1901338. [CrossRef]
3. El-Sayed, M.A.; Essa, K.; Ghazy, M.; Hassanin, H. Design optimization of additively manufactured titanium lattice structures for biomedical implants. *Int. J. Adv. Manuf. Technol.* **2020**, *110*, 2257–2268. [CrossRef]
4. Brambilla, C.R.M.; Okafor-Muo, O.L.; Hassanin, H.; El Shaer, A. 3DP Printing of Oral Solid Formulations: A Systematic Review. *Pharmaceutics* **2021**, *13*, 358. [CrossRef] [PubMed]
5. Hassanin, H.; Zweiri, Y.; Finet, L.; Essa, K.; Qiu, C.; Attallah, M. Laser Powder Bed Fusion of Ti-6Al-2Sn-4Zr-6Mo Alloy and Properties Prediction Using Deep Learning Approaches. *Materials* **2021**, *14*, 2056. [CrossRef]
6. Mohammed, A.; Elshaer, A.; Sareh, P.; Elsayed, M.; Hassanin, H. Additive Manufacturing Technologies for Drug Delivery Applications. *Int. J. Pharm.* **2020**, *580*, 119245. [CrossRef]
7. Delic, M.; Eyers, D.R. The effect of additive manufacturing adoption on supply chain flexibility and performance: An empirical analysis from the automotive industry. *Int. J. Prod. Econ.* **2020**, *228*, 107689. [CrossRef]
8. Hassanin, H.; Abena, A.; Elsayed, M.A.; Essa, K. 4D Printing of NiTi Auxetic Structure with Improved Ballistic Performance. *Micromachines* **2020**, *11*, 745. [CrossRef]
9. Paoletti, I. Mass Customization with Additive Manufacturing: New Perspectives for Multi Performative Building Components in Architecture. *Procedia Eng.* **2017**, *180*, 1150–1159. [CrossRef]
10. Trzepieciński, T.; Oleksik, V.; Pepelnjak, T.; Najm, S.M.; Paniti, I.; Maji, K. Emerging trends in single point incremental sheet forming of lightweight metals. *Metals* **2021**, *11*, 1188. [CrossRef]
11. Essa, K.; Hartley, P. An assessment of various process strategies for improving precision in single point incremental forming. *Int. J. Mater. Form.* **2011**, *4*, 401–412. [CrossRef]
12. Essa, K.; Hartley, P. Investigation on the effects of process parameters on the through-thickness shear strain in single point incremental forming using dual level FE modelling and statistical analysis. *Comput. Methods Mater. Sci.* **2010**, *10*, 238–278.
13. Li, W.; Attallah, M.M.; Essa, K. Experimental and numerical investigations on the process quality and microstructure during induction heating assisted increment forming of Ti-6Al-4V sheet. *J. Mater. Process. Technol.* **2022**, *299*, 117323. [CrossRef]
14. Edward, L. Apparatus and Process for Incremental Dieless Forming. U.S. Patent 3,342,051, 1967. Available online: https://www.freepatentsonline.com/3342051.html (accessed on 15 May 2021).
15. Mohanraj, R.; Elangovan, S. Incremental sheet metal forming of Ti–6Al–4V alloy for aerospace application. *Trans. Can. Soc. Mech. Eng.* **2020**, *44*, 56–64. [CrossRef]
16. Pandre, S.; Morchhale, A.; Kotkunde, N.; Kurra, S. Processing of DP590 steel using single point incremental forming for automotive applications. *Mater. Manuf. Process.* **2021**, *36*, 1658–1666. [CrossRef]
17. Qiu, H.Y.; Yu, H.P.; Qiao, Y.; Shen, Y. Numerical simulation and experimental research on flexible incremental forming method of marine plate. *Suxing Gongcheng Xuebao J. Plast. Eng.* **2018**, *25*, 16–22.
18. Essa, K. *Finite Element Prediction of Deformation Mechanics in Incremental Forming Processes*; University of Birmingham: Birmingham, UK, 2011.
19. Duflou, J.; Tunçkol, Y.; Szekeres, A.; Vanherck, P. Experimental study on force measurements for single point incremental forming. *J. Mater. Process. Technol.* **2007**, *189*, 65–72. [CrossRef]
20. Bambach, M.; Araghi, B.T.; Hirt, G. Strategies to improve the geometric accuracy in asymmetric single point incremental forming. *Prod. Eng.* **2009**, *3*, 145–156. [CrossRef]
21. Dejardin, S.; Thibaud, S.; Gelin, J.; Michel, G. Experimental investigations and numerical analysis for improving knowledge of incremental sheet forming process for sheet metal parts. *J. Mater. Process. Technol.* **2010**, *210*, 363–369. [CrossRef]
22. Azaouzi, M.; Lebaal, N. Tool path optimization for single point incremental sheet forming using response surface method. *Simul. Model. Pract. Theory* **2012**, *24*, 49–58. [CrossRef]
23. Arfa, H.; Bahloul, R.; Salah, H.B.H. Finite element modelling and experimental investigation of single point incremental forming process of aluminum sheets: Influence of process parameters on punch force monitoring and on mechanical and geometrical quality of parts. *Int. J. Mater. Form.* **2013**, *6*, 483–510. [CrossRef]
24. Reese, Z.C.; Ruszkiewicz, B.J.; Nikhare, C.P.; Roth, J.T. Effect of Toolpath on the Springback of 2024-T3 Aluminum during Single Point Incremental Forming. In Proceedings of the ASME 2015 International Manufacturing Science and Engineering Conference, Charlotte, NC, USA, 8–12 June 2015; Volume 1.
25. Maaß, F.; Hahn, M.; Tekkaya, A.E. Interaction of Process Parameters, Forming Mechanisms, and Residual Stresses in Single Point Incremental Forming. *Metals* **2020**, *10*, 656. [CrossRef]
26. Malhotra, R.; Cao, J.; Beltran, M.; Xu, D.; Magargee, J.; Kiridena, V.; Xia, C. Accumulative-DSIF strategy for enhancing process capabilities in incremental forming. *CIRP Ann.* **2012**, *61*, 251–254. [CrossRef]
27. Maidagan, E.; Zettler, J.; Bambach, M.; Rodriguez, P.P.; Hirt, G. A New Incremental Sheet Forming Process Based on a Flexible Supporting Die System. *Key Eng. Mater.* **2007**, *344*, 607–614. [CrossRef]
28. Jung, K.-S.; Yu, J.-H.; Chung, W.; Lee, C.-W. Tool Path Design of the Counter Single Point Incremental Forming Process to Decrease Shape Error. *Materials* **2020**, *13*, 4719. [CrossRef]

29. Gonzalez, M.M.; Lutes, N.A.; Fischer, J.D.; Woodside, M.R.; Bristow, D.A.; Landers, R.G. Analysis of geometric accuracy and thickness reduction in multistage incremental sheet forming using digital image correlation. *Procedia Manuf.* **2019**, *34*, 950–960. [CrossRef]
30. Suresh, K.; Nasih, H.R.; Jasti, N.V.K.; Dwivedy, M. Experimental Studies in Multi Stage Incremental Forming of Steel Sheets. *Mater. Today Proc.* **2017**, *4*, 4116–4122. [CrossRef]
31. Shi, X.; Hussain, G.; Zha, G.; Wu, M.; Kong, F. Study on formability of vertical parts formed by multi-stage incremental forming. *Int. J. Adv. Manuf. Technol.* **2014**, *75*, 1049–1053. [CrossRef]
32. Cocchetti, G.; Pagani, M.; Perego, U. Selective mass scaling for distorted solid-shell elements in explicit dynamics: Optimal scaling factor and stable time step estimate. *Int. J. Numer. Methods Eng.* **2015**, *101*, 700–731. [CrossRef]
33. Wang, J.; Chen, T.; Sleight, D.; Tessler, A. Simulating Nonlinear Deformations of Solar Sail Membranes Using Explicit Time Integration. In Proceedings of the 45th AIAA/ASME/ASCE/AHS/ASC Structures, Structural Dynamics & Materials Conference, Palm Springs, CA, USA, 19–22 April 2004.
34. Moheen, M.; Abdel-Wahab, A.; Hassanin, H.; Essa, K. Reconfigurable Multipoint Forming Using Waffle-Type Elastic Cushion and Variable Loading Profile. *Materials* **2020**, *13*, 4506. [CrossRef]
35. Tolipov, A.; Elghawail, A.; Abosaf, M.; Pham, D.; Hassanin, H.; Essa, K. Multipoint forming using mesh-type elastic cushion: Modelling and experimentation. *Int. J. Adv. Manuf. Technol.* **2019**, *103*, 2079–2090. [CrossRef]
36. Ferreira, S.L.C.; Bruns, R.E.; Ferreira, H.S.; Matos, G.D.; David, J.M.; Brandao, G.C.; da Silva, E.G.P.; Portugal, L.A.; dos Reis, P.S.; Souza, A.S.; et al. Box-Behnken design: An alternative for the optimization of analytical methods. *Anal. Chim. Acta* **2007**, *597*, 179–186. [CrossRef] [PubMed]
37. Hassanin, H.; Modica, F.; El-Sayed, M.A.; Liu, J.; Essa, K. Manufacturing of Ti–6Al–4V micro-implantable parts using hybrid selective laser melting and micro-electrical discharge machining. *Adv. Eng. Mater.* **2016**, *18*, 1544–1549. [CrossRef]
38. Elsayed, M.; Ghazy, M.; Youssef, Y.; Essa, K. Optimization of SLM process parameters for Ti6Al4V medical implants. *Rapid Prototyp. J.* **2019**, *25*, 433–447. [CrossRef]
39. Isidore, B.B.L.; Hussain, G.; Shamchi, S.P.; Khan, W.A. Prediction and control of pillow defect in single point incremental forming using numerical simulations. *J. Mech. Sci. Technol.* **2016**, *30*, 2151–2161. [CrossRef]
40. Micari, F.; Ambrogio, G.; Filice, L. Shape and Dimensional Accuracy in Single Point Incremental Forming: State of the Art and Future Trends. *J. Mater. Process. Technol.* **2007**, *191*, 390–395. [CrossRef]

Article

An Efficient Computational Model for Magnetic Pulse Forming of Thin Structures

Mohamed Mahmoud *,†, **François Bay** † and **Daniel Pino Muñoz** †

MINES Paris-Tech, PSL-Research University, CEMEF—Center for Material Forming, CNRS UMR 7635, BP 207, 1 Rue ClaudeDaunesse, CEDEX, 06904 Sophia Antipolis, France; francois.bay@mines-paristech.fr (F.B.); daniel.pino_munoz@mines-paristech.fr (D.P.M.)

* Correspondence: mohamed.mahmoud@mines-paristech.fr

† These authors contributed equally to this work.

Abstract: Electromagnetic forming (EMF) is one of the most popular high-speed forming processes for sheet metals. However, modeling this process in 3D often requires huge computational time since it deals with a strongly coupled multi-physics problem. The numerical tools that are capable of modeling this process rely either on shell elements-based approaches or on full 3D elements-based approaches. The former leads to reduced computational time at the expense of the accuracy, while the latter favors accuracy over computation time. Herein, a novel approach was developed to reduce CPU time while maintaining reasonable accuracy through building upon a 3D finite element analysis toolbox which was developed in CEMEF. This toolbox was used to solve magnetic pulse forming (MPF) of thin sheets. The problem was simulated under different conditions and the results were analyzed in-depth. Innovative techniques, such as developing a termination criterion and using adaptive re-meshing, were devised to overcome the encountered problems. Moreover, a solid shell element was implemented and tested for thin structure problems and its applicability was verified. The results of this element type were comparable to the results of the standard tetrahedral MINI element but with reduced simulation time.

Keywords: high-speed forming; magnetic pulse forming; computational mechanics; solid-shell finite element

Citation: Mahmoud, M.; Bay, F.; Pino Muñoz, D. An Efficient Computational Model for Magnetic Pulse Forming of Thin Structures. *Materials* **2021**, *14*, 7645. https://doi.org/10.3390/ma14247645

Academic Editors: Mateusz Kopec and Denis Politis

Received: 10 November 2021
Accepted: 8 December 2021
Published: 12 December 2021

1. Introduction

Lightweight structures have many applications, especially in aerospace and automotive industries. The huge demand on lightweight structures requires improvements in the forming processes that would lead to cost reduction while maintaining structural resistance. High-speed forming processes, such as explosive forming discussed by Mynors and Zhang [1], electro-hydraulic forming (EHF) by Rohatgi et al. [2] and electromagnetic forming (EMF) by Zittel [3], have been proposed in the literature to overcome the disadvantages of the conventional forming methods. The study of these problems has been cumbersome for a long period of time. However, studying such problems has become more feasible with the advancements in the computing power and the progress in computational simulations.

Multi-physics simulation is a wide domain that contains various problems, including fluid-structure [4], thermal-structural and electromagnetic-structure interactions [5]. This work is mainly concerned with electromagnetic-structure simulations. Nevertheless, there are two types of this simulation. First, the simulation related to using materials having unique electric-magnetic properties, such as functionally graded piezoelectric material (FGPM) and piezomagnetic (PM) materials [6]. Second, the problems that include external magnetic sources that induce magnetic forces causing the metallic structure to deform, known as electromagnetic forming (EMF) [7].

The electromagnetic forming (EMF) process applies an intense electromagnetic pulse on a metallic component, inducing plastic deformations. It is a high-speed forming process that has strain rates ranging from 10^3 s^{-1} to 10^4 s^{-1}. Psyk et al. [8] discussed, in detail, the advantages of high-speed forming processes over the conventional forming processes. First, there is a lack of contact between the tool and the workpiece, and thus no lubrication is needed. Second, there is improved formability with respect to the conventional forming processes. In this process, an intense magnetic field generated by a coil is applied on an adjacent electrical conductive workpiece. The induced current along with the applied magnetic field produce Lorentz body forces on the workpiece. These forces supply additional momentum and energy to the workpiece, causing deformations [9]. Unger et al. [10] stated that the electromagnetic part of the system is highly dependent on the spatio-temporal evolution of the deformation of the workpiece. Therefore, designing this process remains cumbersome as it deals with strongly coupled multi-physics phenomena.

Recently, many endeavors have been made to simulate this problem. Most of these approaches were mainly restricted to axisymmetric geometries or small deformation problems. Fenton and Daehn [11] tackled the simulation problem of magnetic pulse welding and Imbert et al. [12] worked on the simulation of a corner fill operation process, though both of them worked on axisymmetric geometries. Moreover, Schinnerl et al. [13] tackled the simulation of 3D magnetic pulse forming but in small deformations. However, three-dimensional modeling along with plastic deformations are necessary to simulate a real world metal forming process using electromagnetic forming. The numerical simulation of this problem requires high standards of finite element formulation since the workpieces exhibit large bending and plasticity. Therefore, low order elements are not favorable in such applications as they exhibit a locking effect in bending problems which deteriorates the accuracy of the results. There are many solutions to overcome the locking effect. One possibility is to work with higher order elements [14]. Nonetheless, these formulations require complicated meshing procedures and complicated handling of the contact algorithms. Another solution is to use the method of compatible modes or enhanced assumed strains [15,16]. However, these approaches require more internal element variables to be determined, which leads to a higher computational cost. Recently, Belytschko and Bindeman [17], Liu et al. [18] and Reese et al. [19] have proposed a methodology to combine the assumed enhanced strain method with the reduced integration technique and hourglass stabilization. This approach showed robust deformation behavior along with low numerical cost. A new category of elements are based on the same concept, called solid shell elements [20,21]. Unger et al. [10] developed a model for 3D simulation of electromagnetic forming using a solid shell element [22] for the mechanical solution to reduce the computational cost. However, Unger et al. [10] addressed only the simple problem of a rectangular shape workpiece, which did not represent a real life application problem.

Additionally, many multi-physics commercial software already exist in the market, such as LS-DYNA [23]. The electromagnetic computations in LS-DYNA are based on the coupling between finite elements for solid bodies and boundary elements [24] for the surrounding air [25]; therefore, the air domain is not meshed. This can be more adapted for moving bodies but can significantly increase the computational time due to the fact that the boundary element system leads to solving a small linear system but with a full matrix. On the other hand, this work is complementary to the work of Alves Z and Bay [7], in which the authors use a tetrahedral solid element, called the Nedelec element [26], in which all fields are defined on the solid element's edges instead of its nodes. Thus, all bodies in the system are different, solely by the material parameter, and have the same finite element. This approach can be quite fast since it deals with a single element type in the system. Moreover, since the approach produces a single linear system for the resolution of the magnetic potential, the parallelism is also simplified [27]. Finally, remeshing is more adapted to this approach since tetrahedra are the most suited elements for enabling automatic and adaptive remeshing. It is one of the main advantages of our tool and the work carried out here is aimed at developing an original strategy in order to preserve the efficiency of this strategy.

The main novelty of this work is the study of an application of the EMF processes, which is the magnetic pulse forming (MPF) of thin sheets. This process can be used in either direct forming, used with highly conductive materials, or indirect forming, used with less conductive materials. Although the magnetic pulse forming problem can be simulated using axisymmetric model [28], we have chosen to use a 3D approach to model the problem for some reasons. First, the approaches developed in this paper are generic and are used to solve general 3D problems in the future. Second, the ultimate goal is to include the effect of plastic anisotropy in the future simulations and implementing 3D constitutive equations that are more convenient with most of the anisotropic yield criteria. Herein, the simulations are carried out using different element types: MINI element [29] and solid shell element [20]. This solid shell element was developed and modified to fit magnetic pulse forming applications [30]. Using the solid shell element instead of the MINI element reduces the computational cost of the simulation dramatically.

The paper is divided as follows: Section 2 discusses the modeling of the electromagnetic problem and the mechanical problem, in brief, along with a glimpse of the implementation strategy adopted for these simulations. Section 3 tackles the description of the magnetic pulse forming application, its finite element description, the results and their physical interpretation. Section 4 is an extensive study of the difficulties that have been encountered in the problem and some proposed solutions to overcome these difficulties. Finally, Section 5 states the concluding remarks.

2. Modeling of the Magnetic Pulse Forming Process

Multi-physics simulations, including electromagnetic simulations, can be very computationally expensive. Additionally, design processes that include multiple simulation iterations and optimization processes require high computational power to be carried out. Thus, it is important to select the most appropriate numerical method to solve these problems in a reasonable time while maintaining good accuracy.

A numerical toolbox based on finite elements methods for the electromagnetic forming applications has been developed to solve electromagnetic forming problems [31,32]. This toolbox is a coupling between FORGE—for the mechanical modeling of large deformation—and MATELEC—which solves the electromagnetic wave propagation problem—based on the Maxwell's equations. The following subsections explain the electromagnetic and mechanical models used in the simulation of the magnetic pulse forming process.

2.1. *Electromagnetic Model*

2.1.1. Maxwell'S Equations and the Potential Formulation

The electromagnetic solver is based on the well-known Maxwell's electromagnetic field equations:

$$\vec{\nabla} \times \vec{E} = -\frac{\partial \vec{B}}{\partial t} \tag{1}$$

$$\vec{\nabla} \times \vec{H} = \vec{J} \tag{2}$$

$$\vec{\nabla} \cdot \vec{D} = 0 \tag{3}$$

$$\vec{\nabla} \cdot \vec{B} = 0 \tag{4}$$

where

$\vec{E}$: Electric field intensity	$\vec{D}$: Electric flux intensity
$\vec{H}$: Magnetic field intensity	$\vec{B}$: Magnetic flux intensity
ρ^e : Electric charge density	$\vec{J}$: Electric current density

- Equation (1): (Maxwell Faraday) represents the electric induction due to a varying magnetic field.

- Equation (2): (Maxwell Ampere) represents the creation of a magnetic field due to a passing electric current.
- Equation (3): (Maxwell gauss) represents the conservation of electric charge in the material.
- Equation (4): Represents the conservation of the induced magnetic field in the material.

Although, a reduced form of Maxwell-gauss Equation (3) is used in which $\rho^e = 0$, since there is no fixed electric charge to be considered in this problem. Moreover, a reduced form of Maxwell-Ampere Equation (2) is used in the current applications of metal forming, since electromagnetic wave propagation may be neglected [33]; thus, ($\frac{\partial \vec{D}}{\partial t} = 0$).

Then, the model is completed with the constitutive relations:

$$\vec{B} = \mu_H \vec{H}; \vec{J} = \frac{1}{\rho_E} \vec{E} \tag{5}$$

where μ_H is magnetic permeability and ρ_E the electrical resistivity. These material parameters depend on the temperature and μ_H depends also on the intensity of the magnetic field $||\vec{H}||$.

In many cases, it is more convenient to express this system of equations in potential formulation ($\vec{A}$,ϕ) [34] where $\vec{A}$ is the vector potential function and ϕ is the scalar potential function that can be represented by the following equations:

$$\vec{\nabla} \cdot \vec{B} = 0 \Rightarrow \vec{B} = \vec{\nabla} \times \vec{A} \tag{6}$$

Combining Equation (1) with Equation (6):

$$\begin{aligned} \vec{\nabla} \times \vec{E} = -\frac{\partial \vec{B}}{\partial t} \Rightarrow \vec{\nabla} \times \vec{E} = -\frac{\partial}{\partial t}(\vec{\nabla} \times \vec{A}) \\ \Rightarrow \vec{\nabla} \times \left(\vec{E} + \frac{\partial \vec{A}}{\partial t} \right) = 0 \end{aligned} \tag{7}$$

Since for any scalar function ϕ, $\vec{\nabla} \times (-\vec{\nabla}\phi) = 0$ holds, then

$$\begin{aligned} \Rightarrow \vec{E} + \frac{\partial \vec{A}}{\partial t} = -\vec{\nabla}\phi \\ \Rightarrow \vec{E} = -\vec{\nabla}\phi - \frac{\partial \vec{A}}{\partial t} \end{aligned} \tag{8}$$

Finally, after substitution of $(\vec{A}, \phi)$ in Maxwell's equation and considering law of the charge conservation, the final equations can be written as follows:

$$\begin{aligned} \frac{1}{\rho_E} \frac{\partial \vec{A}}{\partial t} + \vec{\nabla} \times \left(\frac{1}{\mu_H} \vec{\nabla} \times (\vec{A}) \right) = -\frac{1}{\rho_E} \vec{\nabla}(\phi) \\ \vec{\nabla} \cdot (\frac{1}{\rho_E} \vec{\nabla}\phi) + \vec{\nabla} \cdot \left(\frac{1}{\rho_E} \partial_t \vec{A} \right) = 0 \end{aligned} \tag{9}$$

This is a four variables (ϕ, A_x, A_y, A_z) four equations system instead of six variables for a double vector field formulation. Equation (9) is discretized in space by Nedelec elements [26] and $\vec{A}$ is solved at the edges while ϕ is solved at the nodes.

2.1.2. Weak Formulation and Discretization of Electromagnetic Problem

The electromagnetic problem consists of a single domain, as indicated in Figure 1. The domain (Ω) is subdivided into three subdomains: the coil or inductor (Ω_I), the workpiece (Ω_P) and the surrounding air (Ω_a)

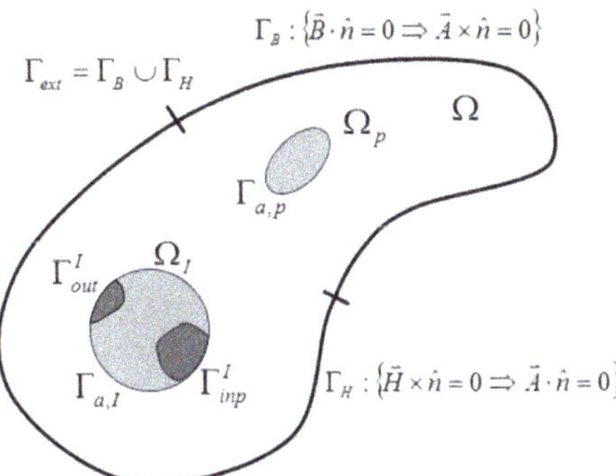

Figure 1. Boundaries of an EMF process. Ω represents the global domain solids + surroundings. Ω_P is the workpiece. Ω_I represents the inductor domain. The electrical input and output connections of the inductor are given by Γ^I_{inp} and Γ^I_{out} [35].

Reese et al. [28] utilized Coloumb gauge condition in order to guarantee uniqueness of solution. This condition indicates that:

$$\vec{\nabla} \cdot \vec{A} = 0 \tag{10}$$

Hence, the second equation of Equation (9) is reduced to:

$$\vec{\nabla} \cdot (\frac{1}{\rho_E} \vec{\nabla} \phi) = 0 \tag{11}$$

Therefore, the weak form of the electromagnetic differential equations in Equation (9) will take the following form:

$$\begin{aligned} \left\langle \vec{\Psi}, \tfrac{1}{\rho_E} \partial_t \vec{A} + \vec{\nabla} \times \tfrac{1}{\mu_H} \vec{\nabla} \times \vec{A} + \tfrac{1}{\rho_E} \vec{\nabla} \phi \right\rangle = 0 \\ \langle \varphi, \vec{\nabla} \cdot (\tfrac{1}{\rho_E} \vec{\nabla} \phi) \rangle = 0 \end{aligned} \tag{12}$$

for all $\vec{\Psi} \in \mathcal{H}^{\text{curl}}$ and $\varphi \in \mathcal{H}^1_0$. Where,

- Space of functions vanishing at the boundary $\mathcal{H}^1_0(\Omega) \subset \mathcal{H}^1(\Omega)$.

$$\mathcal{H}^1_0(\Omega) = \left\{ \varphi \in \mathcal{H}^1(\Omega) / \varphi = 0 \in \partial\Omega \right\} \tag{13}$$

- Space of vector functions with square-integrable curl.

$$\mathcal{H}^{\text{curl}}(\Omega) = \left\{ \vec{\Psi} \in \left(L^2(\Omega) \right)^3 / \vec{\nabla} \times \vec{\Psi} \in \left(L^2(\Omega) \right)^3 \right\} \tag{14}$$

- Inner products: The following notation for the inner product of the spaces will allow simplifying the notation for the weak forms.

$$\int_\Omega f \cdot g d\Omega = \langle f, g \rangle \tag{15}$$

Alves Zapata [27] developed the detailed mathematical model and considered natural conditions to reach the final weak form:

$$\begin{aligned} \left\langle \vec{\Psi}, \sigma \partial_t \vec{A} \right\rangle + \left\langle \vec{\nabla} \times \vec{\Psi}, \tfrac{1}{\mu} \vec{\nabla} \times \vec{A} \right\rangle + \langle \vec{\Psi}, \tfrac{1}{\rho_E} \vec{\nabla} \phi \rangle = 0 \\ \langle \vec{\nabla} \varphi, \tfrac{1}{\rho_E} \vec{\nabla} \phi \rangle = 0 \end{aligned} \tag{16}$$

Afterwards, the approximate fields solutions representing the finite elements discretization is defined as:

$$\begin{aligned} \phi(t,\vec{x}) &\approx \phi^h(t,\vec{x}) = \sum_n \phi_n(t)\varphi_n(\vec{x}) \\ \vec{A}(t,\vec{x}) &\approx \vec{A}^h(t,\vec{x}) = \sum_d a_d(t)\vec{\Psi}_d(\vec{x}) \end{aligned} \tag{17}$$

where $\varphi_n(\vec{x})$ are the nodal shape functions and $\vec{\Psi}_d(\vec{x})$ are the edge shape functions (Nedelec elements). Alves Zapata [27] addresses in more details the interpolation functions and finite element formulation of this problem.

Finally, Lorentz forces can be computed from the potential formulation as follows:

$$\begin{aligned} \vec{F}_{lorentz} &= \vec{J} \times \vec{B} \\ \vec{F}_{lorentz} &= \frac{1}{\rho_E}\left[-\frac{\partial \vec{A}}{\partial t} \times (\vec{\nabla} \times \vec{A})\right] \end{aligned} \tag{18}$$

Lorentz force is only function in $\vec{A}$, which can be computed directly after solving for $\vec{A}$.

2.2. Solid Mechanics Model

The second part of the simulation is related to solid mechanics simulation in which the electromagnetic forces are transferred to the metal part, causing deformation. In this paper, two different formulations are considered in the simulation results: a mixed pressure-velocity element formulation (MINI) element [29] and an enhanced assumed strain element formulation [20]. In the following part, we will present a glimpse of both formulations.

2.2.1. Mini Element Formulation

For the MINI element, the strong form of the mechanical problem is defined by conservation equation along with the boundary conditions. Considering the decomposition of the stress tensor into spherical and deviatoric parts, the mechanical problem is represented on the domain Ω and the external boundary $\Gamma = \partial\Omega$, shown in Figure 2. The boundary is decomposed to several parts depending on the type of loading applied: $\Gamma = \Gamma_{fr} \cup \Gamma_t \cup \Gamma_v \cup \Gamma_c$ where:

- Γ_{fr} : Free surface boundary.
- Γ_t : Imposed external traction boundary.
- Γ_v : Imposed external velocity boundary.
- Γ_c : Contact condition on the boundary with other tools (rigid or deformable).

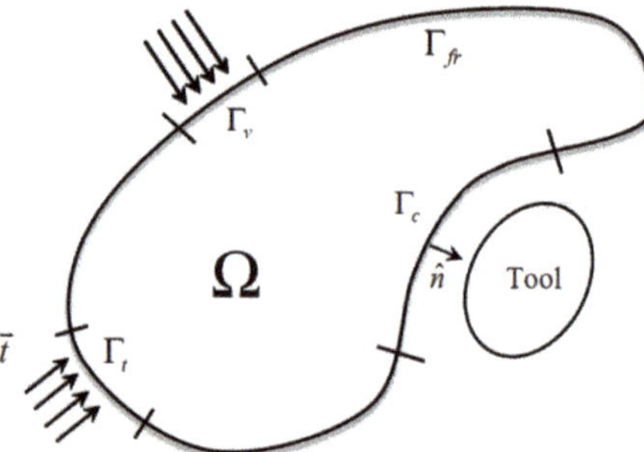

Figure 2. Representation of the domain Ω and the boundary conditions [27].

The system of equations representing the mechanical problem can be summarized as follows:

$$\begin{cases} \vec{\nabla} \cdot s - \vec{\nabla} p = \vec{f}^{ext} \\ \vec{\nabla} \cdot \vec{v} = -\left(\frac{\dot{p}}{\kappa}\right) \\ \vec{v} = \vec{v}_0 \text{ on } \partial\Omega_v \\ \vec{t} = \vec{t}_0 \text{ on } \partial\Omega_t \end{cases} \tag{19}$$

The first two equations represent the conservation of momentum in the system. Although, the representation used here is divided to deviatoric stress and pressure that will help later in developing the mixed formulation element approach.

Third equation represents Dirichlet boundary condition and fourth equation represents Neumann boundary condition.

The weak form is based on a mixed velocity-pressure formulation. The formulation is written for test functions (v^*,p^*) as follows:

$$\begin{cases} \int_\Omega s(\vec{v}) : \dot{\varepsilon}(\vec{v}^*) d\Omega - \int_\Omega p \vec{\nabla} \cdot \vec{v}^* d\Omega - \int_\Omega \vec{F}_{lorentz} \cdot \vec{v}^* d\Omega - \int_{\partial\Omega_t} \vec{t}_0 \cdot \vec{v}^* d\Gamma = 0 \\ \int_\Omega p^* \left(-\vec{\nabla} \cdot \vec{v} - \frac{\dot{p}}{\kappa}\right) d\Omega = 0 \\ \forall(\vec{v}^*, p^*) \in \mathcal{V}^0 \times \mathcal{P} \end{cases} \tag{20}$$

where $s(\vec{v})$ is the deviatoric stress, p is the pressure, $\vec{v}$ is the velocity vector, $\dot{\varepsilon}$ the strain rate and κ is the bulk modulus.

$$\begin{cases} \mathcal{V}^0 = \left\{\vec{v}^* \in \left(H^1(\Omega)\right)^3, \vec{v}^*|_{\partial\Omega_v} = \vec{0} \text{ sur } \Omega\right\} \\ \mathcal{P} = L^2(\Omega) \end{cases} \tag{21}$$

H^1 is the Sobolov space and L^2 is the L^p space of square functions summed on Ω .

This problem has a unique solution. Although, from a numerical point of view, numerical instabilities can arise depending on the choice of the discretization space for v and P. In order to ensure the stability of this approach, the numerical formulation should pass the Brezzi condition [36]. The P1+/P1 discretization allows to pass Brezzi condition leading to a well posed discrete problem.

The element used for this formulation is tetrahedral linear 3D element in which both the pressure and velocity are linearly interpolated. However, the velocity interpolation is enhanced by a dot at the center of the element, called "bubble", as shown in Figure 3.

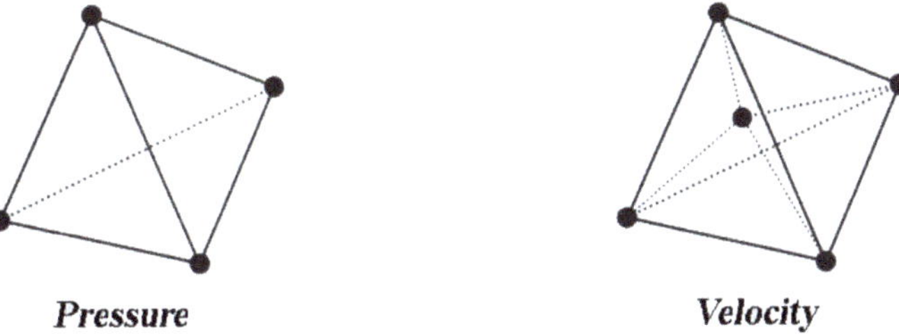

Figure 3. Degrees of freedom for the velocity and pressure for the tetrahedral element P1+/P1.

The following equation presents the interpolation function for both the velocity and pressure. The velocity field is divided into two parts: linear and bubble interpolation functions.

$$\begin{cases} \vec{v}_h(\vec{x}) = \sum_{k=1}^{Nbnode} N_k^l(\vec{x})\vec{V}_k^l + \sum_{j=1}^{Nbelt} N_j^b(\vec{x})\vec{V}_j^b \\ p_h(\vec{x}) = \sum_{k=1}^{Nbnode} N_k^l(\vec{x})P_k \end{cases} \quad (22)$$

where N_k^l , $k = 1 \ldots Nbnode$ are the shape function of the linear interpolation for the velocity and pressure, while $N_j^b, j = 1 \ldots Nbelt$ is the bubble function [37].

2.2.2. Shb Element Formulation

On the other hand, the general variational principle of SHB element is considered for the mechanical problem from Hu–Washizu variational principle:

$$\delta\pi(\vec{v},\dot{\bar{\varepsilon}},\bar{\sigma}) = \int_{\Omega_e} \delta\dot{\varepsilon}^T \cdot \sigma d\Omega + \delta \int_{\Omega_e} \bar{\sigma}^T \cdot (\nabla_s(\vec{v}) - \dot{\bar{\varepsilon}}) d\Omega - \delta v^T \cdot \vec{f}^{ext} = 0 \quad (23)$$

where δ denotes a variation, $\vec{v}$ the velocity field, $\dot{\bar{\varepsilon}}$ the assumed strain rate, $\bar{\sigma}$ the interpolated stress, σ the stress field evaluated by constitutive model, $\vec{v}$ the nodal velocities, $\vec{f}^{ext}$ the external nodal forces (includes Lorentz forces and other external forces in the problem) and $\nabla_s(\vec{v})$ the symmetric part of the velocity gradient.

Then a simplified form of this principle is achieved by considering the interpolated stress orthogonal to the difference between the symmetric part of the velocity gradient and the assumed strain rate, which gives:

$$\delta\pi(\dot{\bar{\varepsilon}}) = \int_{\Omega_e} \delta\dot{\varepsilon}^T \cdot \sigma d\Omega - \delta v^T \cdot \vec{f}^{ext} = 0 \quad (24)$$

In order to avoid a locking problem, a finite element is constructed based on the enhanced assumed strain method and using reduced integration [20].

The element type is different from the one used for the MINI element. It is a linear prism element that interpolates the velocities as the only degrees of freedom. Figure 4 shows the element used for this formulation and the alignment of the integration point along the thickness direction ζ. Trinh et al. [20] discuss, in detail, the element interpolation along with the position of the integration points.

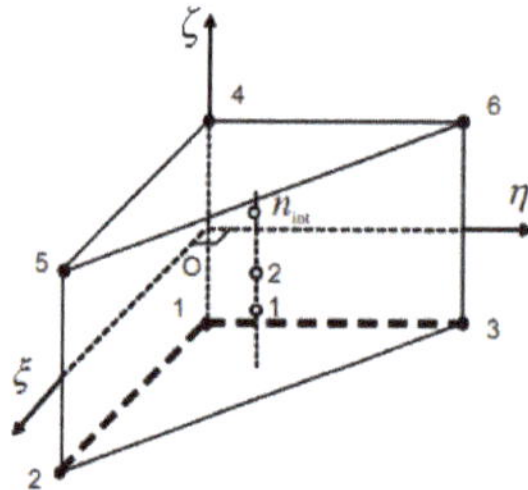

Figure 4. SHB element shape and integration points locations.

The interpolation of the coordinates x_i and the displacements u_i are as follows:

$$x_i = x_{iI}N_I(\xi,\eta,\zeta) = \sum_{I=1}^{n} x_{iI}N_I(\xi,\eta,\zeta) \quad (25)$$

$$v_i = v_{iI}N_I(\xi,\eta,\zeta) = \sum_{I=1}^{n} v_{iI}N_I(\xi,\eta,\zeta) \quad (26)$$

where N_I, v_{iI} and x_{iI} are the shape functions, the nodal velocities and the nodal coordinates, respectively. Moreover, the lowercase subscript i varies from 1 to 3, representing the spatial coordinates, and the uppercase subscript I varies from 1 to n, representing the number of nodes per element [20].

The implementation of the electromagnetic and mechanical solvers is not easy, especially the coupling between them. The following subsection gives a glimpse of the numerical implementation adopted in this work.

2.3. Numerical Implementation

2.3.1. Coupling Algorithm

Figure 5 shows a schematic view of the coupling strategy between the electromagnetic solver and mechanical solver used to solve the MPF problem. A weak coupling is used for the electromagnetic and mechanical problems. Therefore, each solver (MATELEC, FORGE) solves its own physical problem separately, independently of the other solver. Although, after every time increment, the two solvers communicate the corresponding data and variables between each other. This is known as a loosely-coupled scheme [7]. Moreover, this approach allows adapting the mesh separately for each solver which is crucial, especially in the electromagnetic solver in which the air surrounding the moving parts should be remeshed.

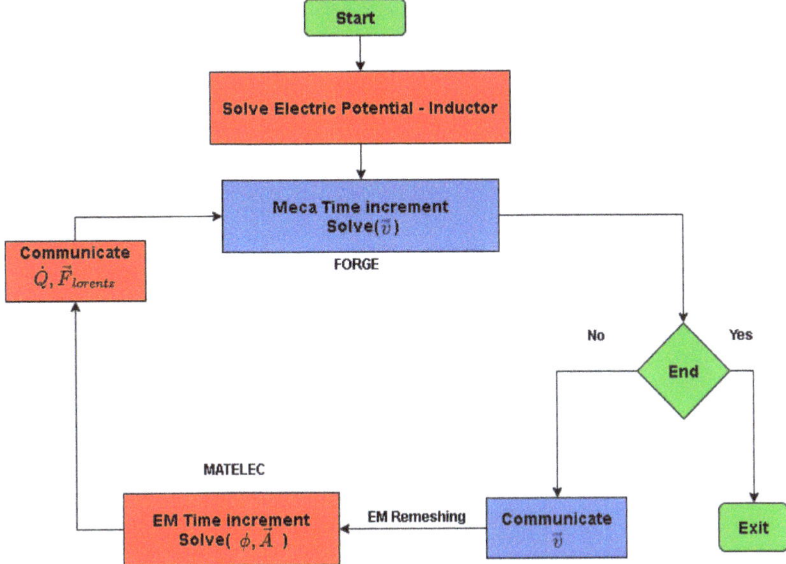

Figure 5. Schematic view of the coupling strategy between mechanical and electromagnetic solvers [7].

2.3.2. Shb Element Implementation Algorithm

The SHB element formulation uses prism elements that have been implemented in a tetrahedral element finite element software. Mahmoud et al. [30] developed a prism division algorithm that ensures that all prism elements are resembled as a set of tetrahedral elements in the software, with minimal changes in the code structure. The main idea was to divide one prism element into six overlapping tetrahedral elements. The overlapping of elements is crucial so that all the components of the original SHB stiffness matrix could be presented in at least one of the generated tetrahedral elements.

Figure 6 shows the corresponding tetrahedral elements generated from the divided prism element. Considering the number of tetra elements sharing each component (node) of the prism stiffness matrix, the new stiffness matrices for the tetra elements can be constructed as follows:

$$element1:\begin{bmatrix} \frac{K_{11}}{4} & \frac{K_{12}}{3} & \frac{K_{13}}{3} & \frac{K_{14}}{2} \\ & \frac{K_{22}}{4} & \frac{K_{23}}{3} & \frac{K_{24}}{2} \\ & & \frac{K_{33}}{4} & \frac{K_{34}}{2} \\ & & & \frac{K_{44}}{4} \end{bmatrix} element2:\begin{bmatrix} \frac{K_{11}}{4} & \frac{K_{12}}{3} & \frac{K_{13}}{3} & \frac{K_{15}}{2} \\ & \frac{K_{22}}{4} & \frac{K_{23}}{3} & \frac{K_{25}}{2} \\ & & \frac{K_{33}}{4} & \frac{K_{35}}{2} \\ & & & \frac{K_{55}}{4} \end{bmatrix} \quad (27)$$

$$element3:\begin{bmatrix} \frac{K_{11}}{4} & \frac{K_{12}}{3} & \frac{K_{13}}{3} & \frac{K_{16}}{2} \\ & \frac{K_{22}}{4} & \frac{K_{23}}{3} & \frac{K_{26}}{2} \\ & & \frac{K_{33}}{4} & \frac{K_{36}}{2} \\ & & & \frac{K_{66}}{4} \end{bmatrix} element4:\begin{bmatrix} \frac{K_{11}}{4} & \frac{K_{14}}{2} & \frac{K_{15}}{2} & \frac{K_{16}}{2} \\ & \frac{K_{44}}{4} & \frac{K_{45}}{3} & \frac{K_{46}}{3} \\ & & \frac{K_{55}}{4} & \frac{K_{56}}{3} \\ & & & \frac{K_{66}}{4} \end{bmatrix} \quad (28)$$

$$element5:\begin{bmatrix} \frac{K_{22}}{4} & \frac{K_{24}}{2} & \frac{K_{25}}{2} & \frac{K_{26}}{2} \\ & \frac{K_{44}}{4} & \frac{K_{45}}{3} & \frac{K_{46}}{3} \\ & & \frac{K_{55}}{4} & \frac{K_{56}}{3} \\ & & & \frac{K_{66}}{4} \end{bmatrix} element6:\begin{bmatrix} \frac{K_{33}}{4} & \frac{K_{34}}{2} & \frac{K_{35}}{2} & \frac{K_{36}}{2} \\ & \frac{K_{44}}{4} & \frac{K_{45}}{3} & \frac{K_{46}}{3} \\ & & \frac{K_{55}}{4} & \frac{K_{56}}{3} \\ & & & \frac{K_{66}}{4} \end{bmatrix} \quad (29)$$

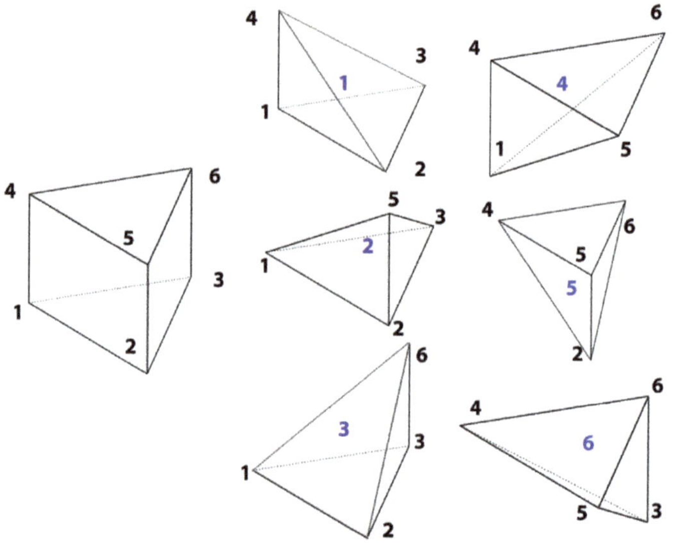

Figure 6. Prism division to six overlapping tetrahedral elements [30].

In this way, the prism element was presented implicitly through the usage of tetrahedral elements and there was no need to profoundly change the code structure used for the simulation.

3. Magnetic Pulse Forming Case Study

Figure 7 shows the schematic view of the free bulging process. A round flat coil is used along with a ring-shaped matrix that blocks the displacement at the circumference of the workpiece. This problem is very convenient for clarifying the coupling between the electromagnetic and mechanical solvers discussed earlier. The following subsections will tackle the details of the simulation setup with respect to the electromagnetic simulation and the mechanical simulation.

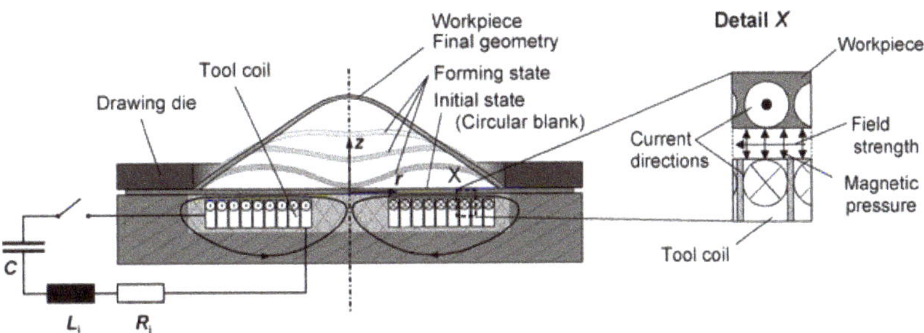

Figure 7. Illustration of magnetic pulse forming setup [38].

3.1. Electromagnetic Simulation Setup

Geometry: Figure 8a shows the geometry of the electromagnetic simulation with the dimensions. The aluminum workpiece has a 0.5 mm thickness and it is located 0.5 mm above the coil.

Mesh: Figure 8b shows the mesh used in the electromagnetic simulation. Three-dimensional global mesh is composed of 43,000 nodes with 260,000 tetrahedral elements. The element size is selected so that the mesh is fine in regions where strong gradients are expected and in the proximity of the metallic sheet.

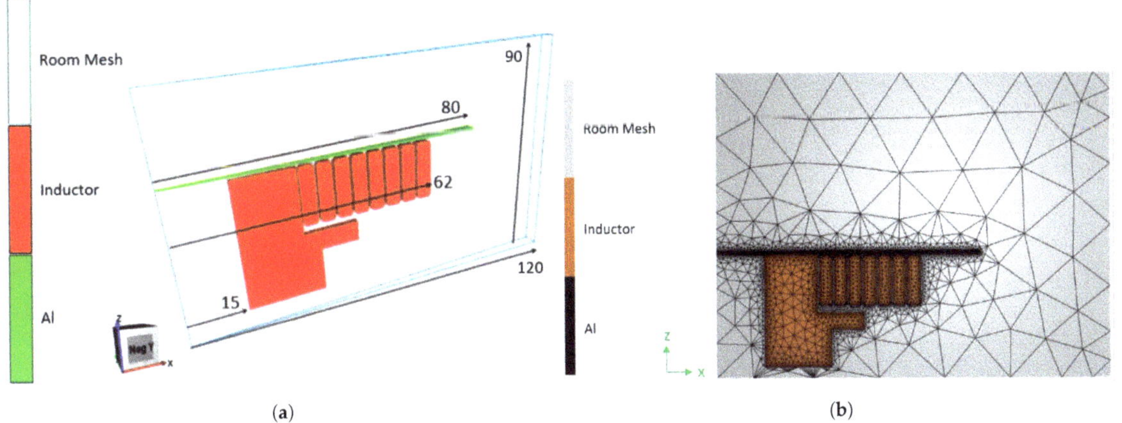

Figure 8. Electromagnetic simulation setup. (**a**) Geometry and dimensions of the electromagnetic simulation; (**b**) mesh configuration of the electromagnetic simulation.

Simulation properties: Table 1 shows the parameters that define the magnetic properties of the materials according to Equation (5). Additionally, it shows the parameters of the machine that produces the electromagnetic field used for the MPF process.

3.2. Mechanical Simulation Setup

Geometry: Figure 9 shows the geometry of the 2° section of the workpiece. The workpiece is fixed in the region of the green ring shown in the figure. The fixed boundary conditions come from the fact that the perimeter of the workpiece is held by a cylindrical clamp that prevent it from moving.

Mesh: Different mesh sizes and element types have been used to simulate the mechanical problem and the results were compared. In the results section, each curve will present the element type and the number of elements used for these results. Mesh study has been carried out to investigate the difference in results using different mesh sizes in Section 4.3 and the most appropriate mesh sizes are used in the results sections.

Table 1. Properties of the electromagnetic simulation.

Property	Value
Electrical resistivity of Al (ρ_{Al})	4 µΩ.cm
Electrical resistivity of Steel (ρ_{Steel})	73 µΩ.cm
Relative magnetic permeability of Al (μ_{Al})	1
Relative magnetic permeability of Steel (μ_{Steel})	1
Magnetic permeability in vacuum (μ_0)	$4\ \pi.10^{-7}$H.m^{-1}
Machine parameters	$U_{max} = 7$ kV $R_m = 1$ mΩ ; $L_m = 3.2$ µH; $C_m = 552$ µF
Time step	1 µs

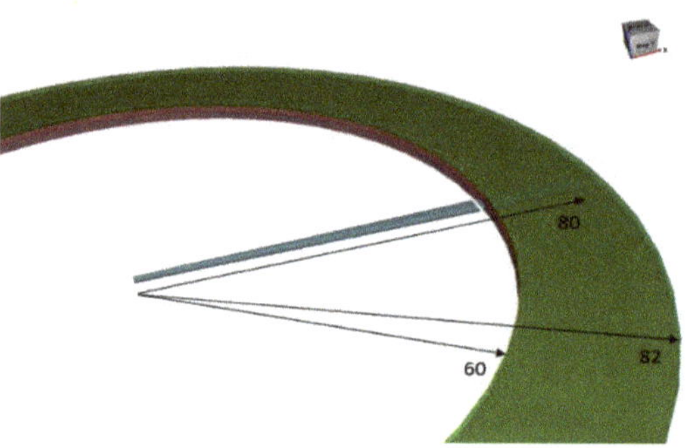

Figure 9. Geometry and dimensions of the mechanical simulation.

Material properties: Material properties of Al (Al1050 [39]) and Steel (AISI 4130 [40]) used in the simulation are shown in Table 2. On the other hand, Johnson–Cook law was used to describe the elastoviscoplastic behavior of the materials. Equation (30) shows the constitutive law used and Table 2 shows the corresponding constants.

$$\sigma_Y = \left(A + B\bar{\epsilon}_{pl}^n\right)\left(1 + C\ln\left(\frac{\dot{\epsilon}_{pl}}{\dot{\epsilon}_0}\right)\right) \quad (30)$$

where σ_Y is the yield stress, $\bar{\epsilon}_{pl}$ is equivalent plastic strain, $\dot{\epsilon}_{pl}$ plastic strain rate and $\dot{\epsilon}_0$ initial plastic strain rate.

Table 2. Material properties and Johnson–Cook law parameters of Al and Steel.

Property	Al	Steel
Elastic modulus (E)	73.1 GPa	200 GPa
Poisson ratio (ν)	0.279	0.3
A	83 MPa	610 MPa
B	426 MPa	750 MPa
C	0.025	0.008
n	0.35	0.25

Boundary conditions: In the direct forming, the Aluminum workpiece is the only part in the mechanical solver. Thus, there is no contact condition added in this simulation. However, there is a bilateral sticking condition between the green ring manipulator shown in Figure 9 and the workpiece. This induces fixed boundary conditions on the circumference of the part. On the other hand, the indirect forming contains two metal parts: Aluminum and Steel. There is a sliding contact condition between Aluminum and Steel so that the friction does not affect the final deformation results. Table 3 summarizes all the boundary conditions adopted in this simulation.

Table 3. Boundary conditions for the indirect forming simulation.

	Al	Steel
Steel	Sliding	
3D manipulator (matrix)	Bilateral sticking	Bilateral sticking

3.3. Results Overview

This section is fully dedicated to discussing the results of the MPF problem in the utmost details possible. The following subsections will tackle two basic types of the MPF process: direct forming and indirect forming. The former, a 160 mm diameter disc-shaped workpiece, is formed by MPF directly, whereas the latter, an Aluminum disc, is placed between the coil and the Steel workpiece since Al has a much higher electrical conductivity than Steel and will help better form the Steel material [7]. Many tests have been carried out either by direct forming or indirect forming.

3.3.1. Direct Forming

In this subsection, the results of the direct forming process are presented. Two different workpieces were tested under direct forming: 0.5 mm thick Al and 1 mm thick Steel. Equivalent strains of both Steel and Al at 7 kV and 3 kV are shown in Figure 10a,b, respectively. The equivalent strain gives some insights on the local deformation of the workpiece. It is obvious that the strain distribution on the Aluminum part is more homogeneous than in the Steel case. This is justified by the fact that Aluminum is a much better electrical conductor than Steel, which enhances the generated Lorentz forces causing distributed deformation, whereas in Steel, the deformation is concentrated at the center as it is the nearest point to the center of the coil.

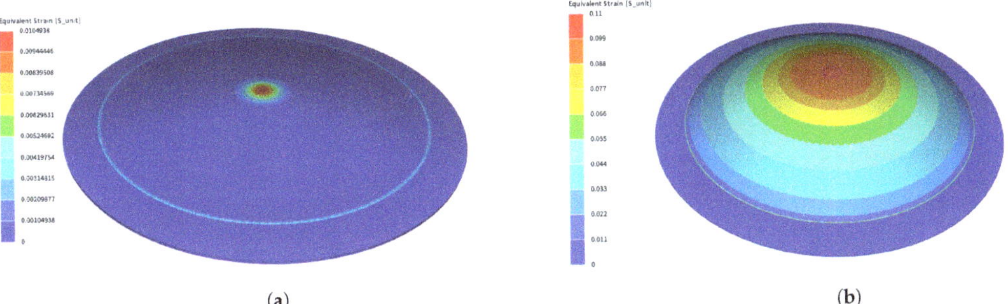

Figure 10. Equivalent strain of Steel and Al separately after direct forming using MINI element. (**a**) Equivalent strain of Steel at 7 kV; (**b**) equivalent strain of Al at 3 kV.

More investigations have been conducted on the deformed shapes. Figure 11a,b shows the deformation profile along the radius of the workpiece at 2 kV and 3 kV, respectively, for Al material. The simulation was conducted using two elements, MINI and SHB. Numerical results show good agreement with each other, even though the number of elements of the SHB element is much smaller than the number of the MINI elements. Likewise, the results of Steel material are presented in Figure 12a,b for 5 kV and 7 kV, respectively. There is an overall agreement among the numerical results for both of the elements.

The final deformation of the Steel material is very small and cannot be shaped through direct forming due to insufficient electrical conductivity and, consequently, not enough deformation range. Therefore, the rest of the work is focused on indirect forming in order to obtain more deformation with the help of an aluminum hammer, which is the same workpiece used in direct forming (0.5 mm thick).

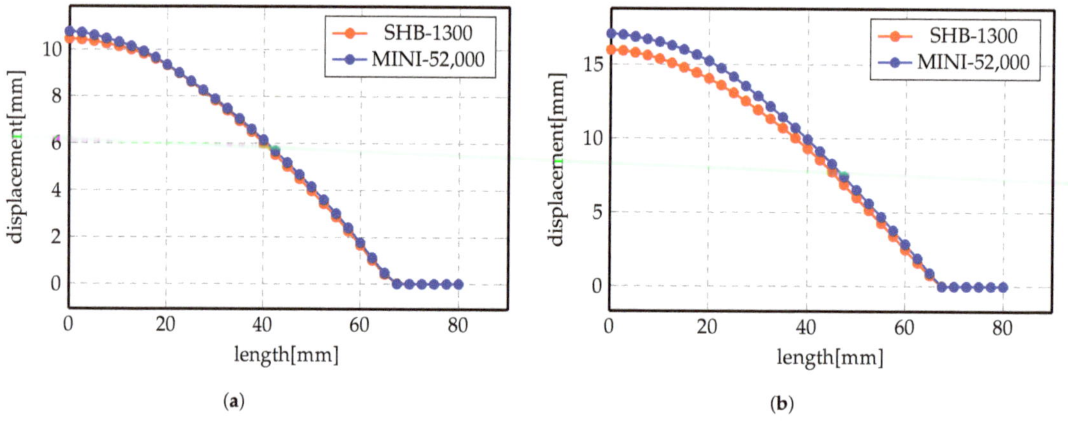

Figure 11. Displacement profile of direct forming process of Al. (**a**) 2 kV; (**b**) 3 kV.

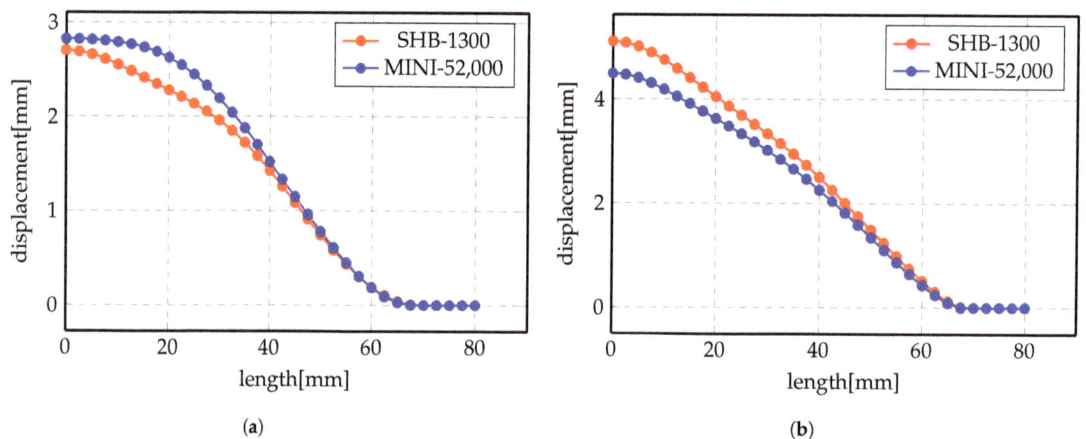

Figure 12. Displacement profile of direct forming process of Steel. (**a**) 5 kV; (**b**) 7 kV.

3.3.2. Indirect Forming

The indirect forming process results will be the main focus of this section. More in-depth investigation will be tackled in this subsection. Figure 13 shows the equivalent strain of the workpiece in the indirect forming case, either for 5 kV or 7 kV. The equivalent strain in the indirect forming is more homogeneous than that shown in the direct forming case due to using the Aluminum workpiece underneath the Steel workpiece to enhance the forming process.

Likewise, numerical simulations are carried out for the indirect case using two elements: MINI and SHB. More deformation has been noticed for the Steel when used with Al since it makes use of the electrical conductivity of Al to gain more force to shape the Steel.

Figures 14 and 15 show the displacement profiles for SHB and MINI elements. The overall conclusion of these results is that the numerical results for both elements are very close even though the mesh sizes are different.

Overall, the results of the recently implemented element SHB showed very good agreement with its counterpart MINI element. These results are very encouraging to use this element in such complicated simulations as it proved its precision and efficiency. Although, more investigation is required to better compare the two elements, which will be introduced in the following section.

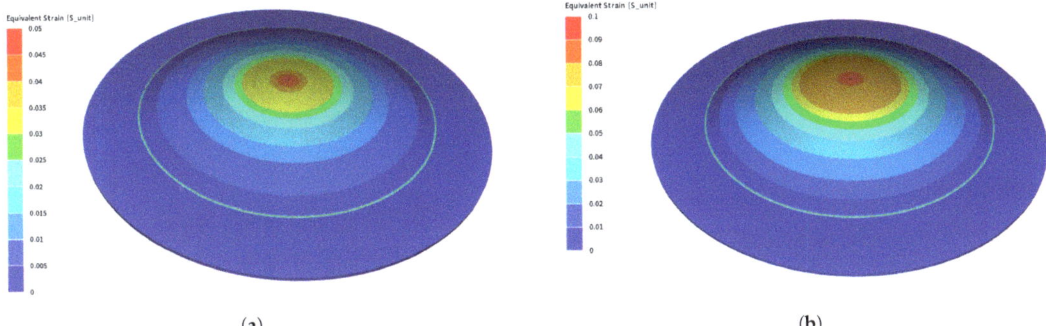

Figure 13. Equivalent strain of Steel and Al after indirect forming using MINI element. (**a**) 5kV; (**b**) 7kV.

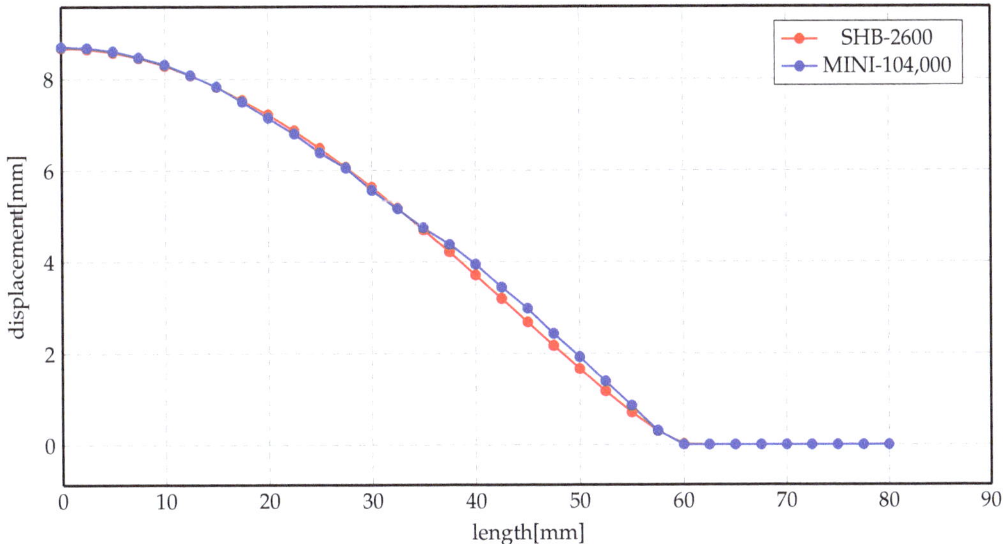

Figure 14. Displacement profile of indirect forming process at 5 kV using different elements.

3.4. Simulation Time

The aim of implementing the new element SHB is to use a special element for bending-dominated problems and obtain accurate results with a low number of elements. The usage of this element was very remarkable as the simulation time was greatly affected. Figure 16 shows a bar chart for the CPU time needed for both of the electromagnetic and mechanical simulations, separately, for indirect forming cases at two different voltages, 5 kV and 7 kV. In contrast to electromagnetic simulation times that were almost identical, the mechanical simulations times were greatly decreased using the SHB element. The simulation time is reduced by almost 10× in the SHB case.

Similarly, the CPU time of the direct forming process represents the same trend. Figure 17 shows the bar chart of the direct forming of Al at two different voltages, 2 kV and 3 kV. Electromagnetic simulation times are almost identical, whereas the mechanical simulation time is much lower, 3× for the SHB element than the MINI element. The MINI element is used with a mesh of 52,000 tetra elements and SHB elements with 1300 elements. This difference in the number of elements causes the simulation time difference highlighted in these figures.

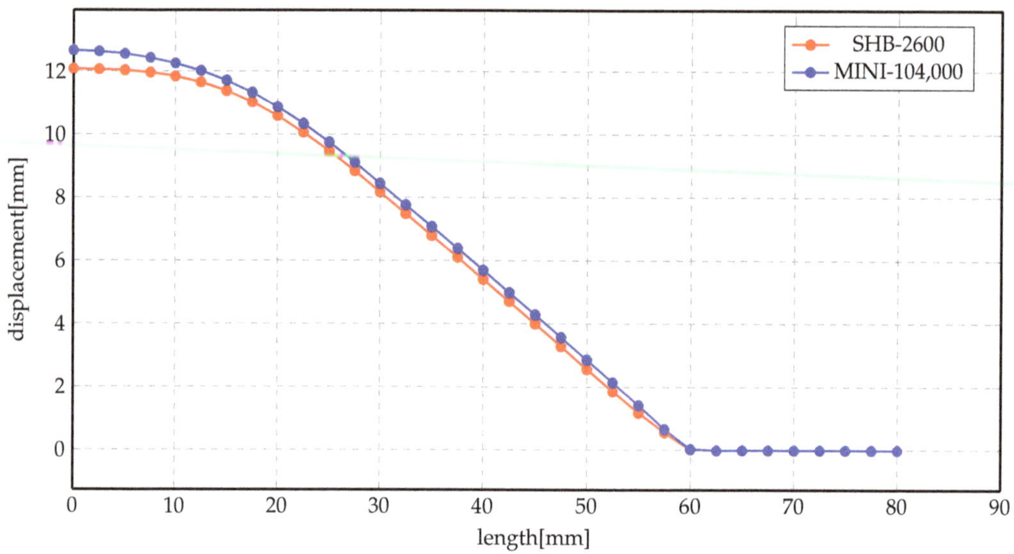

Figure 15. Displacement profile of indirect forming process at 7 kV using different elements.

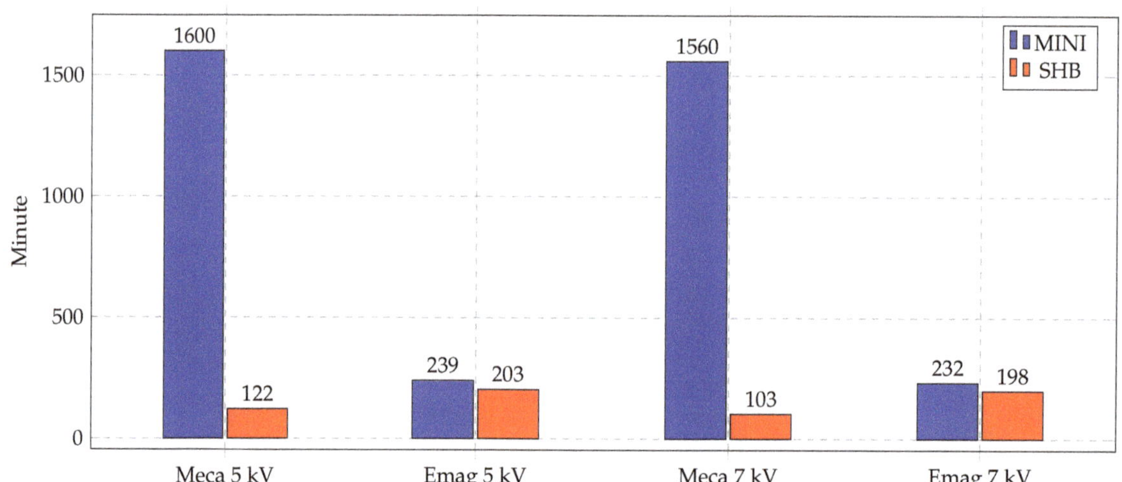

Figure 16. Simulation time of indirect forming process.

Finding a way to reduce the simulation time of the magnetic pulse forming is tremendously important, especially in the optimization processes. Optimization iterations have to be run to optimize the shape of the workpiece, study the effect of the workpiece's thickness and carry out the material parameter identification process. This is considered a crucial milestone in the simulation of MPF processes and similar problems.

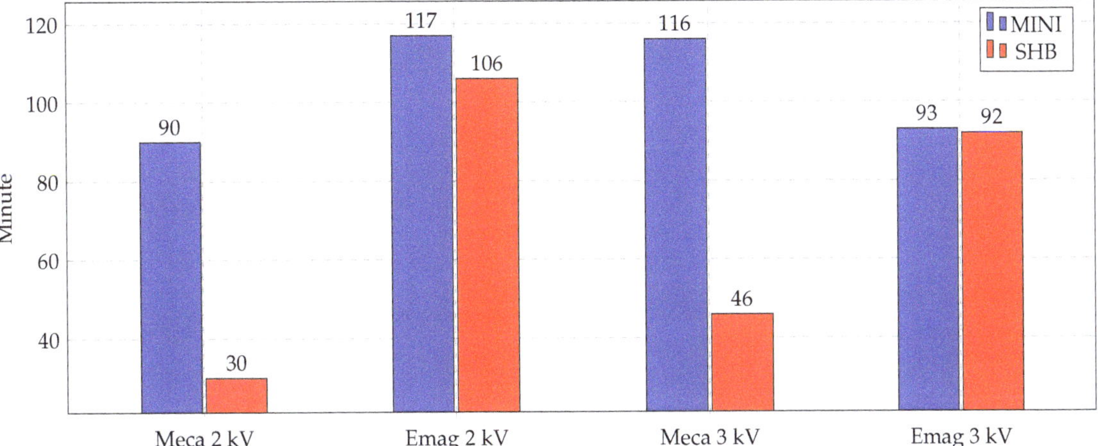

Figure 17. Simulation time of direct forming process of Al.

4. Discussion

This section is dedicated to giving a better insight of the results, along with discussing other results that explain the physical intuition of the process. Nevertheless, some challenges have been encountered while simulating the problem. Therefore, some of these problems are mentioned in the following subsections along with the proposed solutions. These problems include determining the final forming time, remeshing of the electromagnetic domain mesh and the effect of the mechanical mesh on the results. In the following subsections, the energy notion will be used to explain some of the difficulties that have been encountered while solving the problem. Therefore, it is important to explain the energy components that exist in this problem.

The initial energy input in the electromagnetic system is as follows:

$$E_{in} = \frac{1}{2} \cdot C_{ele} \cdot V^2 \tag{31}$$

where E_{in} is the input energy, C_{ele} is the electric capacitance of the coil and V is the voltage applied to the coil.

This energy is equal to the total energy that exists in the system:

$$\begin{aligned} E_{total} &= E_{elec} + E_{therm} + E_{Meca} \\ E_{elec} &= 0 \\ E_{therm} &= 0 \end{aligned} \tag{32}$$

In our case, we are not considering the E_{elec} that is the dissipated energy due to electric resistance of the coil. Moreover, E_{therm} is neglected, which is the dissipated thermal energy in the coil and in the mechanical system. This leaves only E_{Meca} that is represented by :

$$\begin{aligned} E_{Meca} &= E_{el} + E_{pl} + E_{kin} \\ E_{el} &= \int \sigma : \dot{\varepsilon}^{el} dt \\ E_{pl} &= \int \sigma : \dot{\varepsilon}^{pl} dt \\ E_{kin} &= \frac{1}{2} \rho \cdot v(x,t) : v(x,t) \end{aligned} \tag{33}$$

where E_{el}, E_{pl}, E_{kin} are the elastic strain energy, plastic strain energy and kinetic energy, respectively.

4.1. Final Forming Time

One of the main issues with all these models is determining the final forming time. At the beginning, the simulation time was set to 150 μs. However, the exact termination time of the process could not be determined and, by setting a too small value, we obtained an intermediate displacement profile, not the final one. Figure 18 shows a set of transient displacement profiles at different time steps. Therefore, a second stop condition has been added. It is induced from the calculation of a variation of the total energy of the system between the two last time steps. From this point of view, it is possible to set a threshold value at which calculation stops once reached. Otherwise, the recorded deformation will be a transient state and will not represent the real solution.

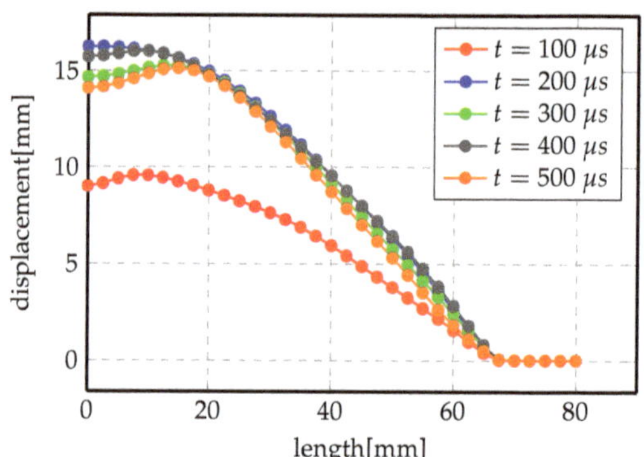

Figure 18. Displacement profile of direct forming process of Al at 3 kV.

To better understand this, Figure 19 shows the evolution of the mechanical energy (strain energy and kinetic energy) of the system in the indirect forming problem. The total energy increases at the beginning of the process in which the electromagnetic energy is maximum (electromagnetic energy represents the total energy in the system, it is then transformed to mechanical energy due to Lorentz force and thermal energy in the system and some loss as electrical energy in the coil). After some time, the system stops acquiring energy from the electromagnetic system and mechanical energy stays steady for the rest of the simulation time (highlighted in red dashed rectangle). The final simulation time is taken once the steady state value has been reached.

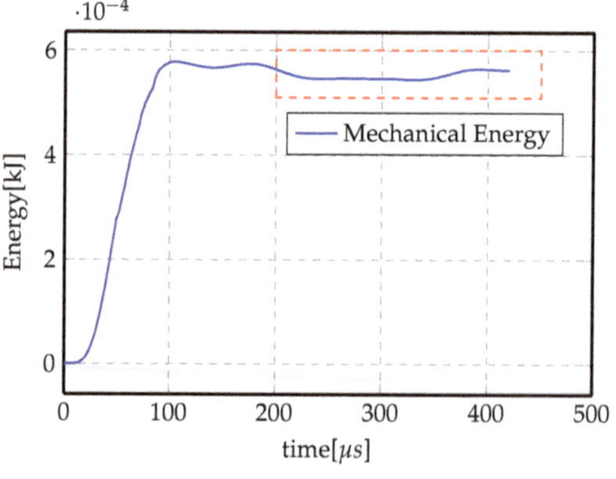

Figure 19. Total mechanical energy evolution with forming time.

4.2. Remeshing of Electromagnetic Domain Mesh

At the beginning, the numerical simulations were carried out using a fixed but fine mesh in the electromagnetic domain, as shown in Figure 20, since it is more convenient and more computationally efficient as remeshing takes more computation time. Although the results were not very satisfactory, by checking the energy transferred between the electromagnetic mesh and the mechanical mesh shown in Figure 21a, it is obvious that at the end, there is energy loss.

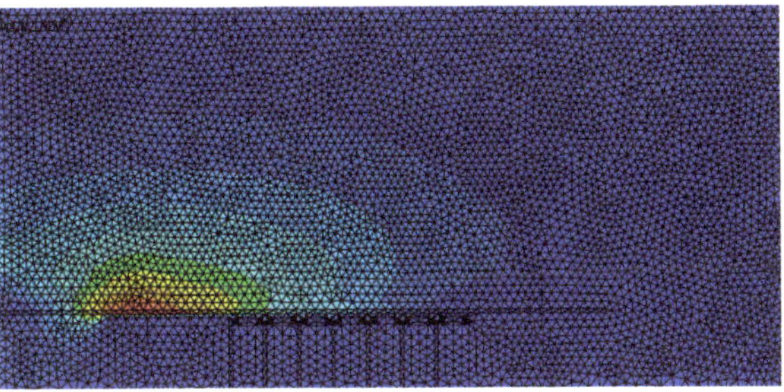

Figure 20. Electromagnetic mesh without remeshing.

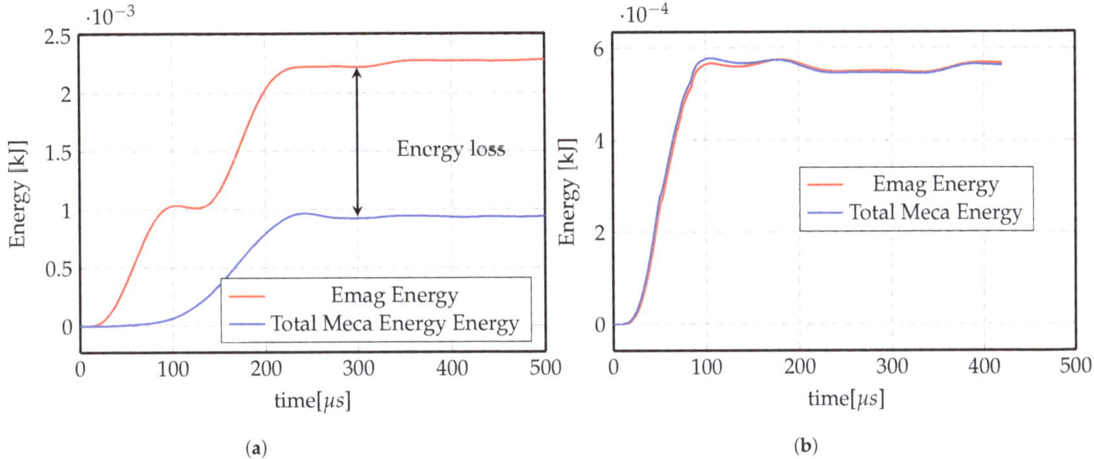

Figure 21. Electromagnetic energy vs. total mechanical energy of Al direct forming at 3 kV. (**a**) No Remeshing. (**b**) Remeshing.

On the other hand, Figure 21b shows the energy transferred with activating remeshing and the energy loss is negligible. The remeshing algorithm checks the deformation of the workpiece in the mechanical simulation and remeshes the surrounding of the workpiece in the electromagnetic simulation. Once the elements around the workpiece are highly deformed, the remesher refines these elements to maintain good mesh quality. This technique ensures that the mesh around the workpiece is always fine and clean, and thus guarantees correct calculations of the electromagnetic field and Lorentz forces, preventing the loss of energy. Figure 22 shows the electromagnetic mesh with activating remeshing before and after remeshing process.

Therefore, all the results in the previous subsections are obtained with remeshing activated. Although this increased the total simulation time, the obtained results can be trusted and a negligible amount of energy is lost.

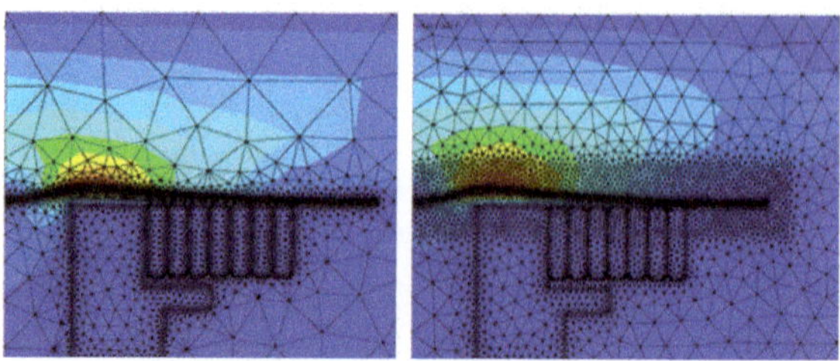

Figure 22. Electromagnetic mesh with remeshing.

4.3. Effect of Mechanical Mesh Refinement

One of the questions that was intriguing while studying this process was the effect of mechanical mesh on the accuracy of the results. Thus, the indirect forming process has been simulated again with coarser mesh of the MINI element and its results have been compared to both the MINI fine mesh and SHB element. Figure 23 shows the displacement profile of the final deformed profiles for the previously mentioned elements and mesh sizes. The overall observation of the results is that displacement profiles are too close to be distinguishable. These results are not very comprehensible, since the MINI element should be stiffer than the SHB element and using a less number of elements should alter the displacement results. Moreover, in order to make sure that we have converged to mesh independent results, the simulation was repeated using more SHB elements, ≈3600 elements, as shown in the black curve. The difference between the red and the black curves are really small, meaning that the results obtained with the lower mesh size can be trusted.

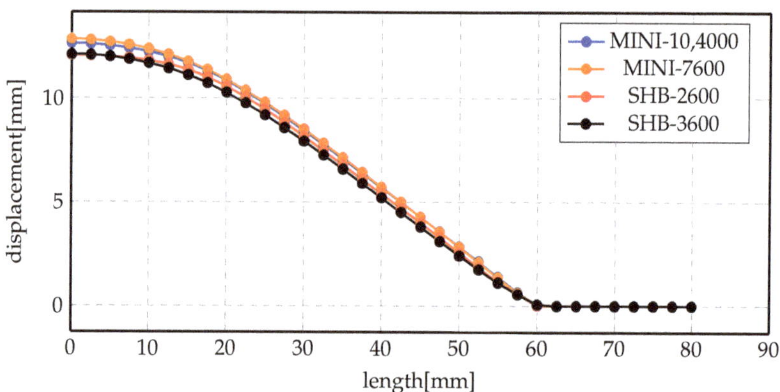

Figure 23. Displacement profile of indirect forming process at 7 kV using different elements and mesh sizes.

However, more insight has been given to the mechanical energy of the three cases. Figure 24 shows the total mechanical energy for the three mesh cases. It is fairly noticeable that the energy of the SHB coarse mesh is the lowest and the MINI element with fine mesh energy is approaching it with minute difference. However, the energy of the MINI coarse mesh has a quite higher value. This means that given almost the same final deformation profile, the energy required is smallest in the case of the SHB element, even with coarse mesh, while it takes higher energy to achieve the same displacement in the case of using the MINI element with coarse mesh, and the energy decreases approaching the energy of the SHB element by refining the mesh. This clearly indicates that the MINI element is stiffer than the SHB element and would need much finer mesh to achieve comparable results to the SHB element.

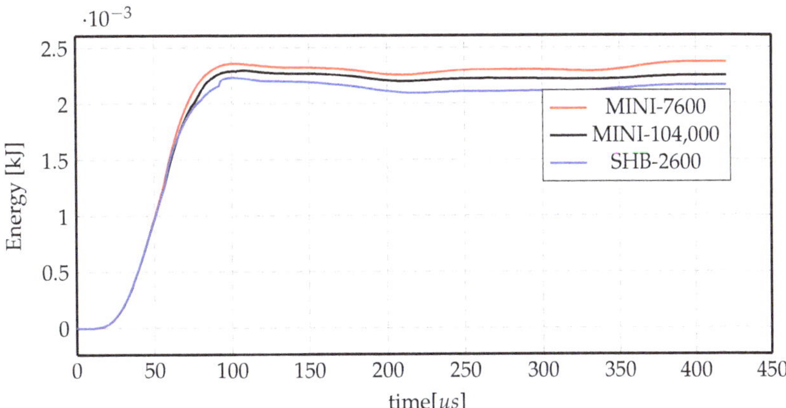

Figure 24. Total mechanical energy comparisons for different elements and mesh sizes for indirect forming at 7 kV.

5. Conclusions

An efficient approach for the simulation of the magnetic pulse forming process of thin sheet metals was developed by combining an electromagnetic solver, relying on Maxwell's equations, with a mechanical solver, based on the conservation of momentum equations. In-depth analysis of the types of the magnetic pulse forming processes was carried out, namely on direct forming and indirect forming. Tetrahedral element (MINI) and solid-shell prism element were employed to solve the mechanical problem and quantitative comparisons were carried out to assess their performance in such applications. The overall results showed that the accuracy obtained with a low resolution SHB approach (low number of elements) was comparable to that of a high resolution MINI element based technique (high number of elements). The SHB element was shown to be less stiff than the MINI element. Finally, a computational cost study was carried out and demonstrated a higher computational efficiency for the SHB element since a smaller number of elements could be used while maintaining comparable accuracy to that of the MINI element. These results are very promising to study the performance of the SHB element, not only in MPF application but also in other applications. Many challenges were encountered during the simulation of this multi-physics problem and some solutions to overcome them were devised. First, the final forming time. It was very challenging to determine the exact final simulation time since the deformation takes place in the order of magnitude of milliseconds. Consequently, criteria based on measuring the change in the mechanical energy variation were adopted to find the termination time at which the energy variation is minimum. Second, the electromagnetic mesh had to be adjusted to consider the new mechanical deformation since the electromagnetic solution is highly dependent on the spatio-temporal evolution of the deformation of the workpiece in the mechanical problem. Hence, a remeshing strategy was utilized for the electromagnetic mesh and the results were compared to those conducted without the remeshing step.

Author Contributions: Conceptualization, M.M. and D.P.M.; formal analysis, M.M. and F.B.; project administration, D.P.M.; software, M.M.; supervision, F.B. and D.P.M.; validation, M.M. and D.P.M.; visualization, M.M.; writing—original draft, M.M.; writing—review and editing, F.B. All authors have read and agreed to the published version of the manuscript.

Funding: The APC was funded by Armines, France.

Institutional Review Board Statement: Not applicable.

Informed Consent Statement: Not applicable.

Conflicts of Interest: The authors declare no conflict of interest.

References

1. Mynors, D.J.; Zhang, B. Applications and capabilities of explosive forming. *J. Mater. Process. Technol.* **2002**, *125*, 1–25. [CrossRef]
2. Rohatgi, A.; Stephens, E.V.; Davies, R.W.; Smith, M.T.; Soulami, A.; Ahzi, S. Electro-hydraulic forming of sheet metals: Free-forming vs. conical-die forming. *J. Mater. Process. Technol.* **2012**, *212*, 1070–1079. [CrossRef]
3. Zittel, G. *A Historical Review of High Speed Metal Forming*; Institute für Forming Technology-Technical Universityät: Dortmund, Germany, 2010.
4. Dowell, E.H.; Hall, K.C. Modeling of fluid-structure interaction. *Annu. Rev. Fluid Mech.* **2001**, *33*, 445–490. [CrossRef]
5. Keyes, D.E.; McInnes, L.C.; Woodward, C.; Gropp, W.; Myra, E.; Pernice, M.; Bell, J.; Brown, J.; Clo, A.; Connors, J.; et al. Multiphysics simulations: Challenges and opportunities. *Int. J. High Perform. Comput. Appl.* **2013**, *27*, 4–83. [CrossRef]
6. Singhal, A.; Sahu, S.A.; Chaudhary, S. Liouville-Green approximation: An analytical approach to study the elastic waves vibrations in composite structure of piezo material. *Compos. Struct.* **2018**, *184*, 714–727. [CrossRef]
7. Alves Z, J.R.; Bay, F. Magnetic pulse forming: Simulation and experiments for high-speed forming processes. *Adv. Mater. Process. Technol.* **2015**, *1*, 560–576. [CrossRef]
8. Psyk, V.; Risch, D.; Kinsey, B.L.; Tekkaya, A.E.; Kleiner, M. Electromagnetic forming—A review. *J. Mater. Process. Technol.* **2011**, *211*, 787–829. [CrossRef]
9. Psyk, V.; Beerwald, C.; Henselek, A.; Homberg, W.; Brosius, A.; Kleiner, M. Integration of electromagnetic calibration into the deep drawing process of an industrial demonstrator part. In *Key Engineering Materials*; Trans Tech Publications Ltd.: Bäch SZ, Switzerland, 2007; Volume 344, pp. 435–442.
10. Unger, J.; Stiemer, M.; Schwarze, M.; Svendsen, B.; Blum, H.; Reese, S. Strategies for 3D simulation of electromagnetic forming processes. *J. Mater. Process. Technol.* **2008**, *199*, 341–362. [CrossRef]
11. Fenton, G.K.; Daehn, G.S. Modeling of electromagnetically formed sheet metal. *J. Mater. Process. Technol.* **1998**, *75*, 6–16. [CrossRef]
12. Imbert, J.; Winkler, S.; Worswick, M.; Oliveira, D.; Golovashchenko, S. Numerical modeling of an electromagnetic corner fill operation. In Proceedings of the NUMIFORM, Columbus, OH, USA, 13–17 June 2004; pp. 1833–1839.
13. Schinnerl, M.; Schoberl, J.; Kaltenbacher, M.; Lerch, R. Multigrid methods for the three-dimensional simulation of nonlinear magnetomechanical systems. *IEEE Trans. Magn.* **2002**, *38*, 1497–1511. [CrossRef]
14. Heyliger, P.; Reddy, J. A higher order beam finite element for bending and vibration problems. *J. Sound Vib.* **1988**, *126*, 309–326. [CrossRef]
15. Simo, J.C.; Armero, F. Geometrically non-linear enhanced strain mixed methods and the method of incompatible modes. *Int. J. Numer. Methods Eng.* **1992**, *33*, 1413–1449. [CrossRef]
16. Simo, J.; Armero, F.; Taylor, R. Improved versions of assumed enhanced strain tri-linear elements for 3D finite deformation problems. *Comput. Methods Appl. Mech. Eng.* **1993**, *110*, 359–386. [CrossRef]
17. Belytschko, T.; Bindeman, L.P. Assumed strain stabilization of the eight node hexahedral element. *Comput. Methods Appl. Mech. Eng.* **1993**, *105*, 225–260. [CrossRef]
18. Liu, W.K.; Guo, Y.; Tang, S.; Belytschko, T. A multiple-quadrature eight-node hexahedral finite element for large deformation elastoplastic analysis. *Comput. Methods Appl. Mech. Eng.* **1998**, *154*, 69–132. [CrossRef]
19. Reese, S.; Wriggers, P.; Reddy, B. A new locking-free brick element technique for large deformation problems in elasticity. *Comput. Struct.* **2000**, *75*, 291–304. [CrossRef]
20. Trinh, V.D.; Abed-Meraim, F.; Combescure, A. A new assumed strain solid-shell formulation "SHB6" for the six-node prismatic finite element. *J. Mech. Sci. Technol.* **2011**, *25*, 2345. [CrossRef]
21. Abed-Meraim, F.; Combescure, A. An improved assumed strain solid–shell element formulation with physical stabilization for geometric non-linear applications and elastic–plastic stability analysis. *Int. J. Numer. Methods Eng.* **2009**, *80*, 1640–1686. [CrossRef]
22. Reese, S. A large deformation solid-shell concept based on reduced integration with hourglass stabilization. *Int. J. Numer. Methods Eng.* **2007**, *69*, 1671–1716. [CrossRef]
23. L'Eplattenier, P.; Çaldichoury, I. Update on the electromagnetism module in LS-DYNA. In Proceedings of the 12th LS-DYNA Users Conference, Detroit, MI, USA, 3–5 June 2012.
24. Brebbia, C.A.; Dominguez, J. *Boundary Elements: An Introductory Course*; WIT Press: The New Forest, UK, 1994.
25. Bahmani, M.A.; Niayesh, K.; Karimi, A. 3D Simulation of magnetic field distribution in electromagnetic forming systems with field-shaper. *J. Mater. Process. Technol.* **2009**, *209*, 2295–2301. [CrossRef]
26. Nédélec, J.C. A new family of mixed finite elements in R^3. *Numer. Math.* **1986**, *50*, 57–81. [CrossRef]
27. Alves Zapata, J. Magnetic Pulse Forming Processes: Computational Modelling and Experimental Validation. Ph.D. Thesis, Ecole Nationale Supérieure des Mines de Paris, Paris, France, 2016.
28. Reese, S.; Svendsen, B.; Stiemer, M.; Unger, J.; Schwarze, M.; Blum, H. On a new finite element technology for electromagnetic metal forming processes. *Arch. Appl. Mech.* **2005**, *74*, 834–845. [CrossRef]
29. Arnold, D.N.; Brezzi, F.; Fortin, M. A stable finite element for the stokes equations. *Calcolo* **1984**, *21*, 337–344. [CrossRef]
30. Mahmoud, M.; Bay, F.; Mũnoz, D.P. An efficient multiphysics solid shell based finite element approach for modeling thin sheet metal forming processes. *Finite Elem. Anal. Des.* **2021**, *198*, 103645. [CrossRef]

31. Bay, F.; Zapata, J.R.A. Computational modelling for electromagnetic forming processes. In Proceedings of the International Scientific Colloquium Modelling for Electromagnetic Processing-MEP 2014, Hannover, Germany, 16–19 September 2014; pp. 259–264.
32. Bay, F.; Jeanson, A.C.; Zapata, J.A. Electromagnetic forming processes: Material behaviour and computational modelling. *Procedia Eng.* **2014**, *81*, 793–800. [CrossRef]
33. Svendsen, B.; Chanda, T. Continuum thermodynamic formulation of models for electromagnetic thermoinelastic solids with application in electromagnetic metal forming. *Contin. Mech. Thermodyn.* **2005**, *17*, 1–16. [CrossRef]
34. Chari, M.; Konrad, A.; Palmo, M.; D'angelo, J. Three-dimensional vector potential analysis for machine field problems. *IEEE Trans. Magn.* **1982**, *18*, 436–446. [CrossRef]
35. Biro, O.; Preis, K. On the use of the magnetic vector potential in the finite-element analysis of three-dimensional eddy currents. *IEEE Trans. Magn.* **1989**, *25*, 3145–3159. [CrossRef]
36. Brezzi, F.; Fortin, M. *Mixed and Hybrid Finite Element Methods*; Springer Science & Business Media: Berlin, Germany, 2012; Volume 15.
37. Fayolle, S. Etude de la Modélisation de la Pose et de la Tenue Mécanique des Assemblages par Déformation Plastique. Ph.D. Thesis, Ecole Nationale Supérieure des Mines de Paris, Paris, France, 2009.
38. Risch, D.; Beerwald, C.; Brosius, A.; Kleiner, M. On the significance of the die design for electromagnetic sheet metal forming. In Proceedings of the 1st International Conference on High Speed Forming, Dortmund, Germany, 31 March–1 April 2004; pp. 191–200.
39. Jeanson, A.C.; Avrillaud, G.; Mazars, G.; Bay, F.; Massoni, E.; Jacques, N.; Arrigoni, M. Identification du comportement mécanique dynamique de tubes d'aluminium par un essai d'expansion électromagnétique. In *CSMA 2013-11ème Colloque National en Calcul des Structures*; HAL: Giens, France, May 2013.
40. Johnson, J.R.; Taber, G.A.; Daehn, G.S. Constitutive relation development through the FIRE test. In Proceedings of the 4th International Conference on High Speed Forming, Columbus, OH, USA, 9–10 March 2010; pp. 295–306.

Article

Spring Back Behavior of Large Multi-Feature Thin-Walled Part in Rigid-Flexible Sequential Loading Forming Process

Yanfeng Zhang [1,2], Lihui Lang [1,*], Yao Wang [3,4,*], Haizhou Chen [2], Jianning Du [5], Zhihui Jiao [1] and Lin Wang [5]

1 School of Mechanical Engineering and Automation, Beihang University, Beijing 100191, China; 18622709176@163.com (Y.Z.); 18740116811@163.com (Z.J.)
2 Tianjin Tianduan Press Co., Ltd., Tianjin 300142, China; chzit@126.com
3 School of Mechanical Engineering, Hebei University of Technology, Tianjin 300130, China
4 State Key Laboratory of Materials Processing and Die & Mould Technology, Huazhong University of Science and Technology, Wuhan 430074, China
5 Shenyang Aircraft Industry (Group) Co., Ltd., Shenyang 110850, China; djn1221@163.com (J.D.); 18722226453@163.com (L.W.)
* Correspondence: lang@buaa.edu.cn (L.L.); bhwy2014@126.com (Y.W.)

Abstract: The spring back behavior of large complex multi-feature parts in the rigid-flexible sequential forming process has certain special characteristics. The hydraulic pressure loading locus has a significant influence on the spring back of small features of the part, and the applicability of the spring back prediction model to the process needs further research. Therefore, this paper takes the large aluminum alloy inner panel of an automobile as the research object, and the spring back model and the influence laws of the hydraulic pressure loading locus are revealed by combining the theoretical analysis and numerical simulation with the process tests. Meanwhile, based on the theoretical prediction and experimental results, the spring back compensation of the complex inner panel is carried out. Results show that the hardening model has a greater impact on the accuracy of spring back prediction than the yield criterion does, and the prediction accuracy of Barlat'89 + Yoshida–Uemori mixed hardening model is the highest. Finally, the optimized loading locus of hydraulic pressure is obtained, and the accuracy results of the compensated parts verify the accuracy of the analysis model.

Keywords: spring back behavior; prediction model; large multi-feature thin-walled part; hydroforming; rigid-flexible sequential loading forming

Citation: Zhang, Y.; Lang, L.; Wang, Y.; Chen, H.; Du, J.; Jiao, Z.; Wang, L. Spring Back Behavior of Large Multi-Feature Thin-Walled Part in Rigid-Flexible Sequential Loading Forming Process. *Materials* **2022**, *15*, 2608. https://doi.org/10.3390/ma15072608

Academic Editors: Mateusz Kopec and Denis Politis

Received: 18 January 2022
Accepted: 11 March 2022
Published: 1 April 2022

1. Introduction

The spring back problem has always been a hotspot in the field of sheet metal forming. Since the last century, scholars have carried out a large number of fruitful and effective studies on the spring back of sheet metal. The research methods include the analysis of spring back theory of typical simple shape, the establishment of spring back prediction model, and the analysis by combing numerical simulation with experiments [1–3]. For the spring back research of ordinary steel sheet parts such as carbon steel, stainless steel, and automobile steel, we have accumulated a wealth of experimental data and have mature research methods. However, for aluminum alloys, especially large, multi-featured, aluminum alloy, thin-walled parts, due to the complexity of deformation process and the particularity of the process, the spring back data have not accumulated in large quantities, and there is a lack of precise spring back behavior research and prediction model analysis for these kinds of material parts [4–7].

In the manufacturing of large, complex, multi-feature, thin-walled parts, the hydroforming technology has great advantages. Due to the uniform load of the high-pressure fluid, the deformation degree and uniformity of the blank are great, which can effectively reduce the spring back and improve the forming performance. However, for the precise hydroforming of complex, small, rounded corners and small features, it is necessary to load a large hydraulic pressure, which will lead to an increase in the required tonnage of the equipment. Meanwhile, it is very difficult for the part of the blank that has been molded to feed into the deformation zone. The blank in the deformation zone can only depend on the thinning of its wall thickness to adhere to the die. It is very easy to cause excessive thinning or even cracks [8–10]. In order to solve this problem, the authors propose a rigidflexible sequential loading forming process, which takes advantage of the dual technology of hydroforming and rigid forming, and sequentially forms the overall features and local small features of parts. The selection of the hydraulic pressure loading locus has an important influence on the spring back of the local features. Meanwhile, the accuracy of the spring back prediction results is also largely determined by the expression of the mechanical properties of the material. The anisotropy and Bauschinger effect are two important aspects that directly affect the accuracy of spring back prediction. The anisotropy of the material is closely related to the yield criterion chosen, while the Bauschinger effect is directly related to the plastic hardening model chosen [11–13]. Therefore, the coupling matching between the yield criterion and the plastic hardening model is of great significance for spring back prediction.

At present, more studies focus on the bending deformation of aluminum alloys and the spring back of small simple parts. There are few studies on the spring back behavior and prediction models of large, complex, multi-feature parts. Appiah et al. [14] studied the U-bending process of aluminum alloy AA6111-T4. The prediction model used was based on the Armstrong–Frederick model, and a new hybrid hardening model was proposed. The research results showed that the spring back is inversely related to the residual stress. Kinzel et al. [15] studied the spring back behavior of aluminum alloys AA6022-T4 and AA6111-T4, and verified the accuracy of the proposed spring back prediction model considering the Bauschinger effect. Tamura et al. [16] studied the spring back behavior of aluminum alloy AA5052-O and AA6016-T4 sheets, and discussed the planar anisotropy, Bauschinger effect, and cyclic hardening properties of the materials under the framework of Yoshida–Uemori hardening model and anisotropic yield criterion. Zhang et al. [17] used the Numisheet'93 U-bending test as an example to calculate the effect of different hardening models on the spring back of aluminum alloy sheets. The results showed that the spring back amount predicted by the isotropic hardening model is too large, and that predicted by the linear kinematic hardening model is small, but the proposed nonlinear mixed hardening model has high accuracy for predicting the spring back of a sheet under complex loading.

The spring back behavior of large, complex, multi-feature parts in the rigid-flexible sequential forming process has certain special characteristics. The hydraulic pressure loading locus has an important influence on the spring back of small features of the part, and the applicability of spring back prediction model to the process needs further research. Therefore, this paper takes the large aluminum alloy inner panel of an automobile as the research object, the influence laws of the hydraulic pressure loading locus and spring back model are revealed by combining the theoretical analysis and numerical simulation with the process tests. Meanwhile, based on the theoretical prediction and experimental results, the spring back compensation of the complex inner panel is carried out. The accuracy results of the compensated parts verify the accuracy of the analysis model.

2. Analysis of Rigid-Flexible Sequential Loading Forming Process

2.1. Part Material and Spring Back Characteristics

The inner panel of the engine hood and the key local features studied in this paper are shown in Figure 1. Its shape dimensions are 1378.41 mm in length, 481.71 mm in width, and 81.05 mm in height. It has the characteristics of large overall size, small local features, and complex shape. The minimum radius of rounded corners is 2.0 mm. For the four key local features shown in Figure 1, their forming method adopted rigid-flexible sequential forming, that is, the local rigid shaping was performed by hydroforming at the mold closing position at the bottom of die. The material of the part is aluminum alloy 5182-O, the thickness is 1.0 mm, and its mechanical property was obtained by the uniaxial tensile test. The sample, with the width D 12.46 mm, was processed according to the shape and size of Figure 2 and relevant requirements of GB/T228.1-2010 and was selected in three directions, which are 0°, 45°, and 90°, respectively, with the rolling direction. In order to avoid the contingency of the test, two specimens were prepared for tensile test in each direction. The tensile strain rate was 0.005 s^{-1}. The test results are in Table 1.

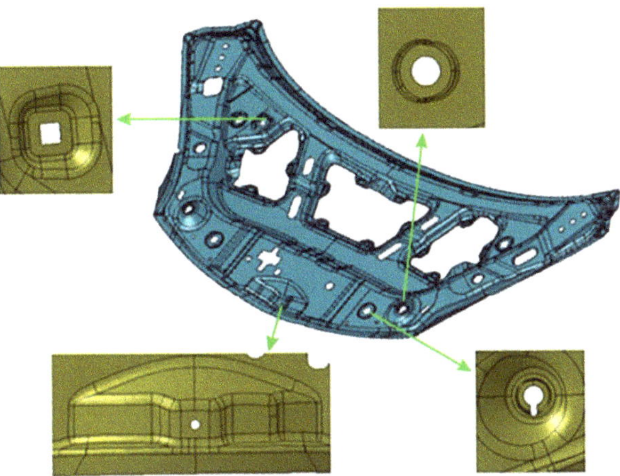

Figure 1. The inner panel of the engine hood and the key local features.

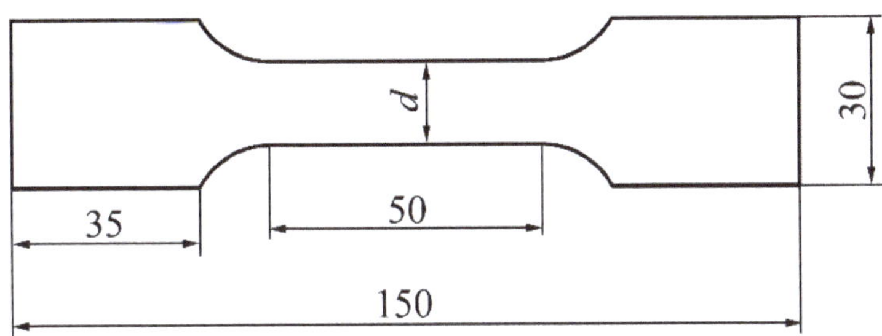

Figure 2. Geometry of tensile specimen.

Table 1. Mechanical properties of aluminum alloy 5182-O.

Parameters	Value
Yield strength, σ_{s0}/MPa	119
Yield strength, σ_{s45}/MPa	118
Yield strength, σ_{s90}/MPa	116
Tensile strength, σ_{b0}/MPa	271
Tensile strength, σ_{b45}/MPa	266
Tensile strength, σ_{b90}/MPa	264
Elongation rate, δ_0/%	28
Elongation rate, δ_{45}/δ_{90}/%	28.5
Work-hardening exponent, n	0.34
Anisotropy coefficient, r_0	0.73
Anisotropy coefficient, r_{45}	0.65
Anisotropy coefficient, r_{90}	0.6

For this complex, multi-feature inner panel, the main detection positions of spring back are at the edge of contour and inner hole, which are the main positions of edge wrapping and assembly, and need to meet tolerance requirements. The edge is close to the punch nose, and the material inflow must undergo bending–reverse bending deformation, which is susceptible to cyclic hardening, and the Bausinger effect is strong. The inner hole edge also undergoes bending–reverse bending deformation in the hydroforming process, and the Bauschinger effect is obvious, the spring back can be reduced by improving the hydraulic pressure loading locus.

2.2. Rigid-Flexible Sequential Loading Forming

Figure 3 shows the principle of the rigid-flexible sequential loading forming process. The sheet metal undergoes hydroforming and precise clamping and shaping of local rigid inserts set at the bottom of the liquid chamber, so that the parts not formed in the hydroforming stage are formed in place, and the sequential precise forming is achieved. The spring back behavior of the part is affected by the hydraulic pressure loading locus and presents complex nonlinear characteristics. The initial hydraulic pressure should not be too large, otherwise the friction between blank and punch will keep a higher level, which is not conducive to material flow, resulting in more storage and serious spring back. Based on this, the initial hydraulic pressure should be within a range, which is represented by $P_{\text{Initial-min}}$ and $P_{\text{Initial-max}}$ in Figure 4. Meanwhile, the hydraulic pressure should not be prematurely loaded. If the loading is too early, the sheet will be intimately attached to the sharp features on the punch, which will result in uneven flow of the blank and a large spring back after cutting. Figure 4 shows that the hydraulic pressure begins to be loaded after the punch stroke reaches S_0. The maximum hydraulic pressure in the later forming stage also plays an important role, too large or too small will have a negative impact on spring back. The range of maximum hydraulic pressure is expressed by $P_{\text{Later-min}}$ and $P_{\text{Later-max}}$ in Figure 4.

2.3. Spring Back Measuring Equipment and Scheme

In order to obtain the original spring back data of aluminum alloy inner panel in the rigid-flexible sequential forming, the spring back compensation was not carried out in advance. The initial values of spring back of the specimen profile were measured and obtained after forming and cutting. Based on the initial measured values and theoretical prediction results of spring back, the specimen was subjected to compensation, the mold was modified, and a new round of forming, cutting, and spring back measurement was performed. By comparison, the accuracy of the theoretical model was verified.

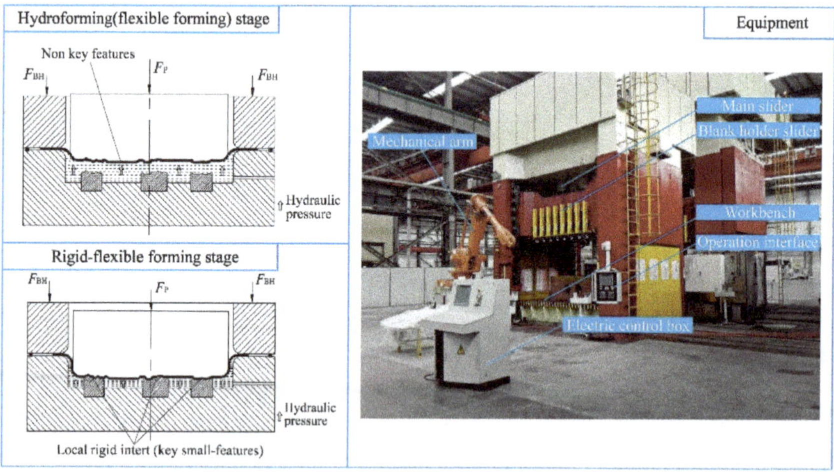

Figure 3. Principle of rigid-flexible sequential loading forming and equipment.

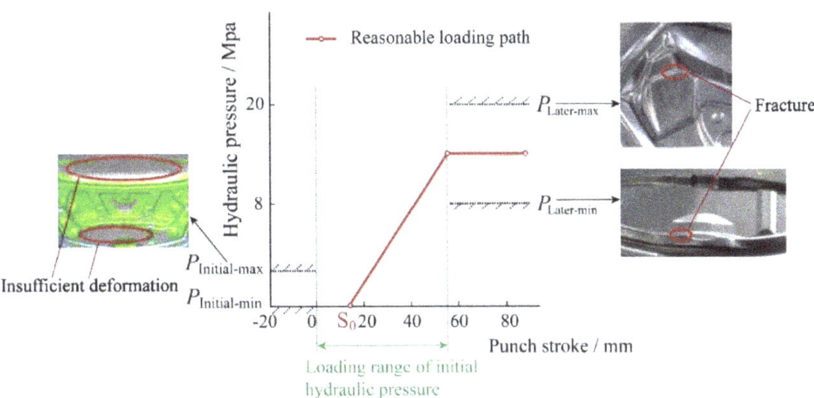

Figure 4. Range of hydraulic pressure loading path.

The measuring device with accuracy of 0.02 mm and reference of spring back is shown in Figure 5. The measurement value was obtained by comparing the cut profile with the checking tool profile. S1, S2, S3, S4, S5, S6, S7, and S8 are reference planes, respectively, with tolerances of 0–0.2 mm. H and h are reference hole positions, and their coordinates are H (−155, 435, 735), h (−416, −260, 659). Both reference holes have the nominal diameters of 20 mm and the diameter tolerances of 0–0.1 mm. The spring back measurement positions of the profile are shown in Figure 6, which are 5 position areas of 4S, 5S, 6S, 7S, and 8S, respectively. Twenty-five points were measured in 4S area, 14 points were measured in 5S area, 25 points were measured in 6S area, 16 points were measured in 7S area, and 8 points were measured in 8S area. Among them, the spring back tolerance of measuring points in 4S and 5S areas was ±0.5 mm, and that of measuring points in 6S, 7S, and 8S areas was ±0.7 mm. There were two specimens measured before and after spring back compensation.

(a)

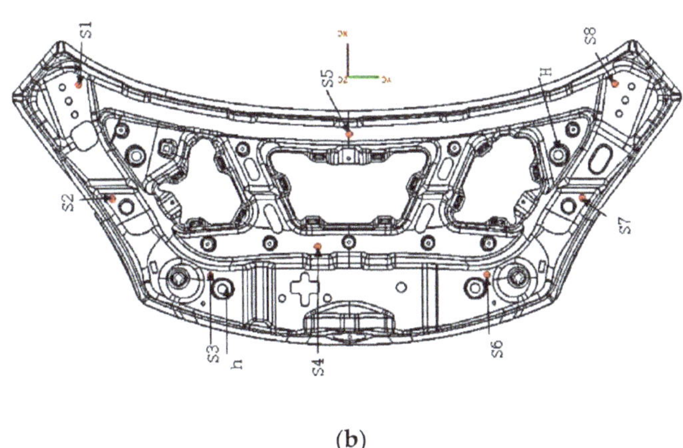

(b)

Figure 5. Measurement device and reference of spring back. (**a**) Three-coordinates measuring machine; (**b**) Measurement reference.

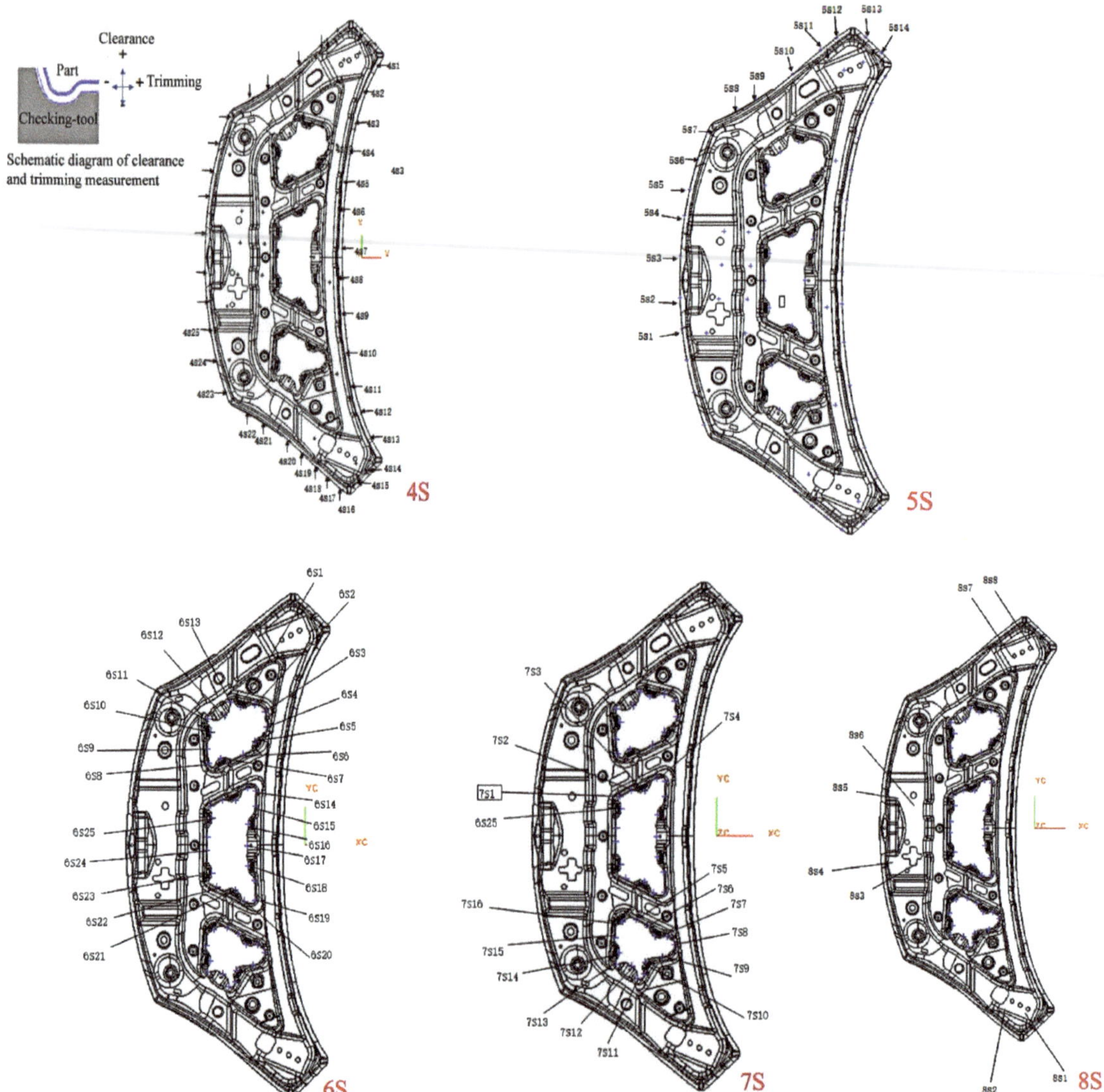

Figure 6. The spring back measurement positions of the profile. The 4S–8S arears are the measurement positions of spring back. Each area measures different points.

3. Spring Back Prediction Model

The accuracy of spring back prediction results for rigid-flexible sequential forming of the large, multi-feature part depends largely on the expression of mechanical properties of materials, except for the key process parameters. The anisotropy and Bauschinger effect are two important aspects that directly affect the accuracy of spring back prediction [18–21]. The aluminum alloy sheets used in the rigid-flexible sequential forming are mostly obtained by repeated rolling and heat treatment. They have obvious fiber structure, crystalline selective orientation, and significant anisotropy. The anisotropy of materials is closely related to the yield criteria selected. Therefore, the accurate selection of yield criteria is of great significance for the numerical simulation of spring back. At present, the commonly used anisotropic yield criteria are Hill'48 anisotropic yield criterion and Barlat'89 anisotropic yield criterion.

Based on the Mises yield criterion, Hill proposed the anisotropic yield condition in 1948 [22]:

$$f = \sqrt{F(\sigma_2 - \sigma_3)^2 + G(\sigma_3 - \sigma_1)^2 + H(\sigma_1 - \sigma_2)^2 + 2L\sigma_{23}^2 + 2M\sigma_{31}^2 + 2N\sigma_{12}^2} = \overline{\sigma} \quad (1)$$

where F, G, H, L, M, and N are the anisotropic parameters of the material.

The Lankford constant is defined as follows:

$$r_0 = \frac{H}{G}, r_{45} = \frac{2N - (F+G)}{2(F+G)}, r_{90} = \frac{H}{F} \quad (2)$$

Therefore, the relationship between the anisotropy parameters and the Lankford constant is as follows:

$$F = \frac{r_0}{(1+r_0)r_{90}}, G = \frac{1}{1+r_0}, H = \frac{r_0}{1+r_0}, N = \frac{(r_0 + r_{90})(1+r_{45})}{2r_{90}(1+r_0)} \quad (3)$$

The L and M cannot be measured, so $L = M = N$ can be considered. The anisotropic parameters in Hill'48 yield function for 5182-O aluminum alloy are $F = 0.703$, $G = 0.578$, $H = 0.422$, and $N = 1.057$.

In 1989, Barlat et al. proposed a new yield criterion based on the in-plane anisotropy of materials. The yield criterion has been widely used because of its good calculation results for the stress–strain curves of aluminum alloy under biaxial tension. The equation is as follows [23]:

$$\overline{\sigma} = \left\{\frac{a}{2}\left[(\sigma_1 - \sigma_3)^m + (-h)^m(\sigma_2 - \sigma_3)^m\right] + \left(1 - \frac{a}{2}\right)(\sigma_1 - \sigma_2 - 2\sigma_3)^m\right\}^{1/m} \quad (4)$$

where m is a parameter related to the crystal structure, $m = 6$ when the material is a body-centered cube, and $m = 8$ when the material is a face-centered cube. a and h are the anisotropic constants. For the 5182-O aluminum alloy, $m = 8$, $a = 1.204$, and $h = 1.061$.

The plastic hardening model is used to describe the variation of the subsequent yield function in the stress space after the material enters the plastic stage. It is related to the stress state, plastic strain, and hardening parameters of the material. For plastic reinforced materials, the commonly used hardening models are the isotropic hardening model, kinematic hardening model, and mixed hardening model.

Another important factor affecting spring back prediction accuracy of rigid-flexible sequential forming of aluminum alloy sheets is the Bauschinger effect. Especially in the forming process of complex, multi-feature inner panel, the material will repeatedly undergo forward loading, reverse loading, unloading, and other states, and the yield strength will be reduced. The Bauschinger effect is obvious. Therefore, the Bauschinger effect should be taken into account when choosing the hardening model. In the mixed hardening model, Yoshida–Uemori model is commonly used in many finite element analysis software programs. The model establishes a two-sided model for cyclic loading conditions, which introduces a boundary surface based on the yield surface, defines the isotropic hardening and kinematic hardening model in the boundary surface, and establishes the anisotropic characteristics of the material. The isotropic hardening model describes the work hardening phenomenon in the deformation process of the material, and the kinematic hardening model describes the reverse softening phenomenon, so the boundary surface is mixed hardening. The yield surface is rigidly translated in the boundary plane, which is the kinematic hardening. The model also effectively describes the Bauschinger effect [24,25].

The relative relationship between the yield surface and the boundary surface of the Yoshida–Uemori hardening model is as follow [26]:

$$\alpha = \alpha^* - \beta \quad (5)$$

$$\alpha = B + R - Y \tag{6}$$

where α is the back stress describing the hardening of the yield surface. β is the back stress describing the hardening of boundary surface. α^* is the difference between back stress α and β (MPa). B is the initial value of boundary surface (MPa). Y is the initial value of yield surface (MPa). R is the isotropic strengthening stress.

In the Yoshida–Uemori model, the boundary surface is mixed hardening. The expressions are as follows:

$$\text{Isotropic hardening}: \ \dot{R} = m(R_{sat} - R)\dot{p} \tag{7}$$

$$\text{Kinematic hardening}: \ \dot{\beta} = m\left(\frac{2}{3}bD^p - \beta\dot{p}\right) \tag{8}$$

where R_{sat} is the saturation value of the isotropic strengthening stress R (when the strain is infinite), m is the material isotropic strengthening rate parameter, and b is the material parameter. D^p and $\dot{p}$ are both the equivalent plastic strain rate.

Yoshida revised the original model to deal with the rapid change of work hardening rate after the initial yield in most cases. The parameters C_1 and C_2 are defined for the difference between the forward and reverse loading. The correction is as follows [27]:

$$R = R_{sat}\left[(C_1 + \bar{\varepsilon}^p)^{C_2} - {C_1}^{C_2}\right] \tag{9}$$

The revised Yoshida–Uemori hardening model has nine parameters: Y, B, C, R_{sat}, b, m, h, C_1, and C_2. For the 5182-O aluminum, alloy, these parameters, obtained by tests, are shown in Table 2.

Table 2. Parameters for the revised Yoshida–Uemori hardening model of aluminum alloy 5182-O.

Y/MPa	B/MPa	C	R_{sat}/MPa	b/MPa	m	h	C_1	C_2
135.62	149.523	432.685	172.853	40.271	10.8	0.25	0.036	0.38

The spring back prediction models were obtained by matching different yield criteria and plastic hardening models. For the large, complex, multi-feature inner panel, four spring back models were studied, including Barlat'89 + Hollomon isotropic hardening model, Hill'48 + Hollomon isotropic hardening model, Hill'48 + Yoshida–Uemori mixed hardening model, and Barlat'89 + Yoshida–Uemori mixed hardening model. Meanwhile, the applicability of spring back models to the rigid-flexible sequential loading forming process was verified.

4. Results and Discussion

4.1. Prediction Accuracy of Spring Back Model

From the perspective of plasticity theory, the complete material model should include the material flow stress relationship, the initial yield criterion, and the subsequent yield hardening model. In the simulation, the associated flow theory was used in the flow relation equation. The combination of different initial yield criteria and hardening models with the true stress–strain relationship of the material obtained by tensile tests formed different material models and different spring back prediction models. The rigid-flexible sequential forming process of aluminum alloy automobile inner panel is a complex plastic deformation process. The applicability of different spring back models to this process needs further study.

The comparisons between the prediction results of the four kinds of spring back models and the test results in different measurement areas are shown in Figure 7. It can be seen that the prediction results of spring back models are larger than the test values. The prediction accuracy of Barlat'89 + Yoshida–Uemori mixed hardening model is the highest and its consistency with the test results is the best. The second is the Hill'48 + Yoshida–Uemori mixed hardening model; the prediction accuracy of the spring back model combined with Hollomon isotropic hardening model is lower. The spring back predicted by isotropic hardening model is larger than that predicted by mixed hardening model, and the difference between the values of isotropic hardening model and test is also larger. When the isotropic hardening model is adopted, the prediction accuracy of the Hill'48 yield criterion is slightly higher than that of the Barlat'89 yield criterion. When the mixed hardening model is adopted, the prediction accuracy of the Barlat'89 yield criterion is significantly higher than that of Hill'48 yield criterion. In general, the hardening model has a greater impact on the accuracy of spring back prediction than the yield criterion does.

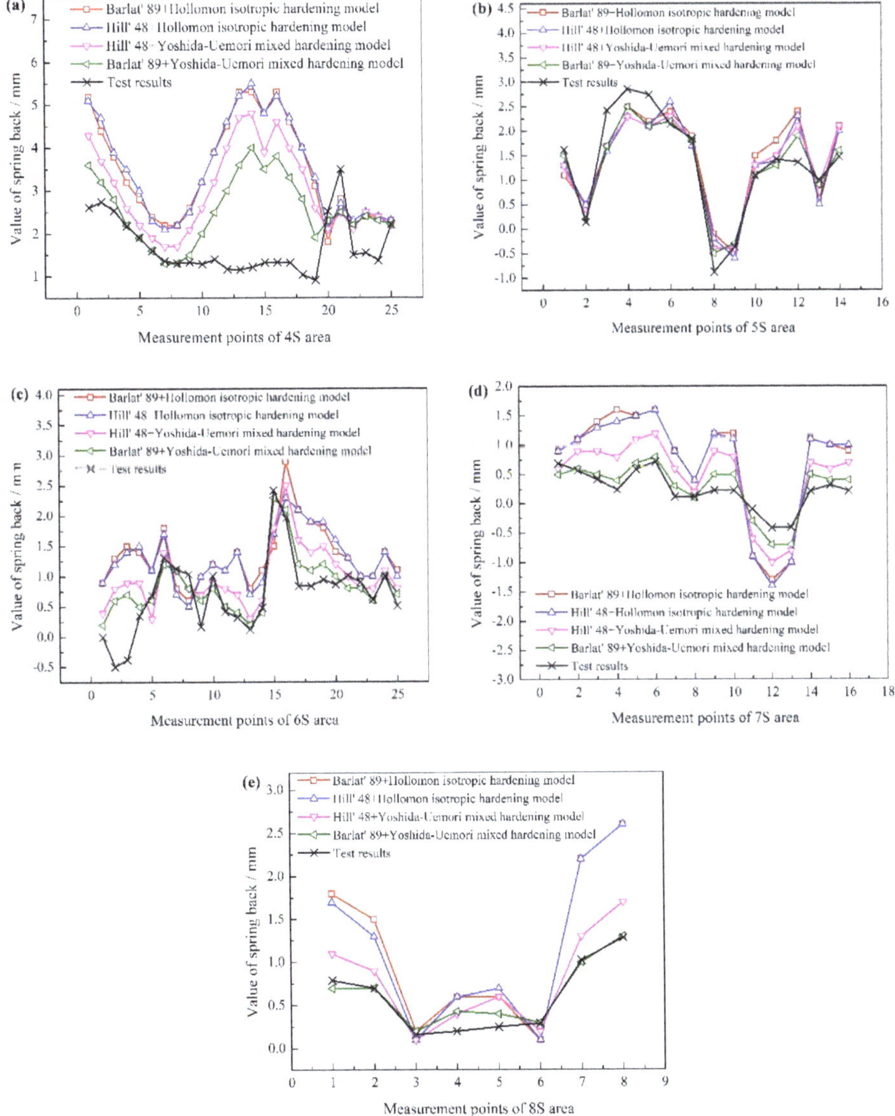

Figure 7. The comparisons between prediction results and the test results: (a) 4S area, (b) 5S area, (c) 6S area, (d) 7S area, and (e) 8S area.

In the 4S measurement area, the predicted trends of spring back model and the test results are basically the same, but the values are quite different. The maximum position of spring back measured by test is point 21, and the maximum value is 3.49 mm. The maximum positions of spring back obtained by the four prediction models are all at point 14. The maximum values are 5.3 mm, 5.5 mm, 4.8 mm, and 4.0 mm, respectively. The experimental value of spring back at point 14 is 1.21 mm, which is closest to the prediction result of Barlat'89 + Yoshida–Uemori mixed hardening model. However, the prediction values of four spring back models at point 21 are 2.8 mm, 2.7 mm, 2.5 mm, and 2.5 mm, respectively, which is closest to the prediction of Barlat'89 + Hollomon isotropic hardening model. The predicted values of spring back models in other measurement areas are in good agreement with the test values, and the prediction accuracy of Barlat'89 + Yoshida–Uemori mixed hardening model is the highest.

4.2. Effect of Hydraulic Pressure

The Barlat'89 + Yoshida–Uemori mixed hardening model was used to investigate the effect of hydraulic pressure on the spring back behavior of the part due to the fact that it has high accuracy for describing the deformation of aluminum alloy and the process of rigid-flexible sequential loading forming. The initial hydraulic pressure paths designed according to the analysis of Section 2.2 and the corresponding spring back values of different areas are depicted in Figure 8. It can be seen that the spring back values in different areas using loading path D are the largest. Because of the late loading of hydraulic pressure, more blank inflow led to inadequate deformation and large spring back after cutting. Meanwhile, the spring backs of specimens obtained by loading paths A and B with big initial hydraulic pressure are also large due to the high friction level between the blank and punch in initial hydroforming stage. In comparison, the loading paths E and F have the low spring back values and uniform wall thickness distribution. The suitable loading stroke in paths E and F means the blank will not be intimately attached to the sharp features on the punch and have the uniform flow. Finally, the optimized path F has the smallest spring back and the best result.

For the spring back of the specimen, the maximum hydraulic pressure also plays a significant role. If it is too large, the concave rounded corner of the part will fracture due to the failure of materials filling during the deformation process. If it is too small, the convex rounded corner of the part will rupture due to the insufficient friction conservation effect. Those will have an important impact on spring back at the same time. Therefore, the designed five loading paths of maximum hydraulic pressure are depicted in Figure 9. The corresponding maximum spring back values are shown in Figure 10. Meanwhile, the unevenness level of wall thickness is introduced for further study on the spring back, which is defined as:

$$\gamma = \frac{t_{\max} - t_{\min}}{t_0} \times 100\% \tag{10}$$

where $t_{\max}$ is the maximum wall thickness and $t_{\min}$ is the minimum wall thickness. The variation of γ is depicted in Figure 10.

It can be seen that when the maximum hydraulic pressure is 12 MPa, the maximum spring back value of the specimen is the smallest, which is 3.8 mm, and the uniformity of wall thickness distribution is the highest. When the maximum hydraulic pressure is 14–18 MPa, the spring back of the specimens varies slightly, but the uniformity of wall thickness distribution varies greatly. Similarly, when the maximum hydraulic pressure is 10 MPa, the wall thickness of the specimen decreases seriously and the distribution uniformity is poor due to the insufficient friction conservation effect during the forming process. Therefore, the selection of the maximum hydraulic pressure needs to satisfy the double conditions of spring back and wall thickness distribution.

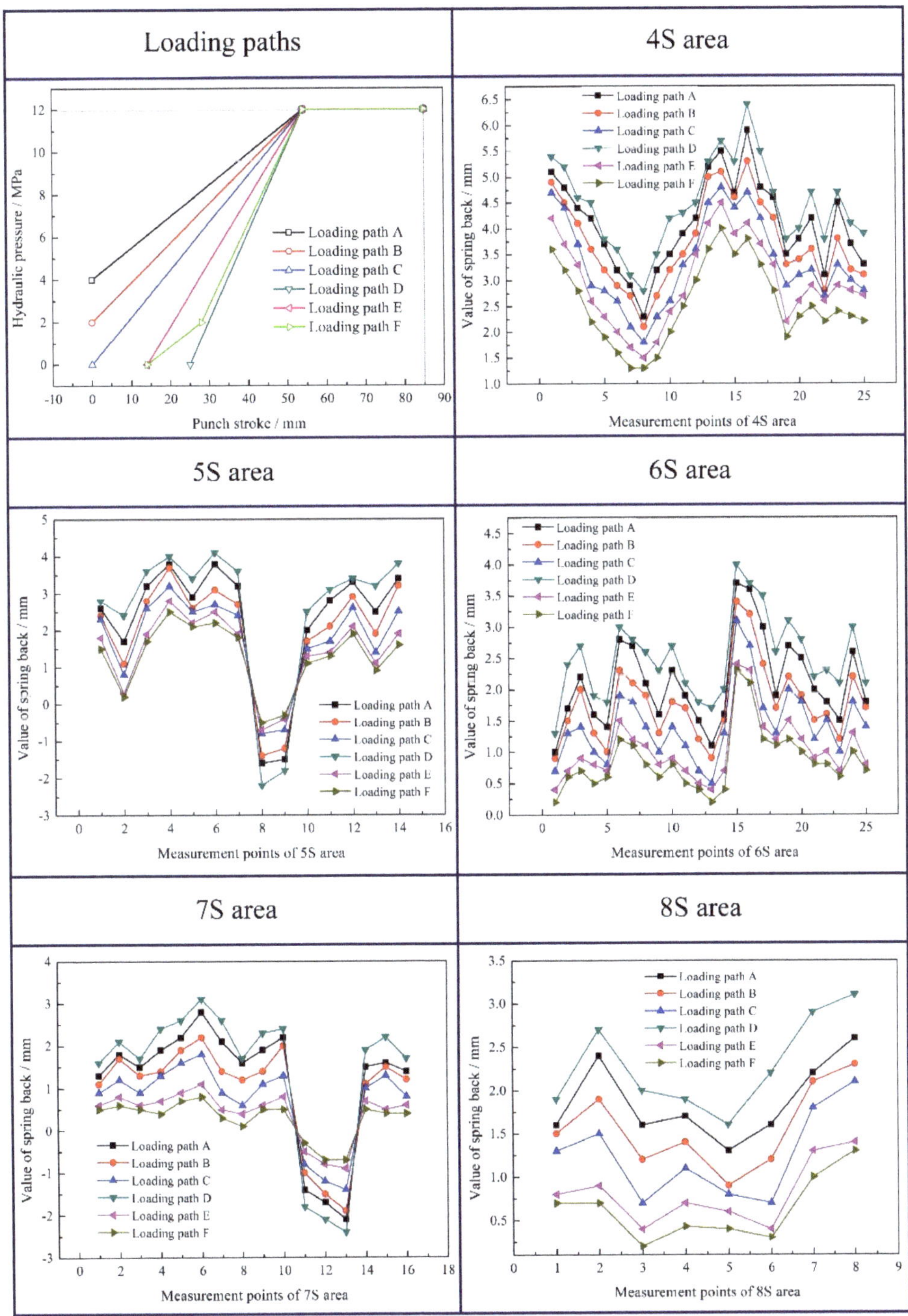

Figure 8. Initial hydraulic pressure paths and the corresponding spring back values of different areas. This figure is the spring back values of different areas (4S–8S areas) under different hydraulic pressure paths. The hydraulic pressure paths are in the first figure of Figure 8.

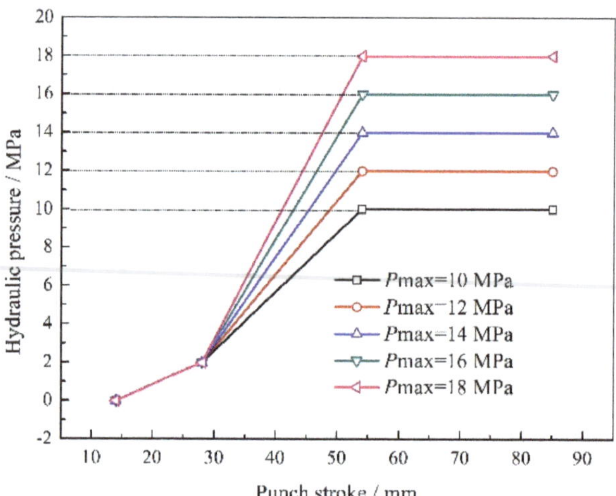

Figure 9. The designed five loading paths of maximum hydraulic pressure.

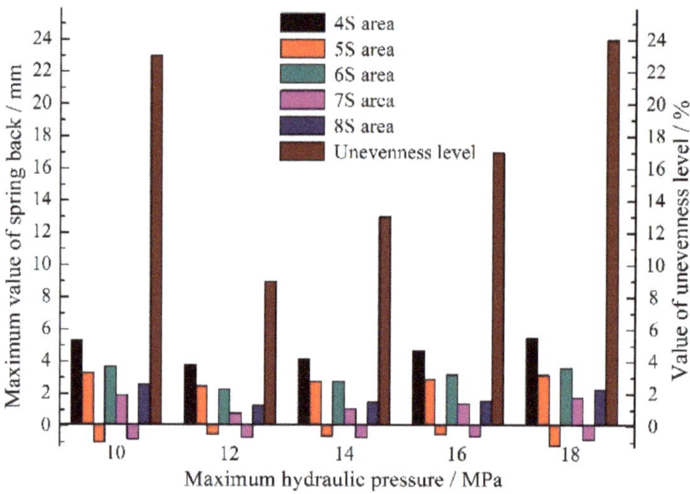

Figure 10. The maximum spring back values and unevenness level of wall thickness with different maximum hydraulic pressures.

4.3. Spring Back Measurement and Compensation Results

The comparison of spring back values of specimens after compensation is shown in Figure 11. Before the spring back compensation, the qualified rate of the two measured specimens is 37.5% and 40.91%, respectively. The consistency of spring back values of each measuring point is good, which shows that the measurement accuracy is high. Meanwhile, it can be seen from the figures that the qualified rate of spring back in 4S area is 0%, and the maximum value of spring back is 3.49 mm. The qualified rate in 5S area is 14.29%, and the maximum value is 2.86 mm. The results of spring back measurement of two specimens in 6S area are slightly different. The qualified rate of specimen 1 is 48%, and that of specimen 2 is 64%, the maximum value of spring back of both specimens is 2.42 mm. The qualified rate in 7S area is higher, close to 100%. The spring back value in one place of specimen 1 is only 0.72 mm, which just exceeds the tolerance requirement. The two specimens in 8S area have good consistency, the qualified rate of spring back is 50%, and the maximum value is 1.28 mm. From the above measurement results, it can be seen that the qualified rate of spring back around the part edge is low and the spring back values of the profile are large. The qualified rate of the edge of inner hole and the key checked profile is high, and the corresponding spring back values are small, which are inseparable from the uniform load of high-pressure liquid in the forming process.

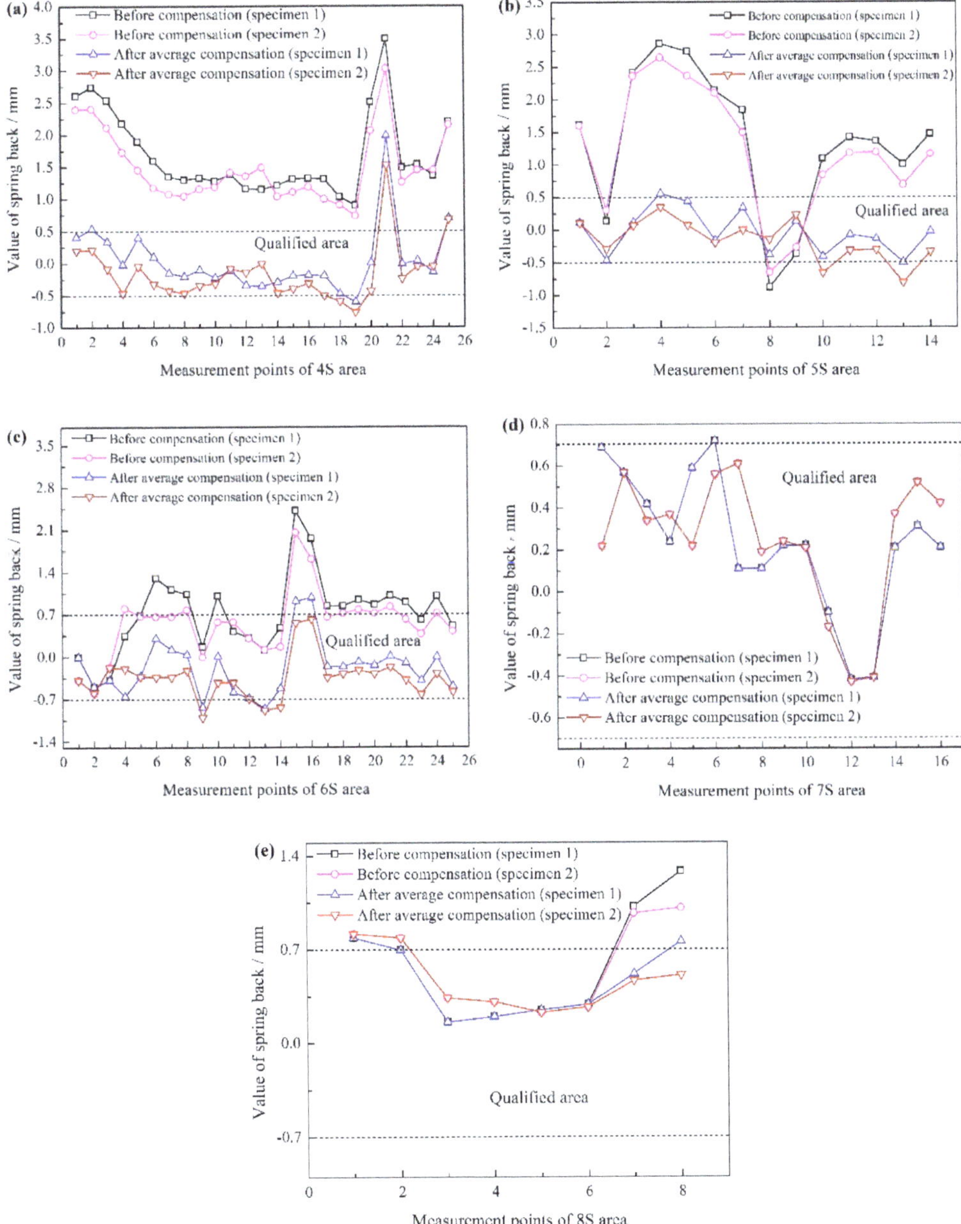

Figure 11. The comparison of spring back values of specimens after compensation: (**a**) 4S area, (**b**) 5S area, (**c**) 6S area, (**d**) 7S area, and (**e**) 8S area.

The qualified rates of spring back of the two measured specimens after the average compensation through the predicted and measured results are 86.36% and 87.5%, respectively, which are 48.86% and 46.59% higher than those before the spring back compensation. The maximum spring back value is also located in the 4S area, which is 1.99 mm and 42.98% lower than that before the compensation. It is indicated that the suitable spring back compensation using the average compensation method can effectively reduce the spring back of specimens and improve the final forming accuracy for the rigid-flexible sequential loading forming process of aluminum alloy inner panel.

5. Conclusions

(1) For the large, complex, multi-feature inner panel, four spring back models are studied, including Barlat'89 + Hollomon isotropic hardening model, Hill'48 + Hollomon isotropic hardening model, Hill'48 + Yoshida–Uemori mixed hardening model, and Barlat'89 + Yoshida–Uemori mixed hardening model. Results show that the hardening model has a greater impact on the accuracy of spring back prediction than the yield criterion does, and the prediction accuracy of Barlat'89 + Yoshida–Uemori mixed hardening model is the highest.

(2) The hydraulic pressure loading path plays a significant role in the specimen spring back. The optimized path F obtains the smallest spring back and uniform wall thickness distribution. Meanwhile, when the maximum hydraulic pressure is 12 MPa, the maximum spring back value of the specimen is the smallest, which is 3.8 mm, and the uniformity of wall thickness distribution is the highest.

(3) The qualified rates of spring back of the measured specimens after the average compensation through the predicted and measured results are improved significantly. The suitable spring back compensation can effectively reduce the spring back of specimens and improve the final forming accuracy for the rigid-flexible sequential loading forming process of aluminum alloy inner panel.

Author Contributions: Conceptualization, L.L. and Y.W.; methodology, Y.W.; software, Z.J.; validation, Y.Z., L.W. and Z.J.; formal analysis, Y.Z.; investigation, L.W.; resources, J.D.; data curation, Y.Z.; writing—original draft preparation, Y.Z.; writing—review and editing, Y.Z. and Y.W.; visualization, H.C.; supervision, L.L.; project administration, J.D.; funding acquisition, Y.W. and H.C. All authors have read and agreed to the published version of the manuscript.

Funding: The authors gratefully acknowledge the financial support from Tianjin "131" innovative talent team—Tianduan skin precision forming intelligent manufacturing technology innovation team, National Natural Science Foundation of China (Grant No. 52005153), local science and technology development fund projects guided by the central government of China (Grant No. 206Z1803G), Natural Science Foundation of Hebei Province of China (Grant Nos. E2019202224), the Project was Supported by State Key Laboratory of Materials Processing and Die & Mould Technology, Huazhong University of Science and Technology (Grant No. P2021-012).

Institutional Review Board Statement: Not applicable.

Informed Consent Statement: Not applicable.

Data Availability Statement: The data in this manuscript cannot be shared at this time as the data also forms part of an ongoing study.

Conflicts of Interest: The authors declare no conflict of interest.

References

1. Wagoner, R.H.; Lim, H.; Lee, M. Advanced Issues in springback. *Int. J. Plast.* **2013**, *45*, 3–20. [CrossRef]
2. Lingbeek, R.; Huétink, J.; Ohnimus, S.; Petzoldtb, M.; Weiherb, J. The development of a finite elements based springback compensation tool for sheet metal products. *J. Mater. Process. Technol.* **2005**, *169*, 115–125. [CrossRef]
3. Joo, G.; Huh, H. Rate-dependent isotropic-kinematic hardening model in tension-compression of TRIP and TWIP steel sheets. *Int. J. Mech. Sci.* **2018**, *146*, 432–444. [CrossRef]

4. Yue, Z.M.; Qi, J.S.; Zhao, X.D.; Badreddine, H.; Gao, J.; Chu, X.R. Springback Prediction of Aluminum Alloy Sheet under Changing Loading Paths with Consideration of the Influence of Kinematic Hardening and Ductile Damage. *Metals* **2018**, *8*, 950. [CrossRef]
5. Geng, L.; Wagoner, R.H. Role of plastic anisotropy and its evolution on springback. *Int. J. Mech. Sci.* **2002**, *44*, 123–148. [CrossRef]
6. Wang, H.; Zhou, J.; Zhao, T.S.; Tao, Y.P. Springback compensation of automotive panel based on three-dimensional scanning and reverse engineering. *Int. J. Adv. Manuf. Technol.* **2015**, *85*, 1187–1193. [CrossRef]
7. Simoes, V.; Laurent, H.; Oliveira, M.; Menezes, L. The influence of warm forming in natural aging and springback of Al-Mg-Si alloys. *Int. J. Mater. Form.* **2019**, *12*, 57–68. [CrossRef]
8. Palumbo, G. Hydroforming a small scale aluminum automotive component using a layered die. *Mater. Des.* **2003**, *44*, 365–373. [CrossRef]
9. Papadia, G.; Prete, A.D.; Anglani, A. Experimental springback evaluation in hydromechanical deep drawing (HDD) of large products. *Prod. Eng.* **2012**, *6*, 117–127. [CrossRef]
10. Kong, D.S.; Lang, L.H.; Sun, Z.Y.; Ruan, S.W.; Gu, S.S. A technology to improve the formability of thin-walled aluminum alloy corrugated sheet components using hydroforming. *Int. J. Adv. Manuf. Technol.* **2016**, *84*, 737–748. [CrossRef]
11. Tang, B.; Zhao, G.; Wang, Z. A mixed hardening rule coupled with Hill'48 yielding function to predict the springback of sheet U-bending. *Int. J. Mater. Form.* **2008**, *1*, 169–175. [CrossRef]
12. Tang, B.; Lu, X.; Wang, Z.; Zhao, Z. Springback investigation of anisotropic aluminum alloy sheet with a mixed hardening rule and Barlat yield criteria in sheet metal forming. *Mater. Des.* **2010**, *31*, 2043–2050. [CrossRef]
13. Xiao, R.; Li, X.X.; Lang, L.H.; Chen, Y.K.; Yang, Y.F. Biaxial tensile testing of cruciform slim superalloy at elevated temperatures. *Mater. Des.* **2016**, *94*, 286–294. [CrossRef]
14. Appiah, E.; Jain, M. A new constitutive model for prediction of springback in sheet metal forming. *AIP Conf. Proc.* **2004**, *712*, 1651.
15. Kinzel, G.L. A new model for springback prediction for aluminum sheet forming. *J. Eng. Mater. Technol.* **2005**, *127*, 279–288.
16. Tamura, S.; Sumikawa, S.; Uemori, T.; Hamasaki, H. Experimental observation of elasto-plasticity behavior of type 5000 and 6000 aluminum alloy sheets. *J. Jpn. Inst. Light Met.* **2011**, *61*, 255–261. [CrossRef]
17. Zhang, D.J.; Cui, Z.S.; Ruan, X.Y.; Li, Y.Q. Sheet springback prediction based on non-linear combined hardening rule and Barlat89′s yielding function. *Comput. Mater. Sci.* **2006**, *1003*, 1218–1225.
18. Lee, M.G.; Kim, D.; Kim, C.; Wenner, M.L.; Wagoner, R.H.; Chung, K. A practical two-surface plasticity model and its application to spring-back prediction. *Int. J. Plast.* **2007**, *23*, 1189–1212. [CrossRef]
19. Zhang, S.L.; Lee, M.G.; Sun, L.; Kim, J.H. Measurement of the Bauschinger behavior of sheet metals by three point bending springback test with pre-strained strips. *Int. J. Plast.* **2014**, *59*, 84–107. [CrossRef]
20. Badreddine, H.; Yue, Z.M.; Saanouni, K. Modeling of the induced plastic anisotropy fully coupled with ductile damage under finite strains. *Int. J. Solids Struct.* **2017**, *180*, 49–62. [CrossRef]
21. Dudescu, C.; Naumann, J.; Stockmann, M.; Steger, H. Investigation of non-linear springback for high strength steel sheet by ESPI. *Strain* **2011**, *47*, 8–18. [CrossRef]
22. Hill, R. A theory of the yielding and plastic flow of anisotropic metals. *Proc. R. Soc. Lond.* **1948**, *193*, 281–297.
23. Barlat, F. Plastic behavior and stretchability of sheet metals. Part I: A yield function for orthotropic sheets under plane stress conditions. *Int. J. Plast.* **1989**, *5*, 51–66. [CrossRef]
24. Yoshida, F.; Urabe, M.; Toropov, V.V. Identification of materials parameters in constitutive model for sheet metals from cyclic bending tests. *Int. J. Mech. Sci.* **1998**, *40*, 237–249. [CrossRef]
25. Yoshida, F.; Uemori, T. A model of large-strain cyclic plasticity describing the Bauschinger effect and workhardening stagnation. *Int. J. Plast.* **2002**, *18*, 661–686. [CrossRef]
26. Yoshida, F.; Uemori, T. A model of large-strain cyclic plasticity and its application to springback simulation. *Int. J. Mech. Sci.* **2003**, *45*, 1687–1702. [CrossRef]
27. Zhang, L.; Chen, J.S.; Chen, J. Springback prediction of advanced high strength steel part based on Yoshida-Uempri hardening model. *Die Mould Technol.* **2012**, *3*, 11–14.

 materials

Article

Development of a Formability Prediction Model for Aluminium Sandwich Panels with Polymer Core

Xiaochuan Liu [1,2,†], Bozhou Di [1,†], Xiangnan Yu [1,*], Heli Liu [1], Saksham Dhawan [1], Denis J. Politis [3], Mateusz Kopec [1,4] and Liliang Wang [1]

1 Department of Mechanical Engineering, Imperial College London, London SW7 2AZ, UK; liuxiaochuan2020@xjtu.edu.cn (X.L.); bozhou.di20@imperial.ac.uk (B.D.); h.liu19@imperial.ac.uk (H.L.); saksham.dhawan12@imperial.ac.uk (S.D.); mkopec@ippt.pan.pl (M.K.); liliang.wang@imperial.ac.uk (L.W.)
2 School of Mechanical Engineering, Xi'an Jiao Tong University, Xi'an 710049, China
3 Department of Mechanical and Manufacturing Engineering, University of Cyprus, Nicosia 1678, Cyprus; politis.denis@ucy.ac.cy
4 Institute of Fundamental Technological Research Polish Academy of Sciences, Pawinskiego 5B, 02-106 Warsaw, Poland
* Correspondence: x.yu19@imperial.ac.uk
† These authors contributed equally to this work.

Abstract: In the present work, the compatibility relationship on the failure criteria between aluminium and polymer was established, and a mechanics-based model for a three-layered sandwich panel was developed based on the M-K model to predict its Forming Limit Diagram (FLD). A case study for a sandwich panel consisting of face layers from AA5754 aluminium alloy and a core layer from polyvinylidene difluoride (PVDF) was subsequently conducted, suggesting that the loading path of aluminium was linear and independent of the punch radius, while the risk for failure of PVDF increased with a decreasing radius and an increasing strain ratio. Therefore, the developed formability model would be conducive to the safety evaluation on the plastic forming and critical failure of composite sandwich panels.

Keywords: formability; M-K model; failure criteria; composite sandwich panel

Citation: Liu, X.; Di, B.; Yu, X.; Liu, H.; Dhawan, S.; Politis, D.J.; Kopec, M.; Wang, L. Development of a Formability Prediction Model for Aluminium Sandwich Panels with Polymer Core. *Materials* **2022**, *15*, 4140. https://doi.org/10.3390/ma15124140

Academic Editor: Alexander Yu Churyumov

Received: 18 May 2022
Accepted: 8 June 2022
Published: 10 June 2022

1. Introduction

The Forming Limit Diagram (FLD) is a mature method to describe the formability of sheet metals. The right side of the FLD, representing positive minor and major strain, was proposed by Keeler and Backofen [1] and subsequently extended by Goodwin [2] with the addition of the negative minor strain. As the metals exhibit different behaviour at various strain rates and temperatures, multiple Forming Limit Curves (FLCs) are generally presented on a single FLD, of which biaxial strain states could represent necking and subsequent failure.

The Nakajima test is one of the most commonly used methods to determine the FLD, and has been defined in standards such as ISO 12004-2 [3] and BSI 12004-2 [4], in which an upper and lower die with draw beads are applied to ensure the flow of the sheet metal along the perimeter of the hole, and a hemispherical punch is used to plastically deform the material. Sheet metal samples are designed with different widths, which leads to a varied strain ratio between the lateral direction (minor strain) and longitudinal direction (major strain). In order to demonstrate a linear strain ratio evolution (loading path), a lubricant is used to minimise the friction. Consequently, the samples with different strain ratios are able to capture linear loading paths, and their critical strains at the verge of necking could be recorded on FLCs. The FLD is subsequently generated by obtaining multiple critical strains from different sample designs.

Theoretical FLD prediction models have been the subject of significant research, with the development of models including the Swift's diffuse necking model [5], Hill's localized necking model [6], the bifurcation analysis-based models [7] and the Marciniake-Kuczynski (M-K) model [8]. Especially, the M-K model enabled the prediction of forming limits for various metals, such as 2060 aluminium alloy [9], 7075 aluminium alloy [10], DP1180 steel [11] and DC01 low-carbon steel [12] at different loading paths, strain rates and elevated temperatures. When combined with the rate-dependent plasticity, polycrystal, self-consistent (VPSC) model, the model was able to realistically predict the fracture behaviour of metals [13]. The fracture angle and local inhomogeneous deformation bands could also be predicted by considering the effect of plastic anisotropy into the model [14]. In addition to the FLD, a generalised forming limit diagram (GFLD) allowing proportional loading with all six independent stress components could be obtained by using the M-K model [15], which has been applied to the prediction of the mesoscaled deformation of clad foils [16].

In recent years, sandwich panels bonded by a polymer as the core layer and metals as the skin layers are being increasingly applied to the construction [17–19], aircraft [20,21] and automobile sectors [22] due to the excellent strength/stiffness-to-weight ratio, thermal insulation, bearing capacity and cost-effectiveness [23]. Among them, sandwich panels made from aluminium alloys and polyvinylidene difluoride (PVDF) have found particular applications on lightweight sealing and insulating components due to PVDF's strong chemical resistance, oxidation resistance and shock resistance. Moreover, the range of applications increased by the inclusion of multiwalled carbon nanotubes (MWCNT) reinforced honeycomb structures to enhance the damping effect [24,25]. At present, a great challenge on the application of the sandwich panels is that the core polymer layer is found to fail earlier than the skin metal layer due to its low resistance to normal pressure. In order to increase the formability of polymer layers, polyethylene [26], polymethyl-methacrylate (PMMA) [27] and polypropylene-polyethylene [28] were applied as adhesion layers. In addition, a density-based topology optimization method was designed and integrated with a multistage algorithm (GA) to optimise the formability of the carbon-fibre-reinforced thermoplastics as the core material of the sandwich panels [29]. To this end, although great efforts have been made to improve the performance of these metal-polymer sandwich panels, the fracture mode and crack propagation of polymers is different to those of metals, leading to lower FLCs and thus earlier failure of the polymer layers [30]. However, the traditional FLD model is not capable of predicting the formability of composite materials, resulting in inaccurate failure prediction of the sandwich panels.

In order to overcome this limitation, the compatibility relationship on the failure criteria between aluminium and polymer layers was established in the present work, and a mechanics-based model for a three-layered sandwich panel was subsequently developed based on the M-K model to predict its formability. Consequently, the critical failures of the sandwich panel at various strain ratios and curvature radii were predicted. Furthermore, the analytical failure solution of the sandwich panel consisting of two AA5754 aluminium alloy layers and a polyvinylidene difluoride (PVDF) layer was presented as a case study, demonstrating the curvature-radius-dependent FLD.

2. The Mechanics-Based Analysis of the Nakajima Test with Sandwich Panels

2.1. Principles of Plastic Deformation

The Logan-Hosford yield criterion [31] was applied as the principle to describe the anisotropic behaviour of metal layers under the plane stress state:

$$R_2\sigma_1^l + R_1\sigma_2^l + R_1R_2(\sigma_1 - \sigma_2)^l = R_2(R_1 + 1)\overline{\sigma}^l \tag{1}$$

$$R_1(2d\varepsilon_2 + d\varepsilon_1)^l + R_2(2d\varepsilon_1 + d\varepsilon_2)^l + R_1R_2(d\varepsilon_1 - d\varepsilon_2)^l = R_2(R_1 + 1)d\overline{\varepsilon}^l \tag{2}$$

where R_1 and R_2 were the r-values in the first and second principal directions, and l was a material constant. If a stress ratio was defined as $\alpha = \sigma_2/\sigma_1$, Equation (1) could be rewritten as:

$$[R_2 + R_1\alpha^l + R_1R_2(1 - \alpha)^l]\sigma_1^l = R_2(R_1 + 1)\overline{\sigma}^l \tag{3}$$

The external loading path was assumed as linear, leading to a constant strain ratio $\beta = d\varepsilon_2/d\varepsilon_1 = \varepsilon_2/\varepsilon_1$. Consequently, the equivalent strain was expressed as:

$$[R_1(2\beta+1)^l + R_2(\beta+2)^l + R_1R_2(1-\beta)^l]\varepsilon_1^l = R_2(R_1+1)\bar{\varepsilon}^l \tag{4}$$

The constitutive relationship of the sheet metal was described as a power law between the flow stress and plastic deformation, where K was the strength coefficient and n was the strain-hardening exponent:

$$\bar{\sigma} = K\bar{\varepsilon}^n \tag{5}$$

The associate flow rule was used to describe the plastic flow behavior as Equation (6) by defining a positive scaler $d\lambda$, in order to establish the relationship between $d\varepsilon_{ij}$ and $\partial\bar{\sigma}/\partial\sigma_{ij}$. As a result, the strain ratio β could be expressed as a function of the stress ratio α, as shown in Equation (8).

$$d\varepsilon_{ij} = d\lambda\frac{\partial\bar{\sigma}}{\partial\sigma_{ij}} \rightarrow d\lambda = d\varepsilon_{ij}/\frac{\partial\bar{\sigma}}{\partial\sigma_{ij}} \tag{6}$$

$$\frac{d\varepsilon_1}{R_2\sigma_1^{l-1} + R_1R_2(\sigma_1-\sigma_2)^{l-1}} = \frac{d\varepsilon_2}{R_1\sigma_2^{l-1} - R_1R_2(\sigma_1-\sigma_2)^{l-1}} \tag{7}$$

$$\beta = \frac{R_1\alpha - R_1R_2(1-\alpha)^{l-1}}{R_2 + R_1R_2(1-\alpha)^{l-1}} \tag{8}$$

2.2. Strain and Stress Analysis of Sandwich Panels

In the Nakajima tests, the friction was generally ignored in the derivation under the lubricated conditions. Therefore, the failure would occur at the apex of the dome in the frictionless case. Figure 1 shows the stress state at an infinitesimal apex element in a single metal layer, of which the stress equilibrium along the thickness direction was derived as:

$$2\sigma_1 tRd\theta\sin\frac{d\varphi}{2} + 2tRd\varphi\sin\frac{d\theta}{2} = pR^2d\theta d\varphi \tag{9}$$

where R was the radius of the punch, t was the thickness at the deformed stage, and p was the contact pressure between the sheet and forming tools on the element. Considering that $\sin\frac{d\varphi}{2} \approx \frac{d\varphi}{2}$ and $\sin\frac{d\theta}{2} \approx \frac{d\theta}{2}$ for the infinitesimal elements, the stress equilibrium equation could be simplified as:

$$(\sigma_1 + \alpha\sigma_1)t = pR \tag{10}$$

The stress equilibrium equation was modified for the three-layered sandwich panel, consisting of two skin metal layers (layer I and III) and one core polymer layer (layer II), as shown in Figure 2. For each layer, Equations (9) and (10) remained valid as long as the contact pressure was amended to the pressure difference between the top and bottom surfaces, considering that the contact pressure decreased from the inner face to the outer face. Thus, the stress equilibrium was developed as Equations (11)–(13):

$$(\sigma_1^I + \alpha^I\sigma_1^I)t^I = (p_1 - p_2)R \tag{11}$$

$$(\sigma_1^{II} + \alpha^{II}\sigma_1^{II})t^{II} = (p_2 - p_3)R \tag{12}$$

$$(\sigma_1^{III} + \alpha^{III}\sigma_1^{III})t^{III} = (p_3 - p_4)R \tag{13}$$

where σ_1^I, σ_1^{II} and σ_1^{III} were the first principal stresses on each layer, t^I, t^{II} and t^{III} were the thicknesses of each layer at the deformation stage, and p_1, p_2, p_3 and p_4 were the contact pressures on each contact surface.

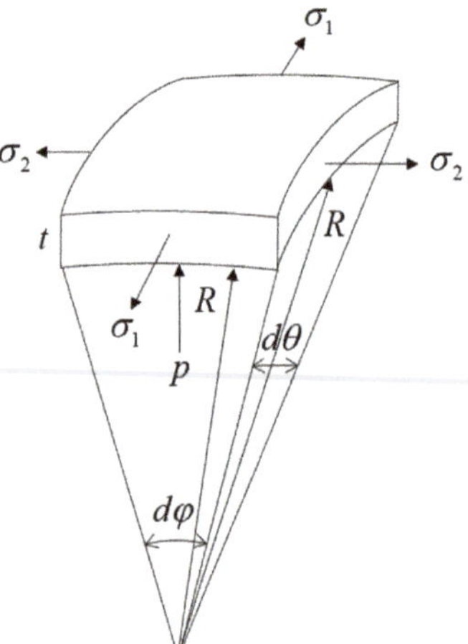

Figure 1. Stress at an infinitesimal apex element in a single metal layer.

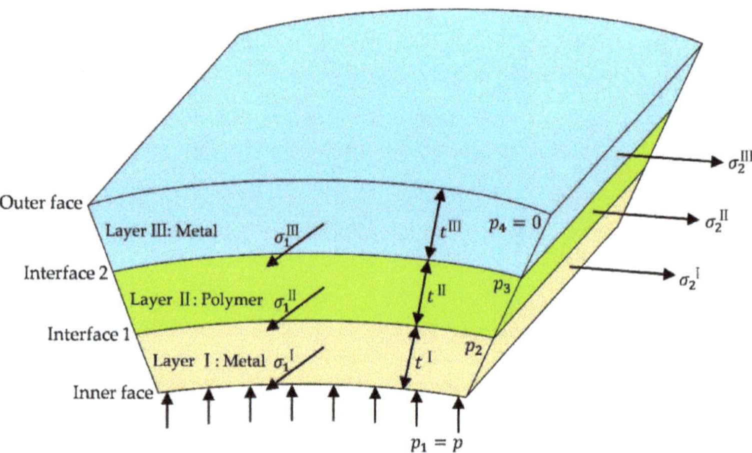

Figure 2. Structure of the three-layered sandwich panel.

Based on the volume constancy, the relationship between the initial thickness and the first principal strain, ε_1 could be derived as:

$$\ln \frac{t^{I/II/III}}{t_0^{I/II/III}} = \varepsilon_3 = -\varepsilon_1 - \varepsilon_2 = -(1+\beta)\varepsilon_1 \tag{14}$$

where $t_0^{I/II/III}$ was the initial thickness of each layer. Note that the strain components and strain ratio β in the three layers were the same, due to the compatibility.

By eliminating t^I, t^{II} and t^{III} in Equations (11)–(14), the relationship of strains was simplified as:

$$R \cdot \exp[(1+\beta)\varepsilon_1] = \frac{t_0^I \sigma_1^I (1+\alpha^I)}{p_1 - p_2} = \frac{t_0^{II} \sigma_1^{II} (1+\alpha^{II})}{p_2 - p_3} = \frac{t_0^{II} \sigma_1^{III} (1+\alpha^{III})}{p_3 - p_4} \tag{15}$$

2.3. *Failure Criteria of Polymer Layer*

A critical normal pressure p_{cp} was defined as the failure criteria of the polymer layer. Thus, the boundary conditions at failure were the contact pressure $p_2 = p_{cp}$ applied on the polymer layer and $p_4 = 0$ applied on the skin layer. As a result, Equation (15) was rewritten as:

$$R \cdot \exp[(1+\beta)\varepsilon_1] = \frac{t_0^I \sigma_1^I (1+\alpha^I)}{p_1 - p_{cp}} = \frac{t_0^{II} \sigma_1^{II} (1+\alpha^{II})}{p_{cp} - p_3} = \frac{t_0^{III} \sigma_1^{III} (1+\alpha^{III})}{p_3} \quad (16)$$

Combining Equations (3) and (4), the power law of Equation (5) was modified as below:

$$\left[\frac{R_2 + R_1\alpha^l + R_1R_2(\alpha-1)^l}{R_2(R_1+1)}\right]^{1/l} \sigma_1 = K\left[\frac{R_2(\beta+2)^l + R_1(2\beta+1)^l + R_1R_2(\beta-1)^l}{R_2(R_1+1)}\right]^{n/l} \varepsilon_1^n \quad (17)$$

By eliminating $\sigma_1^{I/II/III}$ in Equation (16),

$$\frac{R \cdot \exp[(1+\beta)\varepsilon_1]}{1+\alpha^I} = \frac{t_0^I K^I \varepsilon_1^n \left[\frac{R_2^I(\beta+2)^{l^I} + R_1^I(2\beta+1)^{l^I} + R_1^I R_2^I(\beta-1)^{l^I}}{R_2^I(R_1^I+1)}\right]^{n^I/l^I}}{(p_1 - p_{cp})\left[\frac{R_2^I + R_1^I\alpha^{l^I} + R_1^I R_2^I(\alpha-1)^{l^I}}{R_2^I(R_1^I+1)}\right]^{1/l^I}} \quad (18)$$

$$\frac{R \cdot \exp[(1+\beta)\varepsilon_1]}{1+\alpha^{II}} = \frac{t_0^{II} K^{II} \varepsilon_1^n \left[\frac{R_2^{II}(\beta+2)^{l^{II}} + R_1^{II}(2\beta+1)^{l^{II}} + R_1^{II} R_2^{II}(\beta-1)^{l^{II}}}{R_2^{II}(R_1^{II}+1)}\right]^{n^{II}/l^{II}}}{(p_{cp} - p_3)\left[\frac{R_2^{II} + R_1^{II}\alpha^{l^{II}} + R_1^{II} R_2^{II}(\alpha-1)^{l^{II}}}{R_2^{II}(R_1^{II}+1)}\right]^{1/l^{II}}} \quad (19)$$

$$\frac{R \cdot \exp[(1+\beta)\varepsilon_1]}{1+\alpha^{III}} = \frac{t_0^{III} K^{III} \varepsilon_1^n \left[\frac{R_2^{III}(\beta+2)^{l^{III}} + R_1^{III}(2\beta+1)^{l^{III}} + R_1^{III} R_2^{III}(\beta-1)^{l^{III}}}{R_2^{III}(R_1^{III}+1)}\right]^{n^{III}/l^{III}}}{p_3\left[\frac{R_2^{III} + R_1^{III}\alpha^{l^{III}} + R_1^{III} R_2^{III}(\alpha-1)^{l^{III}}}{R_2^{III}(R_1^{III}+1)}\right]^{1/l^{III}}} \quad (20)$$

By eliminating p_3 in Equations (19) and (20),

$$p_{cp} \cdot R \cdot \exp[(1+\beta)\varepsilon_1] = \frac{t_0^{II} K^{II} \varepsilon_1^n \left[\frac{R_2^{II}(\beta+2)^{l^{II}} + R_1^{II}(2\beta+1)^{l^{II}} + R_1^{II} R_2^{II}(\beta-1)^{l^{II}}}{R_2^{II}(R_1^{II}+1)}\right]^{n^{II}/l^{II}} \cdot (1+\alpha^{II})}{\left[\frac{R_2^{II} + R_1^{II}\alpha^{l^{II}} + R_1^{II} R_2^{II}(\alpha-1)^{l^{II}}}{R_2^{II}(R_1^{II}+1)}\right]^{1/l^{II}}} + \frac{t_0^{III} K^{III} \varepsilon_1^n \left[\frac{R_2^{III}(\beta+2)^{l^{III}} + R_1^{III}(2\beta+1)^{l^{III}} + R_1^{III} R_2^{III}(\beta-1)^{l^{III}}}{R_2^{III}(R_1^{III}+1)}\right]^{n^{III}/l^{III}} \cdot (1+\alpha^{III})}{\left[\frac{R_2^{III} + R_1^{II}\alpha^{l^{III}} + R_1^{III} R_2^{III}(\alpha-1)^{l^{III}}}{R_2^{III}(R_1^{III}+1)}\right]^{1/l^{III}}} \quad (21)$$

If the strain ratio β and other material constants were given, α^{II} and α^{III} could be calculated by Equation (8), resulting in the only unknown factor, ε_1 in Equation (21), which would be solved numerically. It should be noted that the solution might not exist, due to the fact that the term on the left side of Equation (21) (Term 1) increased exponentially with increasing ε_1, while the term on the right side (Term 2) increased polynomially. This indicated that Term 1 increased at a higher rate than that of Term 2 after a given point, where the derivative of each side was equal. Thus, there were four bifurcation conditions, as shown in Figure 3:

(a) If Term 1 was always larger than Term 2, no solution existed, suggesting that the sandwich panel would not fail as the failure criteria of the polymer layer was not met;

(b) If Term 1 was larger than Term 2 when $\varepsilon_1 = 0$ and the two terms became equal at a given point, where the derivative of each was equal, a unique solution existed;

(c) If Term 1 was larger than Term 2 when $\varepsilon_1 = 0$ and it became smaller at a given point where the derivative of each was equal, two solutions existed, of which the smaller one was true as the failure occurred and the increase in ε_1 terminated;
(d) If Term 1 was smaller than Term 2 when $\varepsilon_1 = 0$, a unique solution existed.

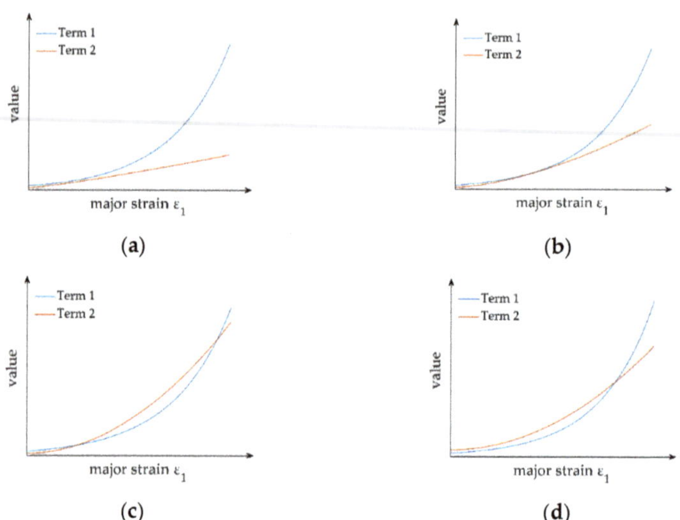

Figure 3. Bifurcation conditions of the failure criteria of the polymer layer: (**a**) Term 1 was larger than Term 2; (**b**) Term 1 was initially larger than Term 2 and became equal at a given point; (**c**) Term 1 was initially larger than Term 2 and became smaller at a given point; (**d**) Term 1 was smaller than Term 2.

It was advised to limit the solution domain to be positive with an initial estimate of 0 when solving Equation (21). Once ε_1 was solved, the pressure on each interface could be obtained by using Equations (18)–(20).

2.4. Failure Criteria of Metal Layers

The M-K model [8] was applied to determine the forming limit of the metal layers, assuming that a perfection Zone *a* and an imperfection Zone *b* coexisted in the metal layers, while a perfection Zone *a* existed in the polymer layer, as shown in Figure 4. Once the failure criterion of Equation (22) was met, the metal layers failed.

$$\frac{d\varepsilon_{3b}^{I/III}}{d\varepsilon_{3a}^{I/III}} = \frac{\varepsilon_{3b,i} - \varepsilon_{3b,i-1}}{\varepsilon_{3a,i} - \varepsilon_{3a,i-1}} > 10 \tag{22}$$

In order to activate the failure criteria of the metal layers, an initial geometrical nonhomogeneity along the thickness direction, known as the 'imperfection factor f', was defined as Equation (23), which had evolved from the initial imperfection factor f_0. The upper script I/III was neglected to simplify the derivation of the M-K model.

$$f = \frac{t_b}{t_a}, \quad f_0 = \frac{t_{b0}}{t_{a0}} \tag{23}$$

where t_a and t_b were the instantaneous thicknesses of Zones *a* and *b*, while t_{a0} and t_{b0} were the initial thickness of Zones *a* and *b*. The relationship between the imperfection factor and the initial imperfection factor was derived as:

$$\varepsilon_{3a} = \ln\left(\frac{t_a}{t_{a0}}\right) \Rightarrow t_a = t_{a0} \exp(\varepsilon_{3a}) \tag{24}$$

$$\varepsilon_{3b} = \ln(\frac{t_b}{t_{b0}}) \Rightarrow t_b = t_{b0}\exp(\varepsilon_{3b}) \tag{25}$$

$$f = f_0 \exp(\varepsilon_{3b} - \varepsilon_{3a}) \tag{26}$$

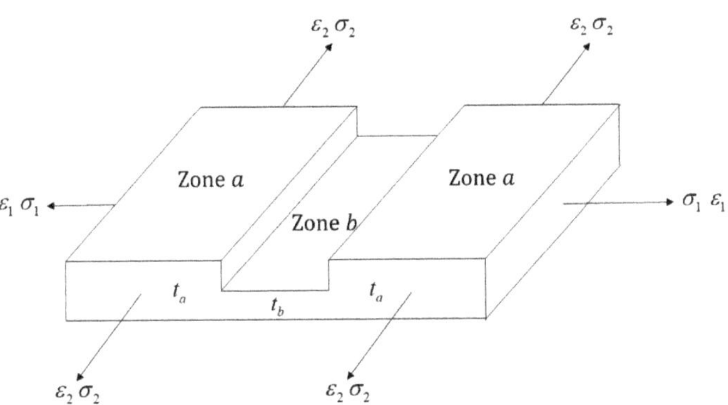

Figure 4. Schematic diagram of Zone *a* and *b* in the M-K model.

The application of the external loading on Zone *a* led to a constant stress ratio α_a and strain ratio β_a. ε_{1a} was assumed to grow by $d\varepsilon$ at any step, ε_{2a} could be calculated based on the linear loading assumption and ε_{3a} could be derived considering the volume constancy:

$$\varepsilon_{1a,i} = \varepsilon_{1a,i-1} + d\varepsilon \tag{27}$$

$$\varepsilon_{2a,i} = \beta_a \varepsilon_{1a,i} \tag{28}$$

$$\varepsilon_{3a,i} = -(\varepsilon_{1a,i} + \varepsilon_{2a,i}) = -(1+\beta_a)\varepsilon_{1a,i} \tag{29}$$

The first principal stress could be calculated from the modified power law in Equation (17):

$$\sigma_{ia,i} = \frac{K[\frac{R_2(\beta_a+2)^l + R_1(2\beta_a+1)^l + R_1R_2(\beta_a-1)^l}{R_2(R_1+1)}]^{n/l}}{[\frac{R_2+R_1\alpha^l+R_1R_2(\alpha-1)^l}{R_2(R_1+1)}]^{1/l}} \cdot \varepsilon_{1a,i}^n \tag{30}$$

Subsequently, σ_{2a} could be expressed by using the stress ratio in Equation (31), and σ_{3a} was assumed to be 0 for metal sheets.

$$\sigma_{2a,i} = \alpha_a \sigma_{1a,i} \tag{31}$$

The stress and strain states on Zone *b* were predicted considering the compatibility of strain and equilibrium of stress between the two zones:

$$\varepsilon_{2b,i} = \varepsilon_{2a,i} \tag{32}$$

$$\sigma_{1b,i} = \frac{1}{f}\sigma_{1a,i} = \frac{\sigma_{1a,i}}{f_0 \exp(\varepsilon_{2b,i}/\beta_{b,i} + \varepsilon_{2b,i} + \varepsilon_{3a,i})} \tag{33}$$

Note that the loading path of Zone *b* was not linear but followed the compatibility relationship with Zone *a*; thus, $\beta_{b,i}$ was changed for a different step *i*.

The plastic deformation of Zone *b* could be modelled based on the modified power law of Equation (17):

$$\begin{aligned}&[\frac{R_2+R_1\alpha_{b,i}^l+R_1R_2(\alpha_{b,i}-1)^l}{R_2(R_1+1)}]^{1/l} \cdot \frac{\sigma_{1a,i}}{f_0\exp(\varepsilon_{2b,i}/\beta_{b,i}+\varepsilon_{2b,i}+\varepsilon_{3a,i})} \\ &+K[\frac{R_2(\beta_{b,i}+2)^l+R_1(2\beta_{b,i}+1)^l+R_1R_2(\beta_{b,i}-1)^l}{R_2(R_1+1)}]^{n/l} \cdot (\frac{\varepsilon_{2b,i}}{\beta_{b,i}})^n = 0\end{aligned} \tag{34}$$

From Equation (8), the strain ratio on Zone b was expressed as:

$$\beta_{b,i} = \frac{R_1\alpha_{b,i} - R_1R_2(1-\alpha_{b,i})^{l-1}}{R_2 + R_1R_2(1-\alpha_{b,i})^{l-1}} \tag{35}$$

By substituting Equation (35) into Equation (34), $\alpha_{b,i}$ remained the only unknown in Equation (34). After solving for $\alpha_{b,i}$ from Equation (34), all the stress and strain components on Zone b could be derived accordingly. In order to numerically solve the model constants, it was necessary to define $\varepsilon_{1c}^{I/II/III}$ as the limit major strain of each layer. Specifically, ε_{1c} and ε_{2c} had to be defined as the limit major and minor strain of the sandwich panel. The flow chart of the analytical model is shown in Figure 5.

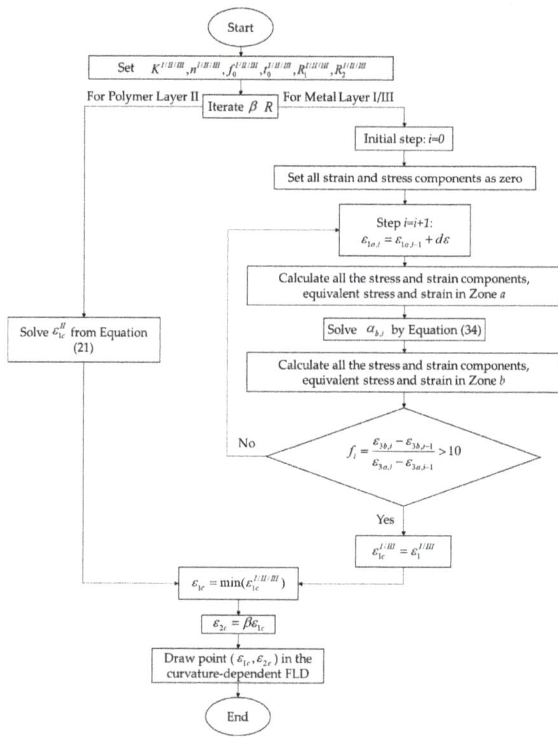

Figure 5. Flow chart of the model.

3. Case Study

In the case study that follows, the model was applied to a sandwich panel, which was produced by combining two skin layers of AA5754 aluminium alloy with a thickness of 1 mm and one core layer of polyvinylidene difluoride (PVDF) with a thickness of 0.1 mm. The properties of AA5754 and PVDF are presented in Table 1 [32,33]. The imperfection factor of AA5754 was set to 0.95. The failure prediction was conducted at a punch radius range between 80 to 180 mm.

Table 1. Mechanical properties of AA5754 and PVDF.

Material	R_1	R_2	σ_0 [MPa]	K [MPa]	n	l
AA5754	0.73	0.69	0	474	0.317	8
PVDF	1	1	0	6.51	0.465	2

Figure 6 shows the FLCs of PVDF and AA5754 under different punch radii. The blue curves represent the critical failure of the AA5754 layers (since the properties of the two aluminium layers were the same and thus the results coincided) and followed a typical V-shape, indicating that the loading path was linear and independent of the punch radius. The green curves represent the critical failure of the PVDF layer, which was significantly dependent on the punch radius. Specifically, when the punch radius was small ($R = 80$ or 100 mm), the FLC of PVDF was much lower than that of AA5754. However, the FLC of PVDF increased with an increasing radius. When the radius reached 120 mm, a part of the FLC was greater than that of AA5754, suggesting that the PVDF layer would not fail before the AA5754 layer. Meanwhile, the critical failure of PVDF was not solved by the model at some low strain-ratio values, indicating that the polymer would not fail under those conditions. When the radius increased to 180 mm, no critical failure of PVDF was predicted at all, suggesting that the entire layer was safe regardless of the strain state. The FLCs of PVDF and AA5754 under different punch radii were combined, as shown in Figure 7, to generate a curvature-radius-dependent FLD. As can be seen, it was found that the risk for failure of PVDF increased with a decreasing radius and an increasing strain ratio.

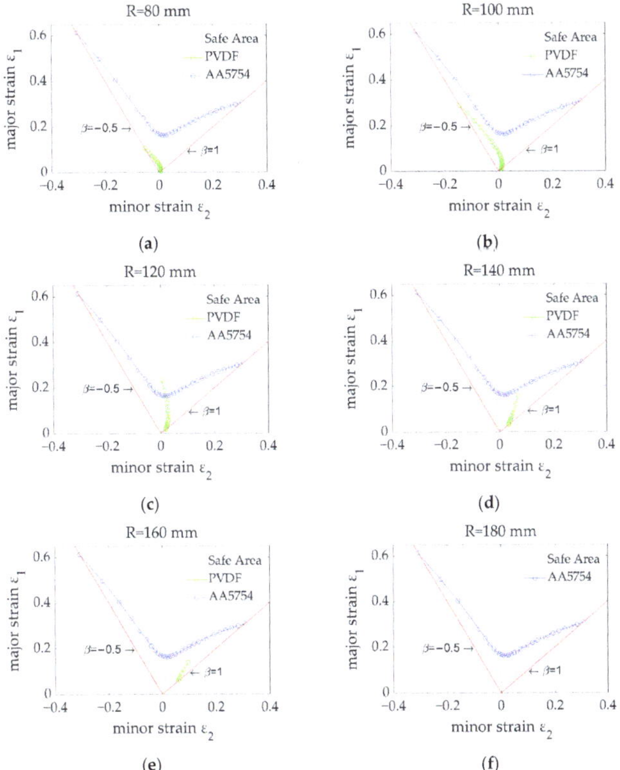

Figure 6. FLCs of PVDF and AA5754 with different punch radii: (**a**) 80 mm; (**b**) 100 mm; (**c**) 120 mm; (**d**) 140 mm; (**e**) 160 mm; (**f**) 180 mm.

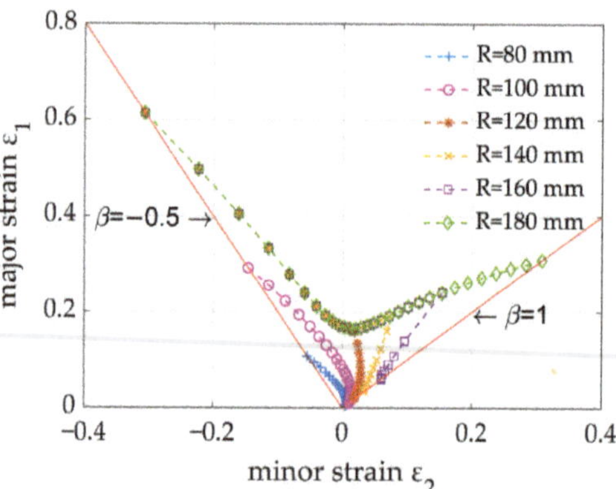

Figure 7. Curvature-radius-dependent FLD for the sandwich panel.

It should be noted that each FLC in Figure 6 could be used to determine the formability of a sandwich panel in a Nakajima test under certain punch radii and strain ratios. However, different regions of a complex-shaped component may withstand various punch radii and strain ratios, which may even vary during forming. Therefore, it is necessary to combine the developed model and a finite element method to predict the evolutionary formability of each element when forming complex-shaped sandwich panels [24,25]. In addition, the developed model has the potential capability of predicting the formability of sandwich panels made from other composite materials with multiple layers by using the proposed compatibility relationship of the failure criteria between polymers and metals.

4. Conclusions

In the present research, an analytical formability model for the sandwich panel has been developed and its capability has been demonstrated by predicting the critical failure of a sandwich panel consisting of two skin AA5754 layers and a core PVDF layer as a case study. The main conclusions from the work were summarised as follows:

- The failure of the polymer layer was curvature-radius dependent. At a small radius, the polymer layer readily failed, while at a large radius, the sandwich panel would be more formable. This feature contributed to the curvature-radius-dependent FLD of the sandwich panel.
- Under the same punch radius/curvature, the polymer layer always failed rapidly at a larger strain ratio. The worst case was the equi-biaxial ($\beta = 1$) loading path.
- The developed FLD model overcame the limitation of traditional FLD models and was capable of predicting the formability of sandwich panels made from composite materials.

The work in this study provided a safety evaluation and theoretical guidance on the plastic forming and critical failure of composite sandwich panels for lightweight sealing and insulating component applications. It is recommended that subsequent studies conduct Nakajima tests and forming trials of complex-shaped components to experimentally verify the formability of the applied composite material and validate the developed model proposed in this study.

Author Contributions: Conceptualization, X.L. and B.D.; Formal analysis, X.L. and X.Y.; Funding acquisition, X.L.; Methodology, B.D. and L.W.; Software, B.D. and H.L.; Supervision, X.Y. and L.W.; Validation, X.L. and B.D.; Visualization, B.D.; Writing—original draft, X.L. and B.D.; Writing—review & editing, X.L., S.D., D.J.P. and M.K. All authors have read and agreed to the published version of the manuscript.

Funding: This work is supported by National Key R&D Program of China (2021YFF0500100).

Institutional Review Board Statement: Not applicable.

Informed Consent Statement: Not applicable.

Data Availability Statement: Data is contained within the article.

Conflicts of Interest: The authors declare no conflict of interest.

References

1. Keeler, S.P.; Backofen, W.A. Plastic instability and fracture in sheets stretched over rigid punches. *ASM Trans.* **1963**, *56*, 25–48.
2. Goodwin, G.M. Application of strain analysis to sheet metal forming problems in the press shop. *SAE Tech. Pap.* **1968**, *77*, 680093. [CrossRef]
3. *ISO 12004-2*; Metallic Materials—Determination of Forming-Limit Curves for Sheet and Strip—Part 2: Determination of Forming-Limit Curves in the Laboratory. ISO: Geneva, Switzerland, 2021.
4. *BSI 12004-2*; Metallic Materials—Determination of Forming-Limit Curves for Sheet and Strip. Determination of Forming-Limit Curves in the Laboratory. BSI: London, UK, 2021.
5. Swift, H.W. Plastic instability under plane stress. *J. Mech. Phys. Solids.* **1952**, *1*, 1–18. [CrossRef]
6. Hill, R. A theory of the yielding and plastic flow of anisotropic metals. *Proc. R. Soc. Lond. A* **1948**, *193*, 281–297. [CrossRef]
7. Stören, S.; Rice, J.R. Localized necking in thin sheets. *J. Mech. Phys. Solids.* **1975**, *23*, 421–441. [CrossRef]
8. Marciniak, Z.; Kuczyński, K. Limit strains in the processes of stretch-forming sheet metal. *Int. J. Mech. Sci.* **1967**, *9*, 609–620. [CrossRef]
9. Gao, H.; El Fakir, O.; Wang, L.; Politis, D.J.; Li, Z. Forming limit prediction for hot stamping processes featuring non-isothermal and complex loading conditions. *Int. J. Mech. Sci.* **2017**, *131*, 792–810. [CrossRef]
10. Zhu, M.; Lim, Y.C.; Liu, X.; Cai, Z.; Dhawan, S.; Gao, H.; Politis, D.J. Numerical forming limit prediction for the optimisation of initial blank shape in hot stamping of AA7075. *Int. J. Light. Mater. Manuf.* **2021**, *4*, 269–280. [CrossRef]
11. Butcher, C.; Khameneh, F.; Abedini, A.; Connolly, D.; Kurukuri, S. On the experimental characterization of sheet metal formability and the consistent calibration of the MK model for biaxial stretching in plane stress. *J. Mater. Process. Tech.* **2021**, *287*, 116887. [CrossRef]
12. Eyckens, P.; Van Bael, A.; Van Houtte, P. An extended Marciniak–Kuczynski model for anisotropic sheet subjected to monotonic strain paths with through-thickness shear. *Int. J. Plast.* **2011**, *27*, 1577–1597. [CrossRef]
13. Jeong, Y.; Jeon, B.; Tome, C.N. Finite element analysis using an incremental elasto-visco-plastic self-consistent polycrystal model: FE simulations on Zr and low-carbon steel subjected to bending, stress-relaxation, and unloading. *Int. J. Plast.* **2021**, *147*, 103110. [CrossRef]
14. Yu, K.; Li, Q.; Wu, Y.; Guo, M.; Li, D.; Liu, C.; Zhuang, L.; Wu, P. An improved Marciniak-Kuczynski approach for predicting sheet metal formability. *Int. J. Mech. Sci.* **2022**, *221*, 107200. [CrossRef]
15. Allwood, J.M.; Shouler, D.R. Generalised forming limit diagrams showing increased forming limits with non-planar stress states. *Int. J. Plast.* **2009**, *25*, 1207–1230. [CrossRef]
16. Wang, C.; Wang, H.; Chen, G.; Zhu, Q.; Zhang, P.; Fu, M.W. Experiment and modeling based studies of the mesoscaled deformation and forming limit of Cu/Ni clad foils using a newly developed damage model. *Int. J. Plast.* **2022**, *149*, 103173. [CrossRef]
17. Joudi, A.; Svedung, H.; Rönnelid, M. Energy efficient surfaces on building sandwich panels-A dynamic simulation model. *Energy Build.* **2011**, *43*, 2462–2467. [CrossRef]
18. Silva, T.; Correia, L.; Dehshirizadeh, M.; Sena-Cruz, J. Flexural creep response of hybrid GFRP-FRC sandwich panels. *Materials* **2022**, *15*, 2536. [CrossRef]
19. Sun, Y.; Li, C.; You, J.; Bu, C.; Yu, L.; Yan, Z.; Liu, X.; Zhang, Y.; Chen, X. An Investigation of the properties of expanded polystyrene concrete with fibers based on an orthogonal experimental design. *Materials* **2022**, *15*, 1228. [CrossRef]
20. Eschen, H.; Harnisch, M.; Schüppstuhl, T. Flexible and automated production of sandwich panels for aircraft interior. *Procedia Manuf.* **2018**, *18*, 35–42. [CrossRef]
21. Chung, Y.; Shrestha, R.; Lee, S.; Kim, W. Binarization mechanism evaluation for water ingress detectability in honeycomb sandwich structure using lock-in thermography. *Materials* **2022**, *15*, 2333. [CrossRef]
22. Lee, H.; Park, H. Study on structural design and manufacturing of sandwich composite floor for automotive structure. *Materials* **2021**, *14*, 1732. [CrossRef]
23. Kee Paik, J.; Thayamballi, A.K.; Sung Kim, G. The strength characteristics of aluminum honeycomb sandwich panels. *Thin Walled Struct.* **1999**, *35*, 205–231. [CrossRef]
24. Maneengam, A.; Siddique, M.J.; Selvaraj, R.; Kakaravada, I.; Arumugam, A.B.; Singh, L.K.; Kumar, N. Influence of multi-walled carbon nanotubes reinforced honeycomb core on vibration and damping responses of carbon fiber composite sandwich shell structures. *Polym. Compos.* **2022**, *43*, 2073–2088. [CrossRef]
25. Kassa, M.K.; Selvaraj, R.; Wube, H.D.; Arumugam, A.B. Investigation of the bending response of carbon nanotubes reinforced laminated tapered spherical composite panels with the influence of waviness, interphase and agglomeration. *Mech. Based Des. Struct. Mach.* **2021**, 1–23. [CrossRef]
26. Liu, J.; Liu, W.; Xue, W. Forming limit diagram prediction of AA5052/polyethylene/AA5052 sandwich sheets. *Mater. Design* **2013**, *46*, 112–120. [CrossRef]

27. Reggente, M.; Harhash, M.; Kriegel, S.; He, W.; Masson, P.; Faerber, J.; Pourroy, G.; Palkowski, H.; Carrado, A. Resin-free three-layered Ti/PMMA/Ti sandwich materials: Adhesion and formability study. *Compos. Struct.* **2019**, *218*, 107–119. [CrossRef]
28. Khardin, M.; Harhash, M.; Chernikov, D.; Glushchenkov, V.; Palkowski, H. Preliminary studies on electromagnetic forming of aluminum/polymer/aluminum sandwich sheets. *Compos. Struct.* **2020**, *252*, 112729. [CrossRef]
29. Zhang, J.; Yanagimoto, J. Density-based topology optimization integrated with genetic algorithm for optimizing formability and bending stiffness of 3D printed CFRP core sandwich sheets. *Compos. B. Eng.* **2021**, *225*, 109428. [CrossRef]
30. Zhang, X.; Chen, Q.; Cai, Z. Prediction and prevention of fracture defect in plastic forming for aluminum foam sandwich panel. *J. Mater. Res. Technol.* **2021**, *15*, 1145–1154. [CrossRef]
31. Logan, R.W.; Hosford, W.F. Upper-bound anisotropic yield locus calculations assuming <111>-pencil glide. *Int. J. Mech. Sci.* **1980**, *22*, 419–430. [CrossRef]
32. Iadicola, M.A.; Foecke, T.; Banovic, S.W. Experimental observations of evolving yield loci in biaxially strained AA5754-O. *Int. J. Plast.* **2008**, *24*, 2084–2101. [CrossRef]
33. Hosford, W.F. *Mechanical Behavior of Materials*, 2nd ed.; Cambridge University Press: Cambridge, UK, 2010; p. 85.

MDPI
St. Alban-Anlage 66
4052 Basel
Switzerland
www.mdpi.com

Materials Editorial Office
E-mail: materials@mdpi.com
www.mdpi.com/journal/materials

www.ingramcontent.com/pod-product-compliance
Lightning Source LLC
LaVergne TN
LVHW080502180726
843515LV00016B/896

* 9 7 8 3 0 3 6 5 8 9 1 6 9 *